A C S S Y M P O S I U M S E R I E S **616**

Light-Activated Pest Control

James R. Heitz, EDITOR
Mississippi State University

Kelsey R. Downum, EDITOR
Florida International University

Developed from a symposium sponsored
by the Division of Agrochemicals
at the 209th National Meeting
of the American Chemical Society,
Anaheim, California,
April 2–6, 1995

American Chemical Society, Washington, DC 1995

Library of Congress Cataloging-in-Publication Data

Light-activated pest control / James R. Heitz, editor, Kelsey R. Downum, editor.

    p.   cm.—(ACS symposium series, ISSN 0097–6156; 616)

    "Developed from a symposium sponsored by the Division of Agrochemicals at the 209th National Meeting of the American Chemical Society, Anaheim, California, April 2–6, 1995."

    Includes bibliographical references and indexes.

    ISBN 0–8412–3334–9

    1. Light-activated pesticides—Congresses.

    I. Heitz, James R., 1941– . II. Downum, Kelsey R., 1952–
III. American Chemical Society. Division of Agrochemicals.
IV. American Chemical Society. Meeting (209th: 1995: Anaheim, Calif.) V. Series.

SB951.145.L54L535   1995
668'.65—dc20
                           95–43870
                                CIP

This book is printed on acid-free, recycled paper.

Foreword

THE ACS SYMPOSIUM SERIES was first published in 1974 to provide a mechanism for publishing symposia quickly in book form. The purpose of this series is to publish comprehensive books developed from symposia, which are usually "snapshots in time" of the current research being done on a topic, plus some review material on the topic. For this reason, it is necessary that the papers be published as quickly as possible.

Before a symposium-based book is put under contract, the proposed table of contents is reviewed for appropriateness to the topic and for comprehensiveness of the collection. Some papers are excluded at this point, and others are added to round out the scope of the volume. In addition, a draft of each paper is peer-reviewed prior to final acceptance or rejection. This anonymous review process is supervised by the organizer(s) of the symposium, who become the editor(s) of the book. The authors then revise their papers according to the recommendations of both the reviewers and the editors, prepare camera-ready copy, and submit the final papers to the editors, who check that all necessary revisions have been made.

As a rule, only original research papers and original review papers are included in the volumes. Verbatim reproductions of previously published papers are not accepted.

Contents

Preface

DEVELOPMENT OF NEW PESTICIDES that are efficacious, environmentally safe, and benign to nontarget organisms continues to be a priority for the agricultural chemistry community in order to protect and increase our food and fiber production. Over the past few decades, new and better toxic strategies have been applied to this problem. We have seen the rise and fall of the organochlorine insecticides due to long-term environmental concerns, and their replacement by the organophosphate and carbamate insecticides. These latter insecticides are under pressure at present as being too toxic to nontarget species. We have seen the development and the eventual difficulties of synthetic pyrethroids and Larvadex due to insect resistance.

The symposium upon which this book is based was concerned with the development of pesticides based on a new chemical mechanism that requires light activation as an integral part of the toxic mechanism. Molecules of broad diversity, from visible dyes to allelochemicals, are capable of functioning in this manner. Serious attempts are now being made to develop and register at least three different light-activated molecules as pesticides. Phloxine B and uranine are being developed as a product called "SureDye" for a variety of insect pests. α-Terthienyl is being developed as a mosquito larvicide. δ-Aminolevulinic acid in the presence of any of several modulators is being developed as both an insecticide and a herbicide. When any of these products goes into commercial use, it will legitimize the entire field.

Chapter Added in Press

A chapter was added to the end of the book, positioned after the indexes as a Supplement. The content of this chapter was presented by Daniel Moreno at the symposium. This chapter, authored by Daniel Moreno and Robert Mangan and entitled "Responses of the Mexican Fruit Fly (Diptera: Tephritidae) to Two Hydrolyzed Proteins and Incorporation of Phloxine B To Kill Adults," was inadvertently left out of the book during the development stage. We sincerely regret that this happened.

Acknowledgments

We thank the ACS Division of Agrochemicals and the SunKist Corporation for generous financial support of the symposium and the Division of

Agrochemicals for sponsoring the forum. The authors are to be commended for their dedication in the preparation of their chapters. The overall quality of the book is also dependent on the efforts of a group of people who often go unnoticed: they are the reviewers who read and made suggestions for the improvements in each of the chapters. They receive our utmost thanks.

JAMES R. HEITZ
Department of Biochemistry
  and Molecular Biology
Mississippi State University
Mississippi State, MS 39762

KELSEY R. DOWNUM
Department of Biological Sciences
Florida International University
Miami, FL 33199

September 21, 1995

Chapter 1

Pesticidal Applications of Photoactivated Molecules

James R. Heitz

Department of Biochemistry and Molecular Biology, Mississippi
Agricultural and Forestry Experiment Station, Mississippi State
University, Mississippi State, MS 39762

Upon illumination, some photoactive molecules have been shown to cause toxic reactions in living cells. This is especially significant when the photoactive chemical is taken up or ingested by arthropods, many of which are agricultural or medical pests. The structures of these molecules are diverse and range from visible dyes to allelochemicals. The common aspect of the toxic reaction involves illumination by photons of light and the presence of oxygen. Over many decades, laboratory and small scale field tests have proven the efficacy of the approach in many pesticidal applications. Most of the dye applications have utilized the halogenated xanthene series of dyes, while the allelochemical applications have utilized a broad search for new chemical structures capable of catalyzing these photochemical reactions. There are current efforts aimed primarily at developing suitable bait technologies for large scale control of commercially important insects.

PHOTOACTIVE DYES

Historical Perspective. The use of light to catalyze toxic reactions in insects began in 1928 (1). Since this original observation by Barbieri, the evolution of this process has slowly gained momentum. Originally, dye molecules were used to absorb the photons of light but; over the years, there has been a growing list of plant chemicals which have also been shown to catalyze these same toxic reactions. The phenomenon observed when an organic molecule is illuminated by light in the presence of oxygen and thereby causes a toxic effect on living cells has been called photodynamic action (2,3). This concept of using certain photoactive dyes to kill insects upon illumination by visible light has not been utilized in the field to any great extent thus far. Over the last 67 years, at least two dozen insect species have been documented to be susceptible to photodynamic action using certain organic dye molecules as the toxic principle (Table I). The development of this area of research has continued to the point where, in 1995, a significant effort is being made to register and use some of these compounds in real life scenarios where pest control is required.

Table I

INSECTS SHOWN TO BE SUSCEPTIBLE TO PHOTODYNAMIC ACTION BY DYES

Common Name	Scientific Name	Reference
Mosquito	*Anopheles* (unspecified)	(1)
Mosquito	*Culex* (unspecified)	(1)
Yellowfever mosquito	*Aedes aegypti*	(4)
Malaria mosquito	*Anopheles maculipennis*	(4)
Mosquito	*Anopheles superpictus*	(4)
Codling moth	*Laspeyresia pomonella*	(5)
House fly	*Musca domestica*	(6)
Yellow mealworm	*Tenebrio molitor*	(7)
Black imported fire ant	*Solenopsis richteri*	(8)
Boll weevil	*Anthonomous grandis grandis*	(9)
Oriental cockroach	*Blatta orientalis*	(10)
American cockroach	*Periplaneta americana*	(10)
Cabbage butterfly	*Pieris brassicae*	(11)
Face fly	*Musca autumnalis*	(12)
Eastern Treehole Mosquito	*Aedes triseriatus*	(13)
Southern house mosquito	*Culex quinquefasciatus*	(13)
Cabbage looper	*Trichoplusia ni*	(14)
Pickleworm	*Diaphania nitidalis*	(14)
Corn earworm	*Heliothis zea*	(14)
Black cutworm	*Agrotis ipsilon*	(15)
Apple maggot	*Rhagoletis pomonella*	(16)
Mediterranean fruit fly	*Ceratitis capitata*	(17)
Oriental fruit fly	*Bacterocera dorsalis*	(17)
Mexican fruit fly	*Anastrepha ludens*	(18)
West Indian fruit fly	*Anastrepha obliqua*	(18)
Serpentine fruit fly	*Anastrepha serpentina*	(18)
Guava fly	*Anastrepha striata*	(18)
Common fruit fly	*Drosophila melanogaster*	(19)

Mechanism. There are several types of dyes which have been shown to cause photodynamic action, but the class of compounds shown most effective as insecticides are the halogenated xanthenes. Rose bengal, erythrosin B, and phloxine B are the most efficient pesticides of this class. The presence of halogens ortho to the wing

A	B	Dye
H	H	Fluorescein
Br	H	Eosin Yellowish
I	H	Erythrosin B
Br	Cl	Phloxine B
I	Cl	Rose Bengal

oxygen atoms allows for spin-orbit coupling which in turn allows for the excited singlet dye to move to the first excited triplet state. The molecule that contains the larger halogen atoms exhibits a more efficient transition to the excited triplet state. This is observed as a molecule which is more phosphorescent. Phosphorescence is defined as the emission of a photon of light as a molecule drops from the excited state to the ground state accompanied by spin inversion. Fluorescence is defined as the emission of a photon of light as a molecule drops from the excited state to the ground state without spin inversion. Therefore, it is defined as phosphorescence when the excited dyes emit a photon of light from the first excited triplet state and drop to the ground singlet state. Thus, the dyes which more efficiently move to the excited triplet state, are more phosphorescent. The phosphorescence quantum yield increases from uranine (0.03) < eosin yellowish (0.30) < erythrosin B (0.60) < rose bengal (0.76) (20). Rose bengal was usually the best insecticide and either erythrosin B or phloxine B was nearly as good, depending on the insect species (9, 12, 21-25).

Photodynamic action has been shown to function by one of two mechanisms (26). In Type I mechanisms, the dye absorbs a photon of light and rises first to the singlet excited state and then drops to the excited triplet state. The energy of the photon is then added to the target substrate molecule, making an activated form of the substrate. This activated molecule then adds to ground state oxygen or other oxygen radicals and becomes oxidized in the process. In Type II mechanisms, the dye again absorbs a photon of light as the first step in the process. The dye raises first to the excited singlet state and then to the excited triplet state. The excited dye molecule then gives the energy to ground state oxygen, thereby raising the oxygen to the excited singlet state. Finally, the excited oxygen adds to the target substrate and oxidizes it. The mechanism of action of the halogenated xanthene dyes is considered to occur by a Type II mechanism.

At this point it must be emphasized that the dye photosensitizer does not enter into the toxic reaction. The photosensitizer is a catalyst, not a participant. A single dye molecule is able to cycle through light absorbance, movement to the excited singlet state, transfer to the excited triplet state, sensitization of the ground state oxygen to the excited singlet state, and eventual return to the ground singlet state of the dye in approximately 10msec, or less. It is then possible that a single dye molecule could be responsible for the generation of thousands of molecules of singlet oxygen per minute. A single dye molecule then is capable of initiating oxidation reactions which would destroy thousands of different target molecules rather than the single target molecule destroyed by a single organophosphate molecule, for instance.

Early Insect Studies. Initial experiments showed that there was a direct concentration dependance between the dye in a food source and the observed toxicity in the insect population being tested (8). In other experiments it was generally shown also that the efficacy of the dye in the food source was fairly well correlated with the phosphorescence quantum yield. In the great majority of cases, rose bengal, which is the halogenated xanthene dye with the highest phosphorescence quantum yield, is the most efficacious dye, while uranine, which is the xanthene dye with the lowest phosphorescence quantum yield, is the least efficacious.

It was also apparent in the early studies that the light source was critical to the success of the dye as an efficacious insect control agent. Sunlight is both intense enough and contains photons of proper wavelength to be readily absorbed by the dyes so that it is a very efficient light source for causing toxicity. Artificial light is another matter. For the halogenated xanthene dyes, maximum absorbance is in the 540-560nm range (21). Cool White fluorescent lights have an emission band which overlaps this region very efficiently (27). Grow Lux lamps, as well as others primarily used to stimulate photosynthesis in plants, may not have an efficient overlap between photon emission by the lamp and photon absorbance by the dye, and are thus to be avoided in these studies. It may be expected that sunlight will be approximately 3.4 times more effective than a pair of Cool White fluorescent lamps 4 inches from the insects in the test due to the larger number of total photons in sunlight relative to the fluorescent lamps (12). However, due to the relative levels of light emitted at the wavelengths that the dyes absorb, fluorescent lights are approximately 2.5 times more efficient than the sun.

Most laboratory studies involved a feeding regimen wherein the insects in the test were exposed to a dye-impregnated food source overnight in the dark so that all insects had an opportunity to feed ad libitum. In this fashion, the population of insects illuminated by light would then respond more uniformly. This dark period of feeding is not required for the dyes to be effective as control agents. In a field setting, insects may be allowed to feed ad libitum in the light where they will begin to die as a function of the time after feeding rather than as a function of the time after light initiation. The homogeneity of the toxicity data will, of course, be affected by this latter experimental design. For the most part, the efficacy data were defined in terms of LT_{50} (the time necessary to kill 50% of the population) and LD_{50} (the concentration necessary to kill 50% of the population) values. There has been a fledgling attempt to quantitate the efficacy data using kinetic rate equations and; that, in the long term, may be the most descriptive method of doing this (16,23). The toxic reaction in insects was described as a 3rd order reaction and included the number of insects, the dye concentration, and the light intensity. The complete reaction, however, must be at least a 4th order reaction, as the rate of the toxic reaction must also be dependent on the oxygen concentration. The rate equation is: v=k_4(insect)(dye)(light)(oxygen).

Light Independent Toxicity. It has also been shown that these dyes are capable of killing insects in the dark by a mechanism that does not require energy from photons of light (8,14). This dark reaction may be caused by the presence of halogen on the dye molecule; but at the present time, the mechanism is unknown. This mechanism may become important if the dyes are utilized to control insects which spend a great deal of their time in the dark, such as imported fire ant queens, cockroaches, and termites.

Latent Toxicity. Over the course of decades of research on many dyes and many species of insect, it has been shown that all developmental forms of insects are susceptible to photodynamic action (25,28-30). House fly eggs, dipped in dye solutions and illuminated by visible light, are killed. Mosquito and house fly larvae, when fed on the dyes and illuminated by visible light, are killed. House fly and face fly larvae, when fed on the dyes and allowed to pupate, show a high incidence of death in the puparium and some examples of incomplete emergence from the puparium. The larval form of all three species, when fed on the dyes and allowed to continue development to the adult stage, show a marked susceptibility to photodynamic action. And, of course, the adult form of the insect, when fed on the dyes and illuminated by visible light, is killed. This may become very important when the concept of photodynamic action moves into the practical arena of field control of important agricultural and medical insects.

Synergism. In an attempt to enhance the efficacy of the halogenated xanthene dyes, it was found that mixtures of the water soluble dyes were able to cause a synergistic toxic effect in insects. A dye such as uranine, which is nearly nontoxic in insects, when added to a toxic dye, like erythrosin B or rose bengal, has been shown to cause a more than additive toxicity in insects (31,32). A United States Patent has been issued to cover this observation (33).

Feeding Stimulation. One of the critical parameters involved in the successful application of this technology is the presentation of the dyes to the insects in such a fashion that they will consume as large a quantity of the dye material as possible. In studies with larval forms of mosquitoes, it was found that enhanced toxicity could be obtained if insoluble forms of the dyes were spread on the surface of the breeding water rather than by dissolving the dyes in the water (34). Using a surfactant to disperse the insoluble dye over the breeding water increased the toxicity even further (35). When the insoluble form of the nontoxic dye fluorescein is added to the insoluble, dispersed form of erythrosin B, an even greater enhancement of the toxicity is observed (36). The use of the insoluble forms of the dyes to use as feeding materials for aquatic insects becomes a viable option in those scenarios.

Feeding Inhibition. Feeding inhibition is a definite concern which must be considered as the dye-bait is developed. Just as one can put too much salt on a steak, one may also include too much pesticide in the bait to the point where the insect is repelled from the bait. In an application aimed at adult house flies, there was an indication that feeding inhibition may be observed at 1.0% in a sucrose solution (22). In some bait formulations, it is possible to mask or minimize the feeding inhibition. In applying the dye technology to the suppression of mosquito larvae, it was found that very small particles of pure dye, sometimes coated with a surfactant to facilitate dispersion over a water surface, were rapidly consumed by larval forms of several species of mosquito. In the future, it is quite probable that the success of further applications of the dye approach to insect control will be directly dependent on the development of baits that utilize strong feeding attractants for specific insect species.

Current Insect Studies. At the present time, there is a concerted effort to develop bait materials with specificity for the various species of commercially important fruit flies. Current research at the United States Department of Agriculture is aimed at developing a mixture of uranine and phloxine B as the active ingredients in a bait to replace malathion in fruit fly suppression programs. The basic material in these baits is a protein-sugar composition to which the dye toxicants and other feeding compounds may be added. The Mediterranean fruit fly, Oriental fruit fly, Mexican fruit fly, West Indian fruit fly, the Guava fly, and the Serpentine fruit fly, have all been shown to be susceptible to this technology. Modifications to the Nulure bait, currently used with malathion, are being evaluated as to enhanced uptake by the target insect species. Although Nulure does attract adult Mediterranean fruit flies, a question arises as to the amount of the bait that is actually consumed by the insect because of the high salt content. This is not critical to the success of the malathion-Nulure bait as malathion is an effective contact poison. The current malathion-Nulure bait is somewhat acidic in order that the pesticide does not hydrolyze too rapidly. This acidic material may be responsible for damages to automobile finishes. The Nulure bait may be made more basic when the pesticides used are phloxine B and uranine because these compounds are stable at neutral pH. This may make the bait less deleterious to automobile finishes and therefore more acceptable to the public. Laboratory studies have now been supported by small scale field studies. An application has been submitted to the United States Environmental Protection Agency for an Experimental Use Permit for large scale field tests to demonstrate the effective field control of fruit flies by photodynamic action caused by the mixture of xanthene dyes.

Resistance. In the eventual use of this approach in a commercial field application, the question of resistance development must be considered. A field collected strain exhibited 48-fold resistance after 32 generations of selection pressure due to increasing concentrations of erythrosin B in the diet and 5 hours of light exposure (37). After selection was removed, the resistance remained relatively stable. It is inherited as a codominant character and is not sex-linked. The erythrosin B-resistant strain of house flies did not exhibit cross resistance to propoxur, DDT, permethrin, and dichlorvos but was resistant to rose bengal and phloxine B. Erythrosin B was toxic to dieldrin-, knockdown-, and diazinon-resistant house fly strains (38). The lack of cross resistance indicates that the practical application of photosensitizing dyes will generate acceptable substitutes in the control of insects where resistance to other modes of action is a major problem. Resistance in the target field insect population may be slow to develop due to the expected multiplicity of biochemical targets in the cell and the observed difficulty in developing resistance in the laboratory setting.

Human and Environmental Safety. Several members of this class of dyes have also been used for years as coloring agents in drugs, cosmetics, and food. An extensive body of literature exists which testifies to the safety of these dyes when higher animals are exposed (39). Because of the defined safety to mammals, nontarget species and the environment, phloxine B has been chosen as the active principle of choice in an expanded interest in developing this mechanism of action as a source for a new form of insect control. These compounds are not metabolized in the body and are excreted rapidly. Intubated rodents have been shown to excrete 65% of phloxine B from the blood stream, unchanged, within 2 hours (40). Neither phloxine B nor uranine is considered to be carcinogenic, mutagenic, or teratogenic.

A comparison may be made between phloxine B and malathion, one of the safest current organophosphate pesticides. The United States Food and Drug Administration has set 1.25 mg/kg/day as the acceptable daily intake for phloxine B as a drug or cosmetic (41). This may be compared to the FAO-WHO acceptable daily intake for malathion of 0.02mg/kg/day (42). The current bait formulation for malathion used against the Mediterranean fruit fly is 10%, while the proposed formulation containing phloxine B is 0.5%. The skin penetrability factor for malathion is calculated to be 87 times greater than that for phloxine B (43). When one considers these 3 parameters, phloxine B may be considered to be greater than 100,000 times safer to humans than malathion via skin exposure to a Mediterranean fruit fly bait material. Overall, it has been estimated that the environment, including all human exposures, would derive on average approximately a 1000-fold increased margin of safety in the event that the photosensitizing dyes replaced malathion in Mediterranean fruit fly suppression (43)

Recently, a fourth grade student in California, Kristopher Kruse, used phloxine B to kill the common fruit fly (*Drosophila melanogaster*) for a science fair project (19). This is probably the first use of this technology at the level of grade school science fairs and was sparked by his interest in finding a replacement for the aerial spraying of malathion bait in residential areas. The new interest in phloxine B as a candidate pesticide is also coupled to the enhanced insect toxicity observed when phloxine B is synergized by the addition of uranine to the pesticidal composition. An impressive list of insects shown to be susceptible to photodynamic action induced by an equimolar mixture of phloxine B and uranine has been compiled. It is safe to say that the question of efficacious susceptibility of problem insects to halogenated xanthene dyes has been put to rest. The main problem associated with the commercial utilization of xanthene dyes as pesticides rests in the development of specific baits for specific insects.

Photodegradation. Another benefit of the photoactive dyes as candidate pesticides is the photolability of the compounds in sunlight. The same photons which activate the dyes and trigger the generation of the singlet oxygen cause the degradation of the dyes. Singlet oxygen, a product of dye photoexcitation, attacks double bonds in target molecules in the cell and can also attack the unsaturation in the dye molecules themselves. When the unsaturation in the dye molecules is destroyed, the ability of the dye to absorb visible light is destroyed, and the ability of the dye to generate singlet oxygen is destroyed. After environmental applications, it is expected that any excess dye may be washed into ecological bodies of water, such as, streams, ponds, and lakes. Depending on the concentration of the dye in the water column and the intensity of the light incident on the water column, the rates of photodegradation are observed to generate half-lives on the order of minutes or hours, rather than weeks or months (44-46).

Herbicidal Activity. An extremely interesting use of the halogenated xanthene dyes is in the control of illegal drug crops (47). Phloxine B was used as a dye stain to mark a crop of contraband marijuana. A week later, when the site was revisited, the marijuana was dead. Investigation of this serendipitous observation indicated that it had rained in the intervening period and the dye had been washed into the soil where it was taken up by the root system. The site of the photodynamic action in the plant tissue has not been ascertained. Further research has shown that within 24 hours, the green plant leaves could be changed into a snow white material with the same consistency as facial tissue. An obvious benefit to law enforcement agencies resides in the fact that drug agents would not be required to be on the ground at risk to the defensive measures put in place by the owner of the crop. Two spherical devices (called "headache balls") have been modified by the United States Department of Agriculture to be suspended from helicopters to spray the dye solution over a specified area from multiple nozzles located on the surface of the sphere. By limiting the on-site presence of the law enforcement agents, the costs associated with the destruction of the marijuana will be lowered enormously. Further, the safety of the dye to humans relative to some other herbicides, the inability of the dye to penetrate human skin to any extent, the lack of an organic carrier solution, and the staining of the skin when the dye contacts the skin, all lead to a control scheme which is very acceptable to law enforcement agencies who have worked with this protocol. A patent has been issued to cover this application as well as the general herbicidal effects of the dyes (48).

It must be noted that plants are not extremely sensitive to the photodynamic effects caused by the xanthene dyes. When the marijuana plants were treated with the dyes, either eosin yellowish or phloxine B, between 1.8 and 3.6 grams of dye were applied to each plant. The solution used contained 6 ounces of dye dissolved in one gallon of water. From that solution, between 40 and 80 ml were applied to the ground at the dripline of the plant depending on the height of the plant. Marijuana plants as tall as 13 feet were killed by this treatment. The most effective procedure involved application of the dye at a single point at the dripline. This will not be a negative factor in the use of the dyes to control commercially important insect species since the proposed control of fruit flies involves the spraying of a bait containing only 2 grams over an entire acre. Early field tests of phloxine B and uranine to control Mediterranean fruit flies and Oriental fruit flies by backpack sprayer used 20 times that amount of dye per acre with no herbicidal effects noted.

PHOTOACTIVE CYCLIC TETRAPYRROLES

Certain porphyrin intermediates (various cyclic tetrapyrroles) have been shown to be photodynamic. All animals require a heme biosynthetic pathway and all plants require a chlorophyll biosynthetic pathway. Therefore, if the heme and chlorophyll biosynthetic pathways could be modulated such that unusually high cellular levels of the photodynamic cyclic tetrapyrroles may be generated within the cells of the animals or plants, one might be able to develop an entirely new mechanism for pest control.

Herbicidal Activity. The herbicidal activity of porphyrin intermediates via the mechanism of photodynamic action was first reported by Rebeiz (49). The basic compound which causes the accumulation of the photodynamically active cyclic tetrapyrroles is delta-aminolevulinic acid, a precursor in the heme biosynthetic pathway. This compound, alone and in mixture with certain modulators such as 2, 2'-dipyridyl and various phenanthroline derivatives, causes the formation of high levels of protoporphyrins in the dark. Upon exposure to light, protoporphyrin catalyzes the formation of singlet oxygen. Singlet oxygen then oxidizes membrane components and other important biomolecules which results in the death of the plant. In some cases, plant injury becomes observable within the first 20 min of light exposure and can become irreversible within 60 min in bright light. Initial symptomology involves isolated bleach spots on green foliage which rapidly coalesce. The bleaching is accompanied by severe dessication. Within 24 hr the green plant becomes a dessicated mass of brown dead tissue (50).

A complete discussion of the monovinyl and divinyl analogs of the protoporphyrin pathways has been reported (50). This review also includes a

R_1	R_2	Mg−Protochlorophyllide
$^-CH=CH_2$	$^-CH_2-CH_3$	Monovinyl
$^-CH=CH_2$	$CH=CH_2$	Divinyl

Table II
ALLELOCHEMICALS SHOWN CAPABLE OF CAUSING PHOTODYNAMIC
ACTION

Phytochemical Class	Specific Structure	Mechanism
Acetophenones	Encecalin	Type I
Acetylenes	Phenylheptatriyne	Type I&II
Benzophenanthrenes	Sanguinarine	Type II
Beta-Carbolines	Harmane	Type II
Coumarins	5,7-Dimethoxycoumarin	Type I
Extended quinones	Cercosporin	Type II
Furanochromones	Khellin	Type I
Furanocoumarins	8-Methoxypsoralen	Type I&II
Furanoquinolines	Dictamnine	Type I
Isoquinolines	Berberine	Type II
Lignans	Nordihydroquaiaretic acid	Unknown
Pterocarpans	Pisatin	Type I
Quinolines	Camptothecin	Type I&II?
Sesquiterpenes	2,7-Dihydroxycadalene	Type I
Thiophenes	Alpha-Terthienyl	Type II

(Adapted from ref. 55.).

comprehensive discussion of the various nitrogen heterocycle derivatives which are capable of functioning as modulators of the delta-aminolevulinic acid effect on the accumulation of cyclic tetrapyrrole intermediates. Variation of the composition, concentration, and treatment times has been shown to allow some specificity in the herbicidal activity (51).

Insecticidal Activity. Delta-aminolevulinic acid and 2,2'-dipyridyl, alone and in mixture, were also shown to induce a toxic photochemical reaction in the larvae of the cabbage looper (52). After the insect larvae were treated with the delta-aminolevulinic acid and the 2, 2'-dipyridyl, they were held in the dark for 17 hrs followed by 3 days illumination by metal halide lamps. Subsequently, it was observed that other modulators could substitute for the 2, 2'-dipyridyl (53). The mechanism appears to function by the enhanced production of protoporphyrin IX during the dark incubation of the delta-aminolevulinic acid and the 2, 2'-dipyridyl in the body of the insect. In fact, either 2, 2'-dipyridyl or 1, 10-phenanthroline, when added to animal or plant cells, causes the induction of protoporphyrin IX. The accumulation of protoporphyrin IX makes the cells susceptible to light-activated generation of toxic singlet oxygen. When the delta-aminolevulinic acid, a natural amino acid, was included with the dipyridyl or phenanthroline, an even higher accumulation of protoporphyrin IX was observed (53). Structure-activity relationships have been studied for the modulation of the conversion of delta-aminolevulinic acid to protoporphyrin IX by a series of phenanthroline derivatives (54).

PHOTOACTIVE ALLELOCHEMICALS

General History. Besides the dyes, there are other compounds shown capable of causing photodynamic action. For the most part, these chemicals are biosynthetic products from plant and fungal tissues (Table II). These allelochemicals are thought to be a response to damage and are produced by the plant to assist in protection. Many plant species have been subjected to extraction procedures in the search for new photosensitizers. In many of the examples, individual molecules have not been studied much beyond the early observation that the molecule is a photosensitizer. There are some cases, however, where there is a considerable amount of study that has been accomplished with a specific compound. Two of the most studied molecules from the extensive list of photodynamically active allelochemicals are the extended quinones (cercosporin) and the thiophenes (alpha-terthienyl).

<u>Cercosporin.</u> Fungi of the genus *Cercospora* have been shown to produce a natural toxin, cercosporin, an extended quinone, which is central to the ability of the fungus to parasitize plant tissue (52). This compound, which is photoinduced, is also photoactivated. Since cercosporin, in the presence of light and molecular oxygen, produces singlet oxygen, the mechanism appears to proceed through a Type II mechanism. Plants, rodents, and other fungi are susceptible to cercosporin-mediated photodynamic action (57). However, *Cercospora* species are resistant to cercosporin-induced photodynamic action, as well as that induced by hematoporphyrin, methylene blue, toluidine blue, and eosin yellowish (58,59). This suggested that the mechanism

of resistance is not based on the photosensitizer, but rather on the toxic product, singlet oxygen. This has provided an excellent model system for investigating the mechanisms by which living cells protect themselves from the damage caused by singlet oxygene via photodynamic action.

Cercosporin

The presence of carotenoids in the cell is the most commonly observed mechanism used by the cell to protect against singlet oxygen (60,61). They are highly efficient quenchers in biological systems and have been shown to quench both the singlet oxygen and the triplet excited state of the photosensitizers (62,63). Carotenoids are present in Cercospora species and may be involved in resistance. However, they are not present in amounts to explain the observed resistance (64). Carotenoid deficient mutant fungal strains are more sensitive than carotenoid producing strains. However, a mutant of *Phycomyces blakesleeanus* which produced 10,000 times as much carotenoid than a deficient mutant was only 3-4 times more resistant.

A more plausible mechanism of resistance has been proposed which involves the reduction of cercosporin, thereby making it unable to generate singlet oxygen. Active, oxidized, cercosporin fluoresces in the red portion of the visible spectrum and exhibits a high singlet oxygen quantum yield, while the inactive, reduced, cercosporin fluoresces in the green portion of the visible spectrum and does not generate singlet oxygen (65). Further, the reduced cercosporin is approximately 80 times more fluorescent than the oxidized cercosporin. This allows the release of the absorbed visible light energy by cercosporin fluorescence rather than by singlet oxygen production. *Cercospora* species, therefore, are postulated to protect themselves from the generation of singlet oxygen by keeping the cercosporin in the inactive, reduced form. When hyphae of *Cercospora* species are killed by any of a variety of methods, the cercosporin rapidly reoxidizes and begins to fluoresce in the red portion of the spectrum (59). A final observation which supports this hypothesis is that rose bengal, a xanthene dye very similar in structure to eosin yellowish, is toxic to *Cercospora* species while eosin yellowish is not. Rose bengal is not reduced by *Cercospora* species while eosin yellowish is reduced. Therefore, the mechanism of resistance in *Cercospora* species does bear directly on the structure of the sensitizer and not on the product of the reaction, singlet oxygen, as was earlier thought.

Thiophenes. A member of the thiophene class of compounds, alpha-terthienyl, has been shown to be photodynamically active against mosquito and blackfly larvae

α − Terthienyl

(66,67). Alpha-terthienyl has been shown to have a small light independent toxicity in insects which can be enhanced greatly upon illumination (67). The mechanism apparently involves a Type II mechanism (68). In an attempt to investigate the mechanism of alpha-terthienyl inside insects, it was shown that larvae of the tobacco hornworm were 70 times more susceptible to alpha-terthienyl induced photodynamic action than were the larvae of the European corn borer (69). This was attributed to the rapid excretion of alpha-terthienyl by the European corn borer (69) and the more rapid metabolism of alpha-terthienyl by the midgut microsomal fraction of the European corn borer (70). Although alpha-terthienyl in the diet of the European corn borer did not cause an induction of the cytochrome P-450, there was a significant induction in cytochrome b5, NADH-cytochrome c reductase, NADPH oxidase, N-demethylase, O-demethylase, glutathione-S-transferase (71).

SUMMARY

The development of photoactive molecules as candidate pesticides is gaining momentum. It is apparent that the mechanism of light activation is not the only criterion to be considered in the understanding of the efficacy of these candidate pesticides. Many molecular structures are capable of absorbing light and catalyzing photodynamic action. The greatly different efficacy observed between these varying structures suggests that different chemical structures cause the molecules to migrate into different areas of the pest and into different subcellular areas. These different structures also affect the ability of the insect to neutralize the compounds, excrete them, or to quench the energy after the photons have been absorbed. This will result in differences in efficacy and, possibly, in the toxic mechanisms inside the cell.

At this time, products are being developed from the xanthene dyes, alpha-terthienyl, and delta-aminolevulinic acid. These products are aimed at a variety of insect and weed pests and several may become registered in the relatively near future. The registration of the first commercially successful light activated pesticide will legitimize this entire field. Thus far, the concept of light activation to initiate toxic reactions in the cells of pests has been considered something of a curiosity. When even one of these products becomes commercially feasible, it will generate a tremendous amount of interest in the area of light activated pesticides.

Eight years ago, when the first ACS symposium on this topic was held, the presentations were fairly focused on the efficacy of the various light activated molecules. At this symposium, the presentations have shown the progress being made in transferring the technology from the laboratory to the field. In the next eight years, we should see several efficacious, environmentally safe, and cost effective, products based on the mechanism of light activation take their place in the arsenal of those charged with the suppression of agricultural pests.

ACKNOWLEDGMENTS

This work was supported in full by the Mississippi Agricultural and Forestry Experiment Station. The author would like to thank Mrs. Cindy Phillips for her assistance in the preparation of the manuscript. MAFES publication number BC-8804.

LITERATURE CITED

1. Barbieri, A. *Riv. Malariol.* **1928**, 7, 456-463.
2. Jodlbauer, A.; von Tappeiner, H. *Muench. Med. Wochenschr.* **1904**, 26, 1139-1141.
3. Spikes, J. D.; Glad, B. W. *Photochem. Photobiol.* **1964**, 3, 471-487.
4. Schildmacher, H. *Biol. Zentralbl.* **1950**, 69, 468-477.
5. Hayes, D. K.;Schecter, M. S. *J. Econ. Entomol.* **1970**, 63, 997.
6. Yoho, T. P.; Butler, L.; Weaver, J. E. *J. Econ. Entomol.* **1971**, 64, 972-973.
7. Graham, K.; Wrangler, E.; Aasen, L.H. *Can. J. Zool.* **1972**, 50, 1625-1629.
8. Broome, J. R.; Callaham, M. F.; Lewis, L. A.; Ladner, C. M.; Heitz, J. R. *Comp. Biochem. Physiol.* **1975**, 51C, 117-121.
9. Callaham, M. F.; Broome, J. R.; Lindig, O. H.; Heitz, J. R. *Environ. Entomol.* **1975**, 4, 837-841.
10. Weaver, J. E.; Butler, L.; Yoho, T. P. *Environ. Entomol.* **1976**, 5, 840.
11. Lavialle, M.; Dumortier, B. *C. R. Hebd. Seances Acad. Sci.* **1978**, 287, 875-878.
12. Fondren, J. E., Jr.; Heitz, J. R. *Environ. Entomol.* **1978**, 7, 843-846.
13. Pimprikar, G. D.; Norment, B. R.; Heitz, J. R. *Environ. Entomol.* **1979**, 9, 856-859.
14. Creighton, C. S.; McFadden, T. L.; Schalk, J. M. *J. Georgia Entomol. Soc.* **1980**, 15, 66-68.
15. Clement, S. L.; Schmidt, R. S.; Szatmari-Goodman, G.; Levine, E. *J. Econ.Entomol.* **1980**, 73, 390-393.
16. Krasnoff, S. B.; Sawyer, A. J.; Chapple, M.; Chock, S.; Reissig, W. H. *Environ. Entomol.* **1994**, 23, 738-743.
17. Liquido, N. J. Personal communication, 1994.
18. Mangan, R. L. Personal communication, 1994.
19. Kruse, K. Personal communication, 1995.
20. Gollnick, K.; Schenck, G. O. *Pure Appl. Chem.* **1964**, 9, 507-525.
21. Broome, J. R.; Callaham, M. F.; Heitz, J. R. *Environ. Entomol.* **1975**, 4, 883-886.
22. Fondren, J. E., Jr.; Norment, B. R.; Heitz, J. R. *Environ. Entomol.* **1978**, 7, 205-208.
23. Fondren, J. E., Jr.; Heitz, J. R. *Environ. Entomol.* **1978**, 7, 891-894.

24. Fondren, J. R., Jr.; Heitz, J .R. *Environ. Entomol.* **1979**, 8, 432-436.
25. Pimprikar, G. D.; Noe, B. L.; Norment, B. R.; Heitz, J. R. *Environ. Entomol.* **1980**, 9, 785-788.
26. Gollnick, K. *Advan. Photochem.* **1968**, 6, 1-122.
27. *Iscotables.* 3rd edition, **1970**, 33-35.
28. Carpenter, T. L.; Heitz, J. R. *Environ. Entomol.* **1980**, 9, 533-537.
29. Fairbrother, T. E.; Essig, H. W.; Combs, R. L.; Heitz, J. R. *Environ. Entomol.* **1981**, 10, 506-510.
30. Sakurai, H.; Heitz, J. R. *Environ. Entomol.* **1982**, 11, 467-470.
31. Carpenter, T. L.; Mundie, T. G.; Ross, J. H.; Heitz, J. R. *Environ. Entomol.* **1981**, 10, 953-955.
32. Carpenter, T. L.; Johnson, L. H.; Mundie, T. G.; Heitz, J. R. *J. Econ. Entomol.* **1984**, 77, 308-312.
33. Crounse, N.; Heitz, J. R. U. S. Patent 4 320 140, 1982.
34. Pimprikar, G. D.; Heitz, J. R. *J. Miss. Acad. Sci.* **1984**, 29, 77-80.
35. Carpenter, T. L.; Respicio, N. C.; Heitz, J. R. *Environ. Entomol.* **1984**, 1366-1370.
36. Respicio, N. C.; Carpenter, T. L.; Heitz, J. R. *J. Econ. Entomol.* **1985**, 78, 30-34.
37. Respicio, N. C.; Heitz, J. R. *J. Econ. Entomol.* **1983**, 76, 1005-1008.
38. Respicio, N. C.; Heitz, J. R. *J. Econ. Entomol.* **1986**, 79, 315-317.
39. Heitz, J. R. In *Insecticide Mode of Action*; Coats, J. R., Ed.; Academic Press, New York, NY., **1982**, 429-457.
40. Webb, J. M., Fonda, M., Brouwer, E. A. *Am. J. Pharmacol. Exptl. Therap.* **1962**, 137, 141-147.
41. *Federal Register*, **1982**, 47, 42566-42569.
42. Duggan, R. E.; Corneliussen, P. E. *Pesticide Monit. J.* **1978**, 5, 331-341.
43. Bergsten, D.A. Personal communication, 1995.
44. Heitz, J. R.; Wilson, W. W. In *Disposal and Decontamination of Pesticides*; Kennedy, M.V., Ed.; ACS Press, Washington, D.C., **1978**, 35-48.
45. Peeples, W. A., II,; Heitz, J. R. In *Biological/Biomedical Applications of Liquid Chromatography II*; Hawk, G. L., Ed.; Marcel Dekker, Inc, New York, NY, **1979**, 437-449.
46. Wilson, W. W.; Heitz, J. R.; *J. Ag. Food Chem.* **1984**, 32, 615-617.
47. Putsche, F. W., Jr. Personal communication, **1993**.
48. Putsche, F. W., Jr. U. S. Patent 5 310 725, **1994**.
49. Rebeiz, C. A.; Montazer-Zouhoor, A.; Hopen, H. J.; Wu, S. M. *Enzyme Microb. Technol.* **1984**, 6, 390-401.
50. Rebeiz, C. A.; Nandihalli, U. B.; Reddy, K. N. In *Topics in Photosynthesis-Volume 10, Herbicides;* Baker, N. R.; Percival, M. P., Eds.; Elsevier, Amsterdam, **1991**, 173-208.
51. Rebeiz, C. A.; Reddy, K. N.; Nandihalli, U. B.; Velu, J. *Photochem.Photobiol.* **1990**, 52, 1099-1117.
52. Rebeiz, C. A.; Juvik, J. A.; Rebeiz, C. C. *Pest. Biochem. Physiol. 1988*, 30, 11-27.

53.   Rebeiz, C. A.; Juvik, J. A.; Rebeiz, C. C.; Bouton, C. E.; Gut, L. J. *Pest. Biochem. Physiol.* **1990**, 36, 201-207.
54.   Gut, L. J.; Lee, K.; Juvik, J. A.; Rebeiz, C. C.; Bouton, C. E.; Rebeiz, C. A. *Pest. Sci.* **1993**, 39, 19-30.
55.   Downum, K. R. New Phytol. **1992**, 122, 401-420.
56.   Daub, M. E. *Phytopathology* **1982**, 72, 370-374.
57.   Daub, M. E.; Ehrenshaft, M. *Physiol. Plant.* **1993**, 89, 227-236.
58.   Daub, M. E. *Phytopathology* **1987**, 77, 1515-1520.
59.   Daub, M. E.; Leisman, G. B.; Clark, R. A.; Bowden, E. F. *Proc. Natl. Acad. Sci. USA* **1992**, 89, 9588-9592.
60.   Foote, C. S. In *Free Radicals in Biology*; Hedin, P.A., Ed.; Vol II, Academic Press, New York, NY., **1976**, 85-133.
61.   Krinsky, N. I. *Pure Appl. Chem* **1979**, 51, 649-660.
62.   Bellus, D. *Adv. Photochem.* **1979**, 11, 105-205.
63.   Truscott, T. G. *Photochem. Photobiol.* **1990**, 6, 359-371.
64.   Daub, M. E.; Payne, G. A. *Phytopathology* **1989**, 79, 180-185.
65.   Leisman, G. B.; Daub, M. E. *Photochem. Photobiol.* **1992**, 55, 373-379.
66.   Wat, C. -K.; Prasad, S. K.; Graham, E. A.; Partington, S.; Arnason, T.; Towers, G. H. N. *Biochem. Syst. and Ecol.* **1981**, 9, 59-62.
67.   Arnason, T.; Swain, T.; Wat, C. -K.; Graham, E. A.; Partington, S.; Towers, G. H. N.; Lam, J. *Biochem. Syst. and Ecol.* **1981**, 9, 63-68.
68.   Reyftmann, J. P.; Kagan, J.; Santus, R.; Morliere, P. *Photochem. Photobiol.* **1985**, 41, 1-7.
69.   Iyengar, S.; Arnason, J. T.; Philogene, B. J. R.; Morand, P.; Werstiuk, N. H.; Timmins, G. *Pest. Biochem. Physiol.* **1987**, 29, 1-9.
70.   Iyengar, S.; Arnason, J. T.; Philogene, B. J. R.; Werstiuk, N. H.; Morand, P. *Pest. Biochem. Physiol.* **1990**, 37, 154-164.
71.   Feng, R.; Houseman, J. G.; Downe, A. E. R.; Arnason, J. T. *J. Chem. Ecol.* **1993**, 19, 2047-2054.

RECEIVED July 19, 1995

Chapter 2

Fullerenes as Photosensitizers

Christopher S. Foote

Department of Chemistry and Biochemistry, University of California, Los Angeles, CA 90095–1569

> Photodynamic action is the action of a sensitizer, light and oxygen on biological materials. It can be mediated by electron or hydrogen-atom transfer or by singlet oxygen. Fullerenes (C_{60} and C_{70}) are excellent photosensitizers for both types of reaction, as are derivatives, dihydrofullerenes (DHFs). A DHF bound to a deoxynucleotide and hybridized to complementary DNA damages guanines near the DHF, but the reaction appears to involve electron transfer rather than singlet oxygen.

Light and oxygen are toxic to all organisms to some extent. Certain anthropogenic or naturally-occurring compounds that absorb light can "sensitize" organisms to photochemical damage by oxygen; this process is called photodynamic action.(1-3) Photodynamic damage is caused by oxidation of biological target molecules, and can lead to membrane lysis by oxidation of unsaturated fatty acids and cholesterol, enzyme deactivation by oxidation of amino acids (methionine, histidine, tryptophan, tyrosine and cyst(e)ine), and oxidative destruction of nucleic acid bases (primarily guanine). Photodynamic effects in humans include photosensitive porphyrias, drug photosensitivity, and photoallergies. Some aspects of aging of sun-exposed skin, cataract induction, and some types of photochemically induced mutations may also result from similar mechanisms.(2-4)

Photodynamic sensitizers are also used in medicine: an example is the use of hematoporphyrin derivatives in tumor phototherapy, which has recently received approval for clinical use in Canada.(5) Second-generation photosensitizers such as phthalocyanines(6) or benzoporphyrins(7) that absorb in the far-red or near-infrared region are receiving increasing attention because of the ability of light of these wavelengths to penetrate tissue. Particular interest has recently been aroused in the use of photodynamic pigments as antiviral agents, for example for selective killing of HIV virus in blood.(8-10) Interest in photodynamic photosensitizers as pesticides and herbicides has also increased in recent years (see other papers in this volume).(4)

Photodynamic Action. Photodynamic action begins when a sensitizer (Sens) absorbs light, giving an excited state (Sens*), often (but not always), the triplet. Sens* can either react directly with the substrate (Type I reaction) or with oxygen (Type II reaction).(11,12) The Type I reaction results in hydrogen atom or electron transfer, yielding radicals or radical ions. The Type II reaction leads mainly to singlet molecular oxygen by energy transfer.

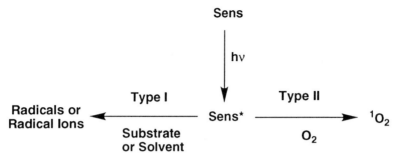

The factors that control the competition between the Type I and Type II processes are now reasonably well understood.(12,13) High oxygen concentrations favor Type II reactions, while high substrate concentrations promote Type I. In biological systems, where binding of sensitizer to substrate is common, Type I reactions are particularly favorable. Electron-rich or hydrogen-atom-donating substrates also favor Type I reactions. Sensitizer characteristics also affect the competition between Type I and Type II processes. For example, ketones favor hydrogen abstraction, while electron-poor sensitizers such as cyanoaromatics and fullerenes are particularly prone to electron transfer.

Singlet oxygen can be produced in high yield in the Type II reaction by energy transfer from Sens*.(14,15) It is an electronically excited state of oxygen, with a lifetime that varies from about 3-4 μs in water to as long as 0.1 s in solvents with no hydrogen atoms.(15-19) In biological lipids and membranes, it probably has a lifetime considerably shorter than that in most organic solvents (50-100 ms) because of quenching by proteins and escape from the membrane into the cytosol.(20)

Reactions of substrates with singlet oxygen include Diels-Alder reaction of dienes to form endoperoxides and ene reaction of alkenes to give allylic hydroperoxides. In addition, singlet oxygen reacts with electron-rich alkenes to form 1,2-dioxetanes, with sulfides to form sulfoxides (via a reactive intermediate with a S-O-O bond) and with electron-rich phenols to form hydroperoxydienones.(21-23)

Singlet oxygen emits at 1270 nm, corresponding to the energy of the singlet-to-triplet oxygen transition, 22 kcal/mol.(24) This "monomol" luminescence at 1270 nm is highly forbidden, and therefore extremely weak, but can be used for direct measurement of both the absolute amount of singlet oxygen produced and its lifetime.*(25)* This luminescence is a direct measure of the amount of singlet oxygen produced, and can lead to confident identification and quantitation.

$$^1O_2 \longrightarrow {}^3O_2 + h\nu \; (1270 \text{ nm})$$

Fullerenes. We have reported the basic photophysical properties of C_{60} and C_{70}.*(26,27)* These studies have since been extended by many other groups.*(28-33)* C_{60} and C_{70} both give very high yields of the triplet state on irradiation. The triplet-triplet absorption spectra of C_{60} and C_{70} were measured and the extinction coefficients estimated; others have since refined them.*(28-30,32,34)* The energy levels of the triplets (E_T) were estimated by triplet-triplet energy transfer to lie near 35 kcal/mole. The quantum yield of singlet oxygen formation is nearly quantitative for C_{60} and slightly lower for C_{70}. These values are also lower limits for the quantum yield of triplet.*(26,27,35-37)*

The reduction potential of $^3C_{60}$ is higher than that of the ground state by the amount of the triplet energy, *(38,39)* and we expected that the triplet should be reduced by electron transfer from electron donors of lower oxidation potential. Quenching of $^3C_{60}$ by electron donors occurs efficiently, by electron transfer, as shown by the production of a prominent transient with maxima at 950 and 1075 nm, which was assigned to the C_{60} radical anion,*(40)* in good agreement with contemporary and subsequent reports*(32,41-44)* as well as the absorption from the donor radical cations.

$$\text{D} + {}^3\text{C}_{60} \longrightarrow \text{D}^{\bullet+} + \text{C}_{60}^{\bullet-}$$

In summary, C_{60} and C_{70} are excellent photosensitizers in both Type II and Type I reactions. They are particularly useful in cases where unreactive substrates are to be oxidized because they quench singlet oxygen at a rate constant one to two orders of magnitude slower than dyes and porphyrins, and they do not react with it at an appreciable rate. They are also only weakly fluorescent, which makes them excellent choices for photophysical studies where small amounts of light are to be detected.

Functionalization. The parent fullerenes are poorly soluble in nonpolar solvents, and not at all in polar solvents, limiting their usefulness as sensitizers in photobiology. Many groups have attached functional groups by various routes which allow greatly improved solubility in various media.*(45-50)*

In particular, Rubin et al. recently demonstrated a simple, high-yield route to dihydrofullerenes (DHFs) via Diels-Alder addition of electron-rich dienes to C_{60}.*(50)* This route allowed the preparation of adducts such as the one below. Dihydrofullerenes all have strong absorption throughout the visible and a weak long-wavelength absorption near 700 nm. Pulsed laser irradiation of both the alcohol and the ketone gave strong transient triplet-triplet absorption spectra. Both spectra are very similar and are also similar to that of the triplet from C_{60}. Both compounds gave singlet oxygen in high yields. The yield of singlet oxygen (which is a lower limit for the quantum yield of formation of the triplet) from the alcohol is $0.84 \pm .04$ at 532 nm and 0.72 ± 0.03 at 355 nm.*(51)*

As mentioned above, triplet excited C_{60} has a reduction potential near +1.6 V, and is readily photoreduced by amines and other donors to C_{60} radical anion and the donor radical cations.(*40*) We expected that this reaction might lead to adducts with covalent bonds. Such adducts are formed with some amines in ground-state chemistry,(*52*) but the photochemical process should be more selective and easily controlled, since only one-electron reduction is possible in the photochemical process. The reduction potential of the triplet is actually high enough that electron-transfer from many donors such as electron-donor-substituted aromatics and alkenes should be possible. This should lead to adducts such as cyclobutanes or other products, depending on the reagents.

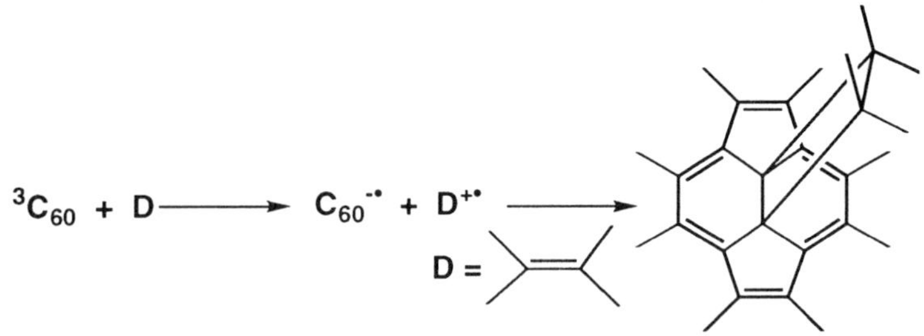

$$^3C_{60} + D \longrightarrow C_{60}^{-\bullet} + D^{+\bullet} \longrightarrow$$

$$D = \text{[structure]}$$

We reacted C_{60} and C_{70} with several different ynamines, which are excellent electron donors. Although the mechanism of the reaction is not yet certain, cyclobutenamine adducts are formed in all cases in > 50% yield.(*53,54*) The enamines are unique in that they also have a photosensitizer in the same molecule, and brief exposure to air and room light or chromatography in the presence of light and oxygen leads to cleavage of the enamine double bond, producing the ketoamides shown below; some of these can undergo interesting further chemical transformations.

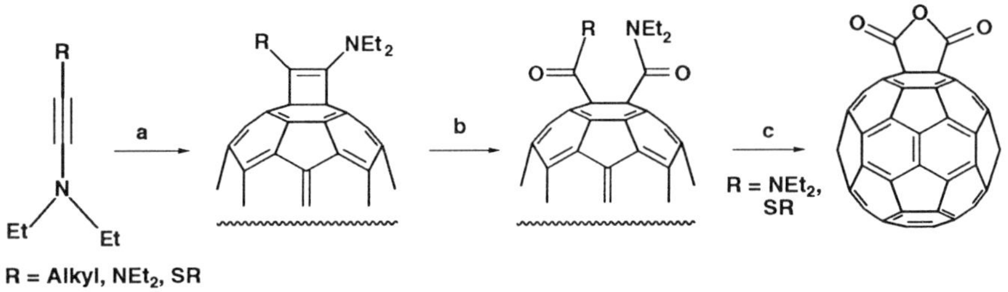

R = Alkyl, NEt$_2$, SR

a. C_{60}, hυ or standing, $C_6H_5CH_3$; b. hυ, O_2, >95%; c. TsOH in $C_6H_5CH_3$

We also used Diels-Alder routes to prepare derivatives in good yield.(*55*)

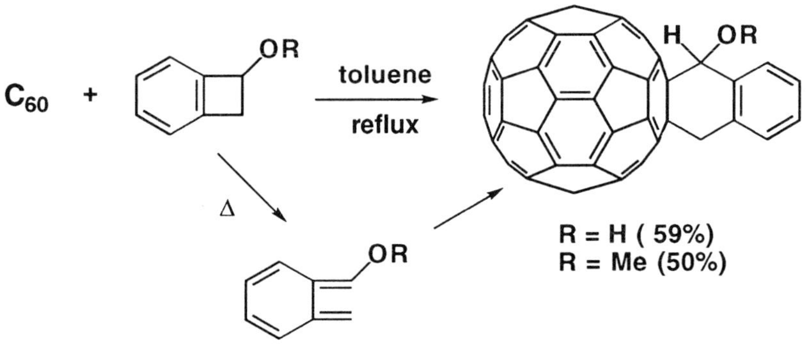

R = H (59%)
R = Me (50%)

Nucleotide linkage. A DHF-linked deoxyoligonucleotide was prepared and hybridized with a DNA strand containing a complementary sequence (see below).(*56*)

This reaction system positions the DHF near both single and double-stranded guanosines and permits comparison of the reactivity of guanosines in these two environments. On irradiation in the presence of air, base-labile sites are created, indicating that guanosines are damaged. Only guanosines in single-stranded regions and only those directly touching or adjacent to the DHF are damaged. The residue which is not directly touching the DHF, but only two bases away is signifiicantly less affected. However, in contrast to suggestions in recent reports,(57,58) the modification does not involve $^1O_2$. Initial evidence suggests that DNA damage is caused by single electron-transfer between guanosine and $^3C_{60}$. This conclusion follows from the following control experiments: a) An eosin linked to the same deoxynucleotide sequence causes similar damage when hybridized and irradiated, but with different specificity (guanosines farther from the sensitizer are damaged, suggesting the intermediate is more diffusible). b) Singlet oxygen quenchers inhibit the reaction with the eosin derivative, but not the DHF. c) Carrying out the reaction in D_2O, which promotes singlet oxygen reactions by increasing its lifetime promotes the eosin reaction but not that with the DHF, which is nearly unaffected by reagents that modify singlet oxygen lifetime.

R = 3'-TCTACGATGTGTTAATCCGAACATGTATAACAGCAATC

0.1 M NaHCO3, 37 °C

cleavage sites
(damage proportional to length of arrow):
5'-AGGGTCTGCTTCAGTAAGCCAGATGCTACACAATTAGGCTTGTACATATTGTCGTTAGAACGCGGCTACAATTAATACAT-3'
3'-TCTACGATGTGTTAATCCGAACATGTATAACAGCAATCT

Summary. These examples of functionalization suggest the rich organic chemistry of the fullerenes that is waiting to be exploited. All of the dihydrofullerene derivatives so far investigated have proven to be excellent photosensitizers, and, since their solubility properties can be controlled by suitable modifications of the functional groups, we believe they may prove to be outstanding photodynamic sensitizers. All fullerenes and dihydrofullerenes investigated share several particular advantages. All give very high yields of singlet oxygen. In addition (and in contrast to most commonly used sensitizing dyes and porphyrins), they have very low rates of quenching of singlet oxygen by the sensitizer and are also almost inert to destruction by singlet oxygen, so that they will be relatively persistent and small amounts can produce relatively large total amounts of singlet oxygen.

Acknowledgments. This work was funded by NSF and NIH grants. The work was carried out by the undergraduate and graduate students and postdoctoral associates credited in the references. The deoxynucleotide work was a collaboration with groups of my colleagues, Yves Rubin and David Sigman.(*56*)

Literature Cited

1. Blum, H. *Photodynamic Action And Diseases Caused By Light*; Reinhold: New York, 1941.
2. Spikes, J. D.; Straight, R. C. In *Light-activated Pesticides*; Heitz, J. R. Downum, K. R., Eds.; American Chemical Society: Washington, D.C., 1987; pp 98-108.
3. Straight, R. C.; Spikes, J. D. In *Singlet O_2*; Frimer, A. A., Ed.; CRC Press: Boca Raton FL, 1985; pp 85-144.
4. *Light-Activated Pesticides;* Heitz, J. R.; Downum, K. R. Eds.; American Chemical Society: Washington, D. C., 1987; Vol. 339.
5. Henderson, B. W.; Dougherty, T. J. *Photochem. Photobiol.* **1992**, *55,* 145-157.
6. Paquette, B.; Boyle, R. W.; Ali, H.; MacLennan, A. H.; Truscott, T. G.; van Lier, J. E. *Photochem. Photobiol.* **1991**, *53,* 323-327.
7. Allison, B. A.; Waterfield, E.; Richter, A. M.; Levy, J. G. *Photochem. Photobiol.* **1991**, *54,* 709-715.
8. North, J.; Neyndorff, H.; Levy, J. G. *J. Photochem. Photobiol. B: Biol.* **1993**, *17,* 99-108.
9. Hudson, J. B.; Harris, L.; Towers, G. H. N. *Antiviral Res.* **1993**, *20,* 173-178.
10. Meruelo, D.; Lavie, G.; Lavie, D. *Proc. Nat. Acad. Sci. (U. S.)* **1988**, *85,* 5230-5234.
11. Foote, C. S. In *Light Activated-Pesticides*; Heitz, J. R. Downum, K. R., Eds.; American Chemical Society: Washington DC, 1987; pp 22-38.
12. Foote, C. S. In *Free Radicals in Biology*; Pryor, W. A., Ed.; Academic Press: New York, 1976; pp 85-133.
13. Foote, C. S. *Accts. Chem. Res.* **1968**, *1,* 104-110.
14. Wilkinson, F.; Helman, W. P.; Ross, A. B. *J. Phys. Chem. Ref. Dat.* **1993**, *22,* 113-262.
15. Gorman, A. A.; Rodgers, M. A. J. In *CRC Handbook of Organic Photochemistry*; Scaiano, J. C., Ed.; CRC Press: Boca Raton, FL, 1989; pp 229-247.
16. Wilkinson, F.; Brummer, J. G. *J. Phys. Chem. Ref. Dat.* **1981**, *10,* 809-1000.
17. Rodgers, M. A. J. *Photochem. Photobiol.* **1983**, *37,* 99-103.
18. Schmidt, R.; Brauer, H. D. *J. Am. Chem. Soc.* **1987**, *109,* 6976-6981.
19. Foote, C. S.; Clennan, E. L. In *Active Oxygen in Chemistry*; Foote, C. S.; Valentine, J. S.; Greenberg, A. Liebman, J. F., Eds.; Chapman and Hall, Inc.: London, 1995; pp 105-140.
20. Kanofsky, J. R. *Photochem. Photobiol.* **1991**, *53,* 93-99.
21. *Singlet O_2*; Frimer, A. A., Ed.; CRC Press: Boca Raton, FL, 1985; Vol. 1.
22. *Singlet Oxygen*; Wasserman, H. H. Murray, R. W., Eds.; Academic Press: New York, 1979.
23. Gollnick, K. *Advan. Photochem.* **1968**, *6,* 1-122.
24. Khan, A. U.; Kasha, M. *Proc. Nat. Acad. Sci.* **1979**, *76,* 6047-6049.
25. Foote, C. S. In *Photosensitisation, NATO ASI Series H: Cell Biology*; Moreno, G.; Pottier, R. H. Truscott, T. G., Eds.; Springer Verlag: Heidelberg, 1988; pp 125-144,.
26. Arbogast, J. W.; Darmanyan, A. O.; Foote, C. S.; Rubin, Y.; Diederich, F. N.; Alvarez, M. M.; Anz, S. J.; Whetten, R. L. *J. Phys. Chem.* **1991**, *95,* 11-12.
27. Arbogast, J.; Foote, C. S. *J. Am. Chem. Soc.* **1991**, *113,* 8886-8889.
28. Ebbesen, T. W.; Tanigaki, K.; Kuroshima, S. *Chem. Phys. Lett.* **1991**, *181,* 501-4.
29. Tanigaki, K.; Ebbesen, T. W.; Kuroshima, S. *Chem. Phys. Lett.* **1991**, *185,* 189-192.
30. Sension, R. J.; Phillips, C. M.; Szarka, A. Z.; Romanow, W. J.; McGhie, A. R.; McCauley Jr., J. P.; Smith III, A. B.; Hochstrasser, R. M. *J. Phys. Chem.* **1991**, *95,* 6075-6078.
31. Kajii, Y.; Nakagawa, T.; Suzuki, S.; Achiba, Y.; Obi, K.; Shibuya, K. *Chem. Phys. Lett.* **1991**, *181,* 100-104.

32. Biczok, L.; Linschitz, H.; Walter, R. I. *Chem. Phys. Lett.* **1992**, *195,* 339-346.
33. Leach, S.; Vervloet, M.; Despres, A.; Bréheret, E.; Hare, J. P.; Dennis, T. J.; Kroto, H. W.; Taylor, R.; Walton, D. R. M. *Chem. Phys.* **1992**, *160,* 451-466.
34. Bensasson, R. V.; Hill, T.; Lambert, C.; Land, E. J.; Leach, S.; Truscott, T. G. *Chem. Phys. Lett.* **1993**, *206,* 197-202.
35. Hung, R. R.; Grabowski, J. J. *Chem. Phys. Lett.* **1992**, *192,* 249-253.
36. Hung, R. R.; Grabowski, J. J. *J. Phys. Chem.* **1991**, *95,* 6073-6075.
37. Terazima, M.; Hirota, N.; Shinohara, H.; Saito, Y. *J. Phys. Chem.* **1991**, *95,* 6490-6495.
38. Rehm, D.; Weller, A. *Z. Phys. Chem. N. F.* **1970**, *69,* 183-200.
39. Mattes, S. L.; Farid, S. In *Organic Photochemistry*; Padwa, A., Ed.; M. Dekker, Inc.: New York, 1983; pp 233-326.
40. Arbogast, J. W.; Foote, C. S.; Kao, M. *J. Am. Chem. Soc.* **1991**, *114,* 2277-2279.
41. Sension, R. J.; Szarka, A. Z.; Smith, G. R.; Hochstrasser, R. M. *Chem. Phys. Lett.* **1991**, *185,* 179-183.
42. Kato, T.; Kodama, T.; Shida, T.; Nakagawa, T.; Matsui, Y.; Suzuki, S.; Shiromaru, H.; Yamauchi, K.; Achiba, Y. *Chem. Phys. Lett.* **1991**, *180,* 446-50.
43. Kato, T.; Kodama, T.; Oyama, M.; Okazaki, S.; Shida, T.; Nakagawa, T.; Matsui, Y.; Suzuki, S.; Shiromaru, H.; Yamauchi, K.; Achiba, Y. *Chem. Phys. Lett.* **1992**, *186,* 35-39.
44. Greaney, M. A.; Gorun, S. M. *J. Phys. Chem.* **1991**, *95,* 7142-4.
45. Wudl, F.; Hirsch, A.; Khemani, K. C.; Suzuki, T.; Allemand, P. M.; Koch, A.; Eckert, H.; Srdanov, G.; Webb, H. M. *ACS Symposium Ser.* **1992**, *481,* 161-75.
46. Wudl, F. *Acc. Chem. Res.* **1992**, *25,* 157-161.
47. Tebbe, F. N.; Harlow, R. L.; Chase, D. B.; Thorn, D. L.; Campbell Jr., G. C.; Calabrese, J. C.; Herron, N.; Young Jr., R. J.; Wasserman, E. *Science* **1992**, *256,* 822-825.
48. Taylor, R.; Langley, J.; Meidine, M. F.; Parsons, J. P.; Abdul-Sada, A. K.; Dennis, T. J.; Hare, J. P.; Kroto, H. W.; Walton, D. R. M. *J. Chem. Soc. Chem. Comm.* **1992**, 667-668.
49. Vasella, A.; Uhlmann, P.; Waldraff, C. A. A.; Diederich, F.; Thilgen, C. *Angew. Chem. Int. Ed.* **1992**, *31,* 1388-1390.
50. An, Y.-Z.; Anderson, J. L.; Rubin, Y. *J. Org. Chem.* **1993**, *58,* 4799-4801.
51. Anderson, J. L.; An, Y.-Z.; Rubin, Y.; Foote, C. S. *J. Am. Chem. Soc.* **1994**, *116,* 9763-9764.
52. Wang, Y. *J. Phys. Chem.* **1992**, *96,* 764-767.
53. Zhang, X.; Romero, A.; Foote, C. S. *J. Am. Chem. Soc.* **1993**, *115,* 11924-11025.
54. Zhang, X.; Foote, C. S. *J. Am. Chem. Soc.* **1995**, *117,* 4271-4275.
55. Zhang, X.; Foote, C. S. *J. Org. Chem.* **1994**, *59,* 5235-5238.
56. An, Y.-Z.; Chen, C.-H. B.; Anderson, J. A.; Sigman, D. S.; Foote, C. S.; Rubin, Y. *Tetrahedron* submitted.
57. Tokuyama, H.; Yamago, S.; Nakamura, E.; Shiraki, T.; Sugiura, Y. *J. Am. Chem. Soc.* **1993**, *115,* 7918-7919.
58. Boutorine, A. S.; Tokuyama, H.; Takasugi, M.; Isobe, H.; Nakamura, E.; Hélène, C. *Angew. Chem. Int. Ed. Engl.* **1994**, *33,* 2462-2465.

RECEIVED August 24, 1995

Chapter 3

Mechanisms of Cellular Photomodification

Dennis Paul Valenzeno and Merrill Tarr

Department of Physiology, University of Kansas Medical Center,
3901 Rainbow Boulevard, Kansas City, KS 66160–7401

An understanding of organismal photosensitization begins with an understanding of sensitized photomodification of cells. Cellular photomodification consists of: 1) transport of the photosensitizer to the site of action, 2) possible binding, aggregation or metabolism, 3) absorption of light at the site, 4) production of energetic intermediate states, 5) reaction with cellular biomolecules, and 6) modification of cellular function. Proteins, lipids, and nucleic acids are susceptible targets, but since many important photosensitizers are lipophilic, membranes are important cellular targets including those of lysosomes and mitochondria. Photosensitized cell death can be either necrotic or apoptotic. An increase in intracellular calcium may trigger either mechanism. We present patch clamp electrophysiological and biochemical results demonstrating that although standard protein channels, including calcium channels, are blocked by photosensitization, a new permeability pathway is induced in cardiac cell membranes. This pathway is selective for cations, shares many properties with the pathway that produces sensitized lysis of erythrocytes, and shows isolated patch currents that suggest that it may be a lipid, rather than a protein, channel.

The responses of living organisms to photosensitized modification are varied, ranging from mild irritation to death. The precise nature of the response depends on many factors, but the majority of cases can be explained based on the changes produced in individual cells. Without question there are tissue-level and organismal responses that depend on extracellular or intercellular events. Nonetheless, an understanding of even these must be rooted in a fundamental understanding of cellular photomodification, the focus of this chapter.

Mechanistic Steps for Effective Cellular Photosensitization

Photosensitized cellular modification, which we refer to as cellular photomodification, is a multi-step process that is schematized in Fig. 1. The figure identifies six steps in the overall process. First, the photosensitizer must get to the target cell (step 1, Fig. 1). This involves introduction into the organism during feeding, via injection or diffusion.

0097–6156/95/0616–0024$12.00/0

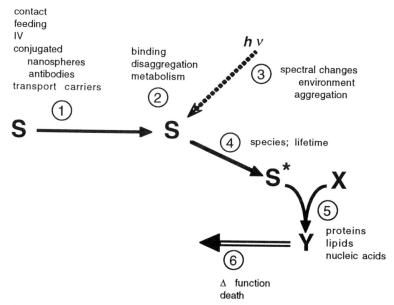

Figure 1. Mechanistic steps leading to photosensitized effects. Six steps are defined starting with contact between the photosensitizer and organism and culminating in altered function or death.

Once in the organism the photosensitizer may be "carried" in combination with proteins, lipoproteins, or other molecules. At the target cell it may need to diffuse through a vascular wall, a cell wall and/or a cell membrane. Ultimately it must localize at or near the site of photomodification. Variations in any of these processes can significantly alter the final effect.

Once the sensitizer has reached its site of action, it may remain free or may be bound to various biomolecules (step 2, Fig. 1). It can also be metabolized to a form that may be more effective or less effective as a photosensitizer. The use of precursors, such as δ-aminolevulinic acid which is metabolized by the organism to a photosensitizing porphyrin, is an example of this possibility *(e.g. 1,2)*.

At the site of action the sensitizer must be able to absorb light (step 3, Fig. 1). This ability can be altered by the environment and physical state of the sensitizer in the organism. Binding, aggregation, metabolism and altered dielectric constant (as in a lipid membrane) all may significantly alter the absorption spectrum of the sensitizer.

After absorbing light, the sensitizer must make use of the energy in a way that effects a change in the cell (step 4, Fig. 1). It may transfer energy to oxygen creating an electronically excited state such as singlet oxygen, which can then react with a cellular structure nearby. Alternately, it can react directly with cellular biomolecules. But, in the cellular environment, it may also be quenched by the impressive array of antioxidants available in cells. It is by no means justifiable to assume that the excited state properties and reactivities of photosensitizers *in vivo* will be the same as they are in simple solution *(3)*.

The next step in the process of cellular photomodification is reaction of either the excited state sensitizer or a reactive intermediate, such as singlet oxygen, with cellular biomolecules which critically impair cell function and/or survival (step 5, Fig. 1). As we shall see below, many biomolecules and cellular structures are photomodifiable, but not all of these affect cells in ways that result in cell death or irreparable damage. For

example, it has been known for many years that cell lipids are peroxidized by photosensitized modification, but a definitive link between that peroxidation and cell killing has remained elusive. Finally, if enough critical cells in an organism are affected, then the foregoing steps will result in death or impairment of function (step 6, Fig. 1).

While each of these steps is critical and can significantly affect the degree of cellular modification, in the following we will focus mainly on step 5, namely the reactions of photosensitizer and light which lead to critical impairment of cell function.

What are the Effects of Photosensitizers and Light on Cells?

Photosensitizers and light can affect a variety of biomolecules and cellular structures *(for a more complete review see 4).* Proteins, lipids and nucleic acids are all susceptible to photomodification. Carbohydrates are much less sensitive. This means that direct effects on cellular energy stores in the form of glucose or glycogen are unlikely to be significant. Conversely, effects on enzymes, nuclear DNA and lipid membranes may be important. In fact there are well documented examples of effects on all of these in the literature. However, it is also apparent from the literature that the photosensitizers used most commonly, and those that are most effective for cellular sensitization, are those that have a high lipid solubility. In fact, many very effective photosensitizers are not able to penetrate to the nucleus in sufficient concentration to sensitize photomodification of DNA. Several sensitizers, are accumulated in lysosomes where they may sensitize swelling and increased permeability of the lysosomal membrane allowing enzymes to be released from the lysosome. Other sensitizers seem to localize in mitochondria where they cause swelling and can affect the function of membrane-bound enzymes involved in energy production for the cell. Finally, a large variety of sensitizers, including many of the porphyrins and xanthene dyes, localize in the plasma membrane (and perhaps other membranes) altering the permeability properties of the surface membrane of the cell when illuminated, either by affecting membrane proteins or lipids.

How do Photosensitizers and Light Kill Cells?

Obviously, even this brief consideration of the effects of photosensitizers and light on cells offers many possible means by which a cell may be killed. Lysosomes may release degradative enzymes into the cytoplasm; mitochondrial production of ATP may be inhibited; plasma membrane barrier function may be compromised; and DNA damage may prevent cell replication and/or impair transcription required for protein synthesis and normal cell function. But, which of these effects is critically related to cell death, and which are merely consequences of the process of cell death?

Two Kinds of Cell Death

It is now recognized that there are two major forms of cell death. One form, referred to as necrotic death or necrosis, is a degenerative cell death resulting from severe cell injury. The second form of cell death, referred to as apoptotic death or apoptosis, is a programmed cell death, a cellularly-controlled process. It can occur under pathological conditions, such as cell injury, but also appears to be the mechanism by which organisms delete certain cell populations when they are no longer needed in the developmental process. A hallmark of apoptosis is a selective cleavage of DNA such that quantized fragments are produced. This results in a characteristic "ladder" pattern of fragments on gels that separate DNA according to molecular weight. By contrast, necrosis produces random DNA degradation having no identifiable pattern on such gels.

Increased Intracellular Calcium Triggers Cell Death

Interestingly, an increase in intracellular calcium concentration has been suggested as the trigger of both forms of cell death, apoptotic and necrotic *(5)*. Even cells which tolerate transient elevation of intracellular calcium concentration as part of their normal function, such as contracting cardiac cells where elevation of calcium serves as the trigger for contraction, cannot tolerate this increased concentration continually. Cardiac cells are well-known to be susceptible to destruction by calcium overload *(6)*.

Recently, Ben-Hur and Dubbelman have reviewed the evidence that cellular photosensitization produces cell death by an apoptotic mechanism *(5)*. They emphasized that although apoptosis is a carefully orchestrated series of reactions within the cell leading ultimately to death, calcium is the trigger for this process. Thus, we should look for mechanisms that can increase intracellular calcium concentrations as the seminal effect of photosensitization. Since cell membranes are responsible for maintaining the very low calcium concentrations inside normal cells, $\approx 10^{-7}$ M, one obvious way that cell calcium concentration can increase is by a change in membrane permeability to calcium. This has led us to study ion movement across the plasma membrane using sensitive electrophysiological techniques to assess the changes in permeability to calcium and other ions.

Electrophysiological Monitoring of Ion Movement

We monitor ion movement across cell membranes by measuring membrane electrical parameters using the patch clamp technique. Figure 2 shows the somewhat deceptively simple concept behind this approach. A glass capillary is drawn to a fine tip, about 1 micrometer in diameter, and filled with an electrolyte solution. When it is pressed against the plasma membrane and a gentle suction is exerted, the pipette seals to the membrane. If suction is maintained, the enclosed patch ruptures, establishing direct electrical connection between the inside of the cell and an electrode in contact with the solution in the shank of the pipette. This allows measurement of membrane potential and currents with respect to a reference electrode in the cell suspension medium. The simplest measurement that can be made with this system is to monitor the membrane potential of the cell. In cardiac cells, this potential is about -70 mV (inside negative), but spontaneously changes periodically to a positive value for a few hundred milliseconds. This change of potential is the cardiac *action potential* which is the signal for the heart to beat. It is produced by currents flowing through the cardiac cell membrane. There are protein channels in the membrane that are specific for different ions, including sodium, calcium and potassium. Studying the changes in the action potential produced by photomodification can provide clues regarding changes in the permeability of these ions through the cell membrane.

Simple Electrophysiology — Action Potentials

Figure 3 shows the effect of the photosensitizer rose bengal and light on the cardiac action potential. The trace labelled "0 sec" is the normal electrical pattern which is repeated at regular intervals, although we show only a single example here. When illumination begins, there is a progressive prolongation of this signal, the trace labelled "120 sec". After a brief time a maximum duration is reached and the action potential begins to shorten again, eventually becoming a brief spike, the trace labelled "180 sec". With continued illumination the action potential disappears entirely *(7)*.

This kind of data provides hints about what may be happening to ion channels. For example, it is well known that the rapid upstroke of the action potential results from the activity of sodium channels. These channels open rapidly and transiently. When they open, the membrane rapidly depolarizes, and the membrane depolarizes more rapidly when more sodium channels are activated. Conversely, it depolarizes more slowly when there are less activated sodium channels. Since the rate of rise of the

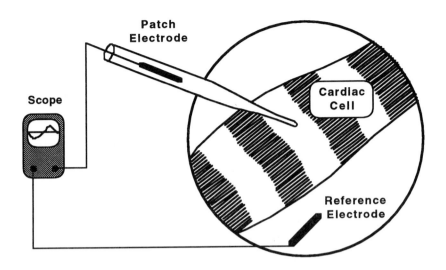

Figure 2. Schematic of a simple patch clamp electrophysiological system used to study membrane potential and ion currents. The scope is an oscilloscope or other device capable of recording rapidly changing voltages.

action potential is reduced by photomodification, as shown in Fig. 3, we might suspect that sodium channels are being blocked. Similarly, it is well known that potassium moving through potassium channels is largely responsible for the return of the membrane potential to its resting level at the end of an action potential. If the number of functional potassium channels is reduced, then the membrane potential returns to its resting level more slowly. Since Fig. 3 shows that photomodification prolongs the action potential, we might suspect that potassium channels are being blocked. But, this kind of measurement of membrane potential does not allow us to make any firm conclusions about what is happening to ion channels or currents. Fortunately, by manipulating the experimental conditions and by electronically controlling the membrane potential across the cell membrane, we can directly monitor specific ionic currents. In this way a more thorough understanding of the effects of photomodification on the action potential can be realized.

Changes in Ion Currents with Photosensitization

Figure 4, upper curve, is a plot of potassium current as a function of illumination time. Zero time is the start of illumination. Potassium current, which was stable in the presence of rose bengal before illumination, began to decrease with light. The loss of current follows an exponential time course and reflects block of potassium channels in the cardiac cell membrane. The time course of this loss of potassium current correlates very well with the time course of prolongation of the action potential, as suspected from the action potential data (8).

Similar studies of sodium and calcium currents (9) reveal that both of these currents are also blocked exponentially by photosensitizers and light with time courses similar to that for potassium (data not shown). In fact, all studies of ion channel photomodification to date, in a variety of cell types, have shown block of channels. Currents mediated by protein channels are always reduced; they never increase.

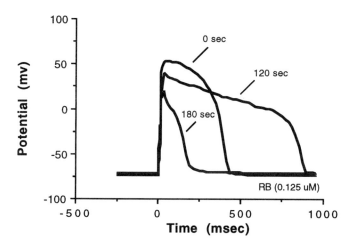

Figure 3. Photomodification of the cardiac action potential recorded from an isolated atrial cell of the frog heart. Cells were illuminated for the time indicated adjacent to each curve in the presence of 0.125 μM rose bengal.

Leak Current Induced by Photosensitizers and Light

On the other hand, there is a current that is not associated with any known membrane protein channel which behaves differently. It may have considerable significance in cell death in many cell types. Here it is significant because its effect is similar to that of potassium current. It acts to terminate action potentials. The lower trace in Figure 4 shows leak current measured as a function of time after illumination. Before light, leak current is small or nonexistent. It begins to increase promptly but relatively slowly with light, but after a time, when most potassium current is blocked, it begins to increase much more rapidly. This rapid phase of increase corresponds to the time when the action potential shortens again. The decrease in action potential duration shown in Fig. 3 at 180 seconds is due to the repolarizing influence of leak current induced by photomodification.

Leak current may reflect a change in membrane permeability that is significant for many cell types. Thus, we've spent a considerable amount of time studying it and trying to correlate it with the development of membrane leakiness in other cell types. Some of our results are summarized below where we've compared the properties of leak current to those of the increase in membrane permeability in red blood cells that leads to cell lysis.

- Although lysis of red cells takes minutes to develop, the permeability change that leads to lysis has been shown to develop rapidly and progressively in light in the presence of a photosensitizer *(10)*. Likewise, we've shown that leak current in cardiac cells begins to increase within 1 msec of the start of illumination and continues to increase in light *(11)*.

- In red cells the initial increase in permeability that leads to lysis is to small molecules only. Hemoglobin cannot penetrate these initial lesions *(12)*. Likewise, leak current is selective for small ions such as sodium and potassium.

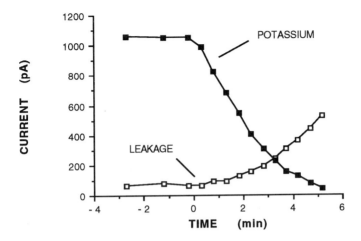

Figure 4. Photomodification of potassium and leakage currents recorded from an isolated atrial cell of the frog heart. Cells were illuminated for the time indicated on the abscissa in the presence of 0.125 μM rose bengal. Illumination intensity was 1.5 mW/cm². Adapted from *(8)* ; see *(8)* for details.

The larger tetraethylammonium ion, for example, is not as permeable through the leak current pathway *(11)*.

• Red cells lose potassium in the early stages of photohemolysis, and later gain sodium as they swell and lyse *(13)*. Similarly, the leak current pathway is initially fairly potassium selective, and later becomes progressively more permeable to sodium *(11)*.

• Finally, the development of lytic lesions in red cells and the development of leak current in cardiac cells are both relatively insensitive to extracellular calcium *(11 and unpublished results)*.

If the leak current that we monitor is a measure of the increase in membrane permeability that leads to photosensitized cell lysis, then it would be of interest to know if leak current is the result of lipid peroxidation or membrane protein modification. Since the other cardiac ionic currents that we measure are mediated by membrane proteins, block of them must ultimately involve modification of those proteins. However, leak current is not known to be mediated by membrane proteins. Could it be the result of lipid peroxidation?

Evidence for the Lipid Pore Theory

If this is to be a tenable hypothesis, we must first be able to show that lipid peroxidation of cardiac cell membranes occurs under the conditions which create leak current. Figure 5 shows HPLC chromatograms of cholesterol oxidation products extracted from cardiac cells. The trace at the left is a set of standards prepared from pure cholesterol. On this silica column cholesterol elutes at about 6.5 minutes. The 5α-hydroperoxide derivative, characteristic of singlet oxygen attack, is the large peak at about 26.5 minutes, and the

7- derivatives, characteristic of radical attack, are at 55 to 60 minutes. The middle trace shows the profile obtained from extracts of unilluminated isolated cardiac cells showing cholesterol, but virtually no oxidation products. The trace at the right is a sample extracted from cardiac cells illuminated in the presence of rose bengal for 10 minutes. There is a prominent peak at 26.5 minutes indicative of significant cholesterol oxidation by rose bengal and light. The peak identified at 37 minutes is likely to be the 6α-hydroperoxide product, although this has not been verified. So, lipid peroxidation does occur in these cells, but can it be responsible for the observed leak current?

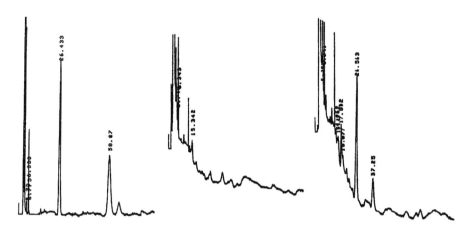

Figure 5. Absorbance at 215 nm of cholesterol standards and frog atrial cell extracts separated by HPLC. The left panel is a combined standard of 5α-, 7α-, and 7ß-cholesterol hydroperoxides. The middle panel is an extract (chloroform extraction) of untreated frog atrial cells. The right panel is an extract of another aliquot of the same preparation of cells that had been exposed to 1 μM rose bengal and illuminated at a light intensity of 0.75 mW/cm².

Isolated Patch Studies of Induced Leak Current

Isolated patch studies can be used to measure leak current flowing through a tiny patch of cardiac cell membrane, just the size of the tip of the patch pipette. This is done by sealing the electrode to the cell, but releasing the suction before the patch ruptures. In this way currents flowing just through the patch can be detected, and often current mediated by a single membrane protein can be detected, a so-called single channel current. If lipid peroxidation is responsible for the observed leak current, then isolated patch studies are not likely to reveal single channel currents with unit conductances that are characteristic of protein channels. Figure 6 shows such recordings. At the left of the figure is an example of a single channel current typical of that from a protein channel. Because the measured current is so small, the trace is quite noisy, but it appears to fluctuate largely between two values with relatively sudden transitions between these two more stable states. The upper level represents the closed state of the channel, the lower open state. With illumination in the presence of rose bengal, this kind of channel behavior disappears from isolated patch records and is replaced by the kind of record seen at the right of the figure which shows a nonquantized current. There are no well defined open and closed states. Rather, it appears that there are a large number of very small current pathways opening and closing rapidly. If many are open at the same time, a large current flows, whereas only a small current is recorded when only a few channels are open.

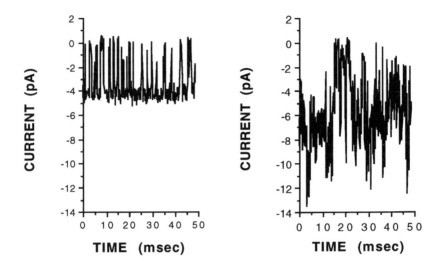

Figure 6. Photomodification of an isolated patch of membrane from an isolated frog atrial cell. The plot at the left shows a typical protein channel that oscillates between open and closed states demonstrating a quantized current. The plot at the right is the same patch after illumination in the presence of 0.125 μM rose bengal. The current pattern is no longer quantized.

What might cause such a pattern of current flow? While there are some examples of extremely complex current recordings from protein channels in membranes *(14)*, it has also been demonstrated that photomodification of simple lipid bilayers can produce a pattern of variable amplitude current spikes *(15)*. Thus, single channel currents are consistent with the hypothesis of a genesis of leak current in lipid peroxidation. Work with lipid peroxidation inhibitors at the single channel level is likely to lead to additional insights into the nature of this photoinduced membrane leak.

Acknowledgements

This work was supported by grants from the National Institutes of Health (NHLBI R01 HL43008), the American Heart Association (890714) and the Kansas Affiliate of the American Heart Association (KS-94-GS-34). HPLC chromatograms were obtained with the assistance of Drs. Edgar Arriaga and Andrey Frolov.

Literature Cited

1. Rebeiz, N.; Rebeiz, C.C.; Arkins, S.; Kelley, K.W.; Rebeiz, C.A. *Photochem. Photobiol.* **1992**, *55*, 431-435.
2. Iinuma, S.; Farshi, S.S.; Ortel, B.; Hasan, T. *Br. J. Cancer* **1994**, *70*, 21-28.
3. Spikes, J.D. In *The Science of Photobiology*; Editor, K.C. Smith; Plenum Press: New York, NY, **1989**; pp. 79-110.
4. Valenzeno, D.P.; Tarr, M. In *Photochemistry and Photophysics*, Editors J.F. Rabek; CRC Press: Boca Raton, FL, **1991**; pp. 137-191.
5. Ben-Hur, E.; Dubbelman, T.M.A.R. *Photochem. Photobiol.* **1993**, *58*, 890-894.

6. Ver Donck, L.; Borgers, M.; Verdonck, F. **1993**, *27*, 349-357.
7. Tarr, M.; Valenzeno, D.P. *J. Mol. Cell. Cardiol.* **1989**, *21*, 539-543.
8. Valenzeno, D.P.; Tarr, M. *Photochem. Photobiol.* **1991**, *53*, 195-201.
9. Tarr, M.; Valenzeno, D.P. *J. Mol. Cell. Cardiol.* **1991**, *23*, 639-649.
10. Pooler, J.P. *Biochim. Biophys. Acta* **1985**, *812*, 199-205.
11. Tarr, M.; Arriaga, E.; Goertz, K.K.; Valenzeno, D.P. *Free Rad. Biol. Med.* **1994**, *16*, 477-484.
12. Deziel, M.R.; Girotti, A.W. *Int. J. Biochem.* **1982**, *14*, 263-266.
13. Schothorst, A.A.; van Steveninck, J.; Went, L.N.: Suurmond, D. *Clin. Chim. Acta* **1970**, *28*, 41-49
14. Coulombe, A.; Lefevre, I.A.; Baro, I.; Coraboeuf, E. *J. Membr. Biol.* **1989**, *111*, 57-67.
15. Mirsky, V.M.; Stozhkova, I.N.; Szitó, T.V. *J. Photochem. Photobiol. B: Biol.* **1991**, *8*, 315-324.

RECEIVED August 18, 1995

Chapter 4

Safety of Xanthene Dyes According to the U.S. Food and Drug Administration

Arthur L. Lipman

Office of Premarket Approval, Center for Food Safety and Applied Nutrition, U.S. Food and Drug Administration, 200 C Street, S.W. (HFS–217), Washington, DC 20204

The Food and Drug Administration (FDA) has published decisions on 32 xanthene dyes. Fourteen were delisted in 1960 because they were not in commercial use in food, drugs, or cosmetics, six were delisted in 1963 because of lack of interest, and two were delisted after 1963. Ten have been listed for use in drugs and cosmetics. FD&C Red No. 3 is listed for use in food and ingested drugs, but was delisted in 1990 for use in cosmetics and external drugs. The reasons for FDA's decisions will be discussed in light of the toxicology data available to the agency at the times of the decisions.

Hyman Gittes, former Toxicologist in the Bureau of Foods, Food and Drug Administration (FDA), once said that the safety of a substance does not change; our perception of the safety changes. That concept places the historical review of color additives by FDA in perspective. FDA's actions on color additives have been based on information available at the time the action occurred in the regulatory framework that existed at the time. This paper will trace the history of the regulation of xanthene dyes by FDA and discuss the reasons for FDA's actions that were taken.

The Food and Drugs Act of 1906 prohibited interstate commerce in misbranded and adulterated foods, drinks, and drugs. However, the burden was placed on FDA's predecessor, the Bureau of Chemistry, United States Department of Agriculture (USDA), to prove that a product containing a poisonous dye (or other poisonous substance) was adulterated. Based on a thorough study (*1*) of some 80 dyes that were in use in food at the time, the USDA published a list of seven dyes considered to be safe for use in food provided that they bear a guaranty from the manufacturer that they are free from subsidiary products and represent the actual substance named (*2*). Erythrosine was the only xanthene dye included in the list of seven. The use in food of a dye, other than the seven listed, would be grounds for prosecution, as would the use of any dye to conceal damage or inferiority. The basis for listing these dyes was that they had been tested physiologically with no unfavorable results (*3*). A system was set up by the USDA to provide for voluntary certification of batches of the seven color additives.

The Federal Food, Drug, and Cosmetic Act of 1938 superseded the 1906 Act and mandated the certification of coal tar colors under the jurisdiction of the FDA. It also broadened the coverage for certifiable colors by creating three categories: colors that could be used in food, drugs, and cosmetics, colors that could be used only in drugs and cosmetics, and colors that could be used only in external drugs and cosmetics. However, the burden of proof of harm still was placed on FDA. Under the 1938 Act and on the basis of public hearings, FDA listed a large number of color additives subject to certification that were believed to be harmless and suitable for use in foods (*4*). FDA also established the "FD&C," "D&C," and "External D&C" nomenclature based on the categories established in the 1938 Act (*4*). This list contained 32 dyes in the xanthene class (Table I). Structures of these dyes are shown in Figure 1. Of these, only 10 remain listed today.

Because the burden of proof of harm under the 1938 law still fell on FDA, the agency conducted its own animal tests during the 1950s and found that several of the color additives caused toxic responses at higher doses (*5-7*). These findings led FDA to propose (*5, 7*) delisting certain colors, including D&C Orange No. 5, D&C Orange No. 6, D&C Orange No. 7, D&C Red No. 19, D&C Red No. 20, D&C Red No. 37, D&C Yellow No. 7, D&C Yellow No. 8, and D&C Yellow No. 9, for general use in drugs and cosmetics, on the basis that they were not harmless when ingested, and listing them under new names for use in external drugs and cosmetics. FDA was unable to set quantity limitations for use of these color additives in ingested drugs and ingested cosmetics, because of the Supreme Court's decision (*8*) that FDA did not have the authority under the 1938 law to do so. Thus, "the Supreme Court decision had rendered the old law obsolete and unworkable and colors were being delisted even though the FDA admitted they were not endangering the public health" (*9*).

Relief for the 1938 Act came through the Color Additive Amendments of 1960. These amendments, which were enacted on July 12, 1960 (*10*), provided for premarket approval by the Secretary of Health, Education, and Welfare of color additives used in food, drugs, and cosmetics, including the decision of whether batch certification was necessary. It also allowed for the Secretary to list color additives for specific uses and to set conditions of use and tolerances on the use, but included a cancer clause (Delaney clause) that precluded listing of a color additive found to induce cancer when ingested by man or animal. The 1960 amendments provided for the continued use of commercially established color additives through the establishment of provisional listings of such color additives, pending completion of scientific investigations to support listing under the Act.

Eighteen xanthene dyes were included on the provisional list established by FDA in 1960 (*11*), leaving 14 that were not included (Table I). Thus, D&C Orange Nos. 6, 7, 9, 12, and 13, D&C Red Nos. 20, 23, 25, and 26, D&C Yellow No. 9, Ext D&C Orange No. 2, and Ext D&C Red Nos. 4, 5, and 6 were delisted by Public Law 86-618, because no batches had ever been certified by FDA to establish that they were in commercial use and eligible for inclusion on the provisional list.

Table I. Xanthene Dyes Listed by FDA in 1939[a]

Color Additive	Common Name	Provisional Listing	Date Delisted	Effective Date for Listing[b]
D&C Yellow No. 9	Uranine K	No	7/12/60	
D&C Orange No. 6	Dibromofluorescein NA	No	7/12/60	
D&C Orange No. 7	Dibromofluorescein K	No	7/12/60	
D&C Orange No. 9	Dichlorofluorescein NA	No	7/12/60	
D&C Orange No. 12	Erythrosine Yellowish K	No	7/12/60	
D&C Orange No. 13	Erythrosine Yellowish NH	No	7/12/60	
D&C Red No. 20	Rhodamine B Acetate	No	7/12/60	
D&C Red No. 23	Eosin YSK	No	7/12/60	
D&C Red No. 25	Tetrachlorofluorescein NA	No	7/12/60	
D&C Red No. 26	Tetrachlorofluorescein K	No	7/12/60	
Ext D&C Orange No. 2	Indelible Orange	No	7/12/60	
Ext D&C Red No. 4	Dichlorotetraiodofluorescein	No	7/12/60	
Ext D&C Red No. 5	Rose Bengale TD	No	7/12/60	
Ext D&C Red No. 6	Rose Bengale TDK	No	7/12/60	
D&C Orange No. 8	Dichlorofluorescein	10/12/60	1/11/63	
D&C Orange No. 14	Orange TR	10/12/60	1/11/63	
D&C Orange No. 16	Diiododibromofluorescein	10/12/60	1/11/63	
D&C Red No. 24	Tetrachlorofluorescein	10/12/60	1/11/63	
D&C Red No. 29	Bluish Orange TR	10/12/60	1/11/63	
Ext D&C Red No. 3	Violamine R	10/12/60	1/11/63	
FD&C Red No. 3	Erythrosine	10/12/60	1/29/90[c]	7/7/69[d]

D&C Yellow No. 7	Fluorescein	10/12/60 (LTD)[c]		12/20/76[f]
D&C Yellow No. 8	Uranine	10/12/60 (LTD)[c]		12/20/76[f]
D&C Orange No. 5	Dibromofluorescein	10/12/60 (LTD)[c]		12/3/82,[g] 5/7/84[f]
D&C Orange No. 10	Diiodofluorescein	10/12/60		4/28/81[f]
D&C Orange No. 11	Erythrosine Yellowish NA	10/12/60		4/28/81[f]
D&C Red No. 19	Rhodamine B	10/12/60 (LTD)[c]	7/15/88[b]	
D&C Red No. 21	Tetrabromofluorescein	10/12/60		1/3/83
D&C Red No. 22	Eosin YS	10/12/60		1/3/83
D&C Red No. 27	Tetrachlorotetrabromofluorescein	10/12/60		10/29/82
D&C Red No. 28	Phloxine B	10/12/60		10/29/82
D&C Red No. 37	Rhodamine B Stearate	10/12/60 (LTD)[c]	6/6/86[h]	

[a] D&C Red No. 37, D&C Orange No. 16, and Ext D&C Orange No. 2 listed 9/16/39 (4 FR 3932); all others listed 5/9/39 (4 FR 1922).

[b] Normally, color additive regulations are effective 31 days after date of publication in the Federal Register unless stayed by objections.

[c] Provisional listing terminated for external drug uses and all cosmetic uses of the straight color additive and all uses of the lake.

[d] Food and ingested drug uses only.

[e] External uses and limited ingested uses with a temporary tolerance.

[f] Externally applied drugs and cosmetics.

[g] Use in lipsticks, drug and cosmetic mouthwashes, and dentifrices.

[h] External drug and cosmetic uses only; ingested uses terminated on 2/4/83.

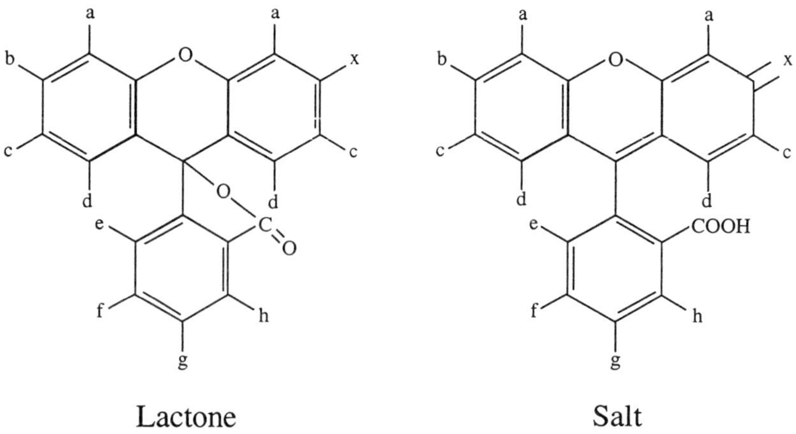

Figure 1. Structures of xanthene dyes considered by FDA.

Xanthene Dye	a	b	c	d	e	f	g	h	x
D&C Yellow No. 7, 8, 9	H	OH	H	H	H	H	H	H	O
D&C Orange No. 5, 6, 7	Br	OH	H	H	H	H	H	H	O
D&C Orange No. 8, 9	Cl	OH	H	H	H	H	H	H	O
D&C Orange No. 10, 11, 12, 13	I	OH	H	H	H	H	H	H	O
D&C Orange No. 14	Br	OH	COOH	H	Br	H	H	H	O
D&C Orange No. 16	Br	OH	I	H	H	H	H	H	O
FD&C Red No. 3	I	OH	I	H	H	H	H	H	O
D&C Red No. 19, 20, 37	H	N(C_2H_5)_2	H	H	H	H	H	H	+N(C_2H_5)_2
D&C Red No. 21, 22, 23	Br	OH	Br	H	H	H	H	H	O
D&C Red No. 24, 25, 26	Cl	OH	Cl	H	H	H	H	H	O
D&C Red No. 27, 28	Br	OH	Br	H	Cl	Cl	Cl	Cl	O
D&C Red No. 29	Br	OH	COOH	Br	Br	H	H	H	O
Ext D&C Red No. 3	H	NNPhCH_3 SO_3Na	H	H	H	H	H	H	+NPhCH_3
Ext D&C Red No. 4, 5, 6	I	OH	I	H	Cl	H	H	Cl	O
Ext D&C Orange No. 2	NO_2	OH	H	H	H	H	H	H	O

Furthermore, a requirement for continued provisional listing of color additives was that studies demonstrating their safety were in progress. Because no such studies were being conducted on D&C Red Nos. 24 and 29, D&C Orange Nos. 8, 14, and 16, and Ext D&C Red No. 3, FDA allowed their provisional listings to terminate on January 11, 1963 (*12*). Thus, by 1963, only 12 of the original 32 xanthene dyes listed for use in 1939 still had commercial value. The provisional listings for D&C Orange No. 5, D&C Red Nos. 19 and 37, and D&C Yellow Nos. 7 and 8 were limited to ingested uses (lipsticks, mouthwashes, and dentifrices) with temporary tolerances and to external drug and cosmetic uses (*11, 13*).

As acceptable safety studies were completed, the provisionally listed color additives were either permanently listed or delisted, depending on the outcome of the safety studies. The permanent listing of FD&C Red No. 3 for use in food and ingested drugs was effective July 7, 1969 (*14*), but the external drug uses, all cosmetic uses, and all uses of lakes of FD&C Red No. 3 remained provisionally listed, while petitioners conducted safety tests appropriate for external use.

On February 4, 1977 (*15*), FDA published revised safety requirements for the color additives then provisionally listed. New chronic toxicity studies were required for 31 provisionally listed colors, including the 12 xanthene dyes still on the provisional list. This requirement was imposed because previous studies were deficient in the following respects: (1) The number of animals studied was too small to permit conclusions to be drawn about the chronic toxicity or carcinogenic potential of the color additive, (2) the number of animals surviving to a meaningful age was inadequate, (3) an insufficient number of animals was reviewed histologically, (4) an insufficient number of tissues was examined for pathology, and (5) lesions or tumors detected under gross examination were not examined microscopically (*16*). Except for D&C Yellow Nos. 7 and 8 and D&C Orange Nos. 10 and 11, which were permanently listed for use in external drugs and cosmetics in 1976 (*17, 18*) and 1981 (*40*), respectively, the results and FDA's interpretations of these new studies were pivotal in the decisions described below for the xanthene dyes that were on the provisional list after 1963.

Before the safety decisions for individual color additives are described, the working definition of safety needs to be discussed. The legislative history of the 1958 Food Additives Amendment to the Federal Food, Drug, and Cosmetic Act explains that "safety requires proof of a reasonable certainty that no harm will result from the proposed use of the additive. It does not--and cannot--require proof beyond any possible doubt that no harm will result under any conceivable circumstance" (*19*). The same concept of safety was invoked for the Color Additive Amendments of 1960. This concept of safety has been incorporated as follows into FDA's color additive regulations: "Safe means that there is convincing evidence that establishes with reasonable certainty that no harm will result from the intended use of the color additive" (*20*). In determining whether a color additive is safe, the agency determines the maximum dose level inducing no adverse effects in each appropriate toxicity study. The agency applies a safety factor to that no-adverse-effect level in the study deemed to be the most appropriate to determine the acceptable daily intake (ADI) for the color additive. Depending on the duration of the study, the safety factor may be 100 for chronic studies or 1000 for shorter-term studies. The agency generally considers a color additive to be safe under its intended conditions of use if the estimated daily intake (exposure) of the additive does not exceed its ADI. In determining the safety of a color additive, the agency also must consider whether the color requires batch certification or should be exempt from that requirement. Batch certification requires that a representative portion of each batch of the color additive be sent to FDA for analysis to certify that it meets the specifications established by regulation. If the sample meets these requirements, FDA issues a certificate to that effect and the color additive may be used legally as allowed by regulation. The certified batch of color must be identified by its official name as listed in the Code of Federal Regulations or by an abbreviated name containing the color and number; however, that name cannot be used for a batch of uncertified color additive. For example Phloxine B may be called D&C Red No. 28 (or Red 28) only if it is from a certified batch; it is a violation of the Federal Food, Drug, and Cosmetic Act to use Phloxine B (from an uncertified batch) in a drug or cosmetic product.

With these concepts in mind, FDA's decisions will be examined for the 12 xanthene dyes remaining on the provisional list after 1963: D&C Yellow Nos. 7 and 8, FD&C Red No. 3, D&C Orange No. 5, D&C Orange Nos. 10 and 11, D&C Red Nos. 19 and 37, D&C Red Nos. 21 and 22, and D&C Red Nos. 27 and 28.

Lakes and Eye Area Use

All of the petitions for these provisionally listed color additives requested that lakes of the color additives be permanently listed along with the straight color additives. (A lake is an insoluble pigment composed of a soluble straight color strongly adsorbed onto an insoluble substratum through use of a precipitant.) However, the agency had proposed permanent listing of all lakes of the certifiable colors in 1965 (21). That proposal was never finalized because the straight color additives were still provisionally listed. In 1979 the lakes proposal was withdrawn (22) because of the long time since the proposal and because several additional questions had arisen. The agency published an advance notice of proposed rulemaking (ANPR) requesting information needed to support permanent listing of the lakes of the certifiable color additives (23). Because the agency was handling the permanent listing of all lakes in a separate action, it deferred the permanent listing of lakes of the xanthene dyes when it permanently listed the straight dyes. Lakes of D&C Red Nos. 19 and 37 were delisted in 1988 and 1986, respectively, when the straight color additives were delisted, and lakes of FD&C Red No. 3 were delisted in 1990.

Although the Cosmetic, Toiletry and Fragrance Association (CTFA) had amended several of its petitions to request listing of FD&C Red No. 3, D&C Red Nos. 27 and 28, and D&C Orange Nos. 10 and 11 for eye area use (24), they did not submit adequate data to support such use. After requesting the information by letter, FDA considered those portions of the petitions to be withdrawn without prejudice to a future filing, because the information was not supplied.

D&C Yellow Nos. 7 and 8

The color additive petition (CAP 5C0034) submitted by the Toilet Goods Association (now the CTFA), the Pharmaceutical Manufacturers Association, and the Certified

Color Industry Committee (later named the Certified Color Manufacturers Association (CCMA) and now the International Association of Color Manufacturers) in 1965 and filed by FDA in 1968 (*25*) requested permanent listing of D&C Yellow Nos. 7 and 8, for use only in externally applied drugs and cosmetics. Thus, FDA's review and final determination of safety of these two dyes did not consider ingested uses. The studies available to FDA, listed in Table II, included acute feeding studies in rats and dogs, a subacute feeding study in rats, a dermal irritation study in rabbits, and a lifetime skin-painting study in mice. A published metabolism study in rats (*26*) indicated that the fluorescein moiety was not degraded, but was excreted in the urine as the glucuronide. All studies were conducted using D&C Yellow No. 7. However, because D&C Yellow No. 8 is the sodium salt of D&C Yellow No. 7, FDA judged that the tests on D&C Yellow No. 7 were applicable to D&C Yellow No. 8. FDA also took into consideration evidence presented by the petitioner on human use experience, because of the low number of complaints by consumers about the types of products in which these two color additives might be used. The final orders were published separately for D&C Yellow No. 7 and D&C Yellow No. 8 (*17, 18*). FDA confirmed the effective date of December 20, 1976, for these permanent listings and the removal of D&C Yellow Nos. 7 and 8 from the provisional list on February 25, 1977 (*27, 28*). Although FDA did not calculate an ADI for D&C Yellow No. 7 when the color additive was listed for external drug and cosmetic use, the author has calculated an ADI, using the no-effect level from the subchronic rat study and a 1000-fold safety factor, of 0.25 mg/kg (Table III).

FD&C Red No. 3

On the basis of studies available at the time in CAP No. 8C0067, which was filed by CCMA on July 2, 1968 (*29*), FDA published a final order permanently listing FD&C Red No. 3 for use in food and ingested drugs on May 8, 1969 (*30*), and subsequently confirmed the effective date of July 7, 1969, for that final order (*31*). These studies included acute oral studies in rats and gerbils, acute injection studies in rats and mice, subacute feeding studies in rats, chronic feeding studies in rats, dogs, and mice, a metabolism study in rats, and a dermal irritation study in rabbits (Table II). On the basis of a 0.5% no-effect level in rats and a 100-fold safety factor, FDA estimated the ADI to be 2.5 mg/kg (150 mg/60 kg person). Table III also lists an ADI established in 1991 by the Joint Expert Committees on Food Additives of the World Health Organization (JECFA). JECFA's ADI, which is based on a no-effect level for thyroid effects in humans and a 10-fold safety factor (*32*), will be used for comparison. The estimated daily intake combining food and ingested drug use was 34 mg/day, well below the ADI set by FDA.

External drug and all cosmetic uses of FD&C Red No. 3 and all uses of FD&C Red No. 3 lakes remained on the provisional list, while tests appropriate to external use were completed for the straight color additive, and the agency completed action on the permanent listing of lakes of the certifiable color additives as proposed in 1965 (*21*).

CTFA, which filed CAP 9C0096 for the external uses of FD&C Red No. 3 in 1973 (*33*), sponsored the new chronic studies required (*15, 16*) for continued provisional listing of the color additive. Although the new chronic mouse-feeding study did not

Table II. Available Toxicity Studies on Provisionally Listed Xanthene Dyes

Type of Study	D&C Yellow No. 7	D&C Orange No. 5	FD&C Red No. 3	D&C Orange No. 10	D&C Red No. 19	D&C Red No. 21	D&C Red No. 27
Acute Oral	Rats, Dogs	Rats, Dogs	Rats, Dogs, Gerbils	Rats, Dogs	Rats, Mice, Dogs	Rats, Dogs	Rats, Dogs
Oral Subchronic	Rats[a]	Rats, Dogs	Rats, Dogs	Rats, Dogs	Rats, Dogs	Rats, Dogs	Rats, Dogs
Chronic Oral		Old - Rats, Dogs New - Rats, Mice	Old - Rats,[a] Dogs, Mice New - Rats, Mice	Old - Rats, Dogs	Old - Rats,[a] Dogs New - Rats, Mice	Old - Rats, Dogs New - Rats,[a] Mice	Old - Rats, Dogs New - Rats,[a] Mice
Reproductive		Rats			Rats	Rats	Rats
Teratology		Rats, Rabbits[a]		Rats, Rabbits[a]	Rats, Rabbits	Rats, Rabbits	
Metabolism	Rats	Rats	Rats	Rats	Rats, Dogs, Rabbits	Rats	Rats
Mutagenicity			Ames, Mammalian, Yeast, Drosophila		Ames, Mammalian, Yeast, Drosophila	Ames	Ames
Subchronic Dermal	Rabbits	Rabbits	Rabbits	Rabbits	Rabbits	Rabbits	Rabbits, Mice
Chronic Dermal	Mice	Mice	Mice	Mice	Mice	Mice	Mice
Human	Use, Complaint	Use, Complaint	Use, Complaint	Use, Complaint	Use, Complaint	Use, Complaint	Use, Complaint
Acute Injection			Rats, Mice		Rats		
Skin Penetration		Yes			Yes		

[a]Study used to set the ADI (Table III).

Table III.   ADIs of Selected Xanthene Dyes

Dye	Study Used	Highest No Effect Level (mg/kg/day)	Safety Factor	ADI (mg/kg)	Molecular Weight	Molar ADI (μmoles/kg)
D&C Yellow No. 7	Subchronic Rat	250	1000	0.25	332.31	0.7
D&C Orange No. 5	Rabbit Teratology	50	1000	0.05	490.1	0.1
FD&C Red No. 3 1969 Value	Human Clinical Old Chronic Rat	1 250	10 100	0.1[a] 2.5[b]	879.66	0.1 2.8
D&C Orange No. 10	Rabbit Teratology	50	1000	0.05	584.1	0.09
D&C Red No. 21	New Chronic Rat	500	100	5	647.9	7.7
D&C Red No. 27	New Chronic Rat	125	100	1.25	785.68	1.6

[a]JECFA, 1991 (*32*).

[b]FDA, 1969 (*30*).

show a carcinogenic response, the new study in Sprague-Dawley rats did show an association between ingestion of the color additive and an increased incidence of combined thyroid follicular cell adenomas and carcinomas in male rats fed at the highest level (4.0%). FDA concluded that FD&C Red No. 3 is an animal carcinogen on the basis of the rat study and other data, and in 1990 terminated the provisional listing for the external drug use and all cosmetic uses of FD&C Red No. 3, as well as all uses of the lakes of the color additive (34). The agency provided a detailed discussion of the basis for its conclusion on the carcinogenicity of FD&C Red No. 3 in a document denying the petition filed by CTFA (35). The agency's conclusion about the carcinogenicity of FD&C Red No. 3 was corroborated by the National Toxicology Program Board of Scientific Counselors, Technical Reports Review Subcommittee (NTP Subcommittee), which met at FDA's request (36) to consider the results of the new chronic studies on FD&C Red No. 3. The proponents of the use of FD&C Red No. 3 did not refute this conclusion, but instead contended that the color additive causes the carcinogenic response through a secondary mechanism. Under the secondary mechanism hypothesis, the color additive causes a physiological response involving an imbalance of the thyroid hormones that would have a threshold below which the imbalance would not occur; consequently the carcinogenic response also would not occur. The proponents submitted the results of several studies that they claimed supported their secondary mechanism hypothesis, but FDA did not find the data or arguments convincing (35). In addition, the design of the previous studies did not effectively eliminate the possibility that the thyroid tumors could have been due to a primary (direct) rather than a secondary (indirect) carcinogenic response. Therefore, under the Delaney clause, FDA terminated the provisional listing of FD&C Red No. 3 and its lakes, and denied the petition.

D&C Orange No. 5

As in the case of FD&C Red No. 3, FDA considered ingested uses and external uses in separate actions, but in this case both actions were based on the same petition. CTFA's petition (CAP 6C0041), requesting permanent listing of D&C Orange No. 5 for general use in drugs and cosmetics, was filed by FDA in 1973 (33). FDA permanently listed the color additive for use in lipsticks or other lip cosmetics, in amounts not exceeding 5%, and in drug and cosmetic mouthwashes and dentifrices, but requested information on the skin penetration of the color additive before it would consider listing the requested external uses (37). As mentioned above, D&C Orange No. 5 had been provisionally listed with a temporary tolerance, because studies conducted in the 1950s had shown that the color additive was a toxic substance when ingested, and thus safe only with restricted use in drugs and cosmetics (11).

On the basis of studies done under the requirements for continued provisional listing, FDA established newer temporary tolerances for D&C Orange No. 5 and deleted ingested drug uses in 1979 (38). These studies included new information on use of the color additive, reports on the chronic feeding studies that were in process, and results of teratology and multigeneration reproduction studies that were conducted for color additives requiring temporary tolerances. The agency concluded that D&C Orange No. 5 is not carcinogenic on the basis of the new chronic studies in rats and mice. Other studies available to the agency were chronic toxicity studies in dogs and rats, a teratology study in rats, a three-generation reproduction study in rats, a short-term feeding study in rabbits, an 18-month skin-painting study in mice, and a dermal study in rabbits (see Table II). The agency used the feeding study in rabbits to establish the acceptable daily intake for D&C Orange No. 5, based on a 1000-fold safety factor for the highest no-effect level in that study of 50 mg/kg/day. Thus, the ADI is 0.05 mg/kg/day or 3 mg/day for a 60-kg person. The agency used a 1000-fold safety factor instead of the normal 100-fold factor (21 CFR 70.40), because of the short duration of the study. FDA estimated the exposure to the color additive for the various potential uses as shown in Table IV.

Table IV. Exposure to D&C Orange No. 5

Use	1982 Exposure (mg/day)	1984 Exposure (mg/day)
Ingested drugs	24	NA
External drugs	2.4	0.5
Lipsticks (6% max)	3	2.1
Mouthwashes/dentifrices	0.2	0.2
Topical cosmetics	2	0.1

On the basis of a comparison of the ADI with the estimated exposure levels, FDA concluded that it could list D&C Orange No. 5 for use in lipsticks only with the tolerance lowered to 5%, because people using more than one product containing the color additive could very easily exceed the ADI. The agency called for new information on use, such as information on the skin penetration of D&C Orange No. 5 from topical uses (37).

In 1984, on the basis of data from a skin penetration study run in its own laboratories, FDA permanently listed D&C Orange No. 5 for use in externally applied drugs and cosmetics (39). The agency concluded that D&C Orange No. 5 is absorbed from topical products in amounts not exceeding 0.5% of the amount applied to intact skin, and revised its exposure estimates accordingly as indicated in Table IV. The agency also adjusted the lipstick calculation because the 1982 calculation assumed that 100% of the color additive in the lipstick is ingested. The 1984 calculation uses a factor of 0.85 of the applied dose in lipstick for short-term exposure to the color additive. For use of the color in topical drugs, the agency considered that penetration would be higher because the product is applied to damaged skin. To calculate the daily exposure to topical drug products, the agency estimated that 10% of the applied color would penetrate damaged skin. The agency established a 5 mg per daily dose limitation for use of D&C Orange No. 5 in topical drug products to ensure that the daily exposure did not exceed the ADI (39).

D&C Orange Nos. 10 and 11

The agency permanently listed D&C Orange No. 10 and D&C Orange No. 11 for use in externally applied drugs and cosmetics in 1981 (40). The action was in response to CAP 6C0042, filed by CTFA. FDA published the notice of filing of this petition on August 6, 1973 (33), and confirmed the effective date of April 28, 1981, for the final order on July 7, 1981 (41). Because the petitioner decided not to do the new chronic feeding studies needed for continued provisional listing of color additives for ingested uses (15), the color additives were permanently listed only for external drug and cosmetic uses. The provisional listing for ingested uses of D&C Orange Nos. 10 and 11 was terminated, because there was no longer a petition to support the continued provisional listing of the colors. Studies available to the agency included acute feeding studies in rats and dogs, subchronic feeding studies in rats and dogs, chronic feeding studies in rats and dogs, dermal studies in rabbits, a chronic skin-painting study in mice, teratology studies in rats and rabbits, a metabolism study in rats, and human use data (Table II). However, the available chronic feeding studies did not meet current criteria (15, 16). The ADI for D&C Orange No. 10, calculated by FDA toxicologists in 1973, was based on a 100-fold safety factor applied to the no-effect level for the rabbit teratology study. For consistency with current policy, the ADI presented in Table III was based on a 1000-fold safety factor because of the short duration of the study.

D&C Red Nos. 19 and 37

D&C Red No. 19 was provisionally listed for external drug and cosmetic uses and for lipsticks with a temporary tolerance of 6% and for ingested drugs with a tolerance of 0.75 mg/day dosage, because studies by FDA had shown that the color additive was unsafe for more extensive drug and cosmetic use (11). Similarly, D&C Red No. 37 was provisionally listed for external drug and cosmetic use and for ingested drug use with a temporary tolerance of 0.75 mg/day (13). As with D&C Orange No. 5, the temporary tolerances were revised in 1979 for both D&C Red No. 19 and D&C Red No. 37 to 1.3% in lipsticks, 0.002% in dentifrices, 0.005% in mouthwashes, and 0.025% in other ingested drugs with unlimited external drug and cosmetic uses (38). The petition supporting the provisional listing of these color additives was submitted by CTFA and filed by FDA in 1973 (33). On the basis of the results of the new chronic studies that were required by FDA (15), the petitioner withdrew that portion of its petition that pertained to the ingested uses of D&C Red Nos. 19 and 37, because D&C Red No. 19 caused a carcinogenic response in animals. Because D&C Red No. 19 and D&C Red No. 37 are closely related structurally (D&C Red No. 37 is the stearate salt and D&C Red No. 19 is the chloride salt), the agency considers them to be toxicologically equivalent on ingestion. Therefore, studies on D&C Red No. 19 are applicable to D&C Red No. 37. Specifically, the agency concluded that D&C Red No. 19 is an animal carcinogen when it is administered in the diet because of an increased incidence of hepatocellular neoplasms (carcinomas or adenomas) in mice and an increased incidence of thyroid and parathyroid neoplasms (malignant and benign follicular cell tumors of the thyroid, adenomas in the parathyroid) in rats. FDA's termination of the provisional listing for the ingested uses of these color additives was effective February 4, 1983 (42).

The color additives remained on the provisional list for external drug and cosmetic uses, while the agency evaluated submissions, including skin penetration data, in support of listing the color additives for those uses. However, by letter of April 29, 1986, CTFA withdrew the portion of its petition that pertained to the listing of D&C Red No. 37 for use in external drugs and cosmetics. Consequently, FDA terminated the provisional listing of the color additive for those uses (43), while D&C Red No. 19 remained provisionally listed for use in external drugs and cosmetics.

On August 7, 1986, FDA published a final order permanently listing D&C Red No. 19 for use in externally applied drugs and cosmetics (44), based on the conclusion that the carcinogenic risk from the external uses of D&C Red No. 19 is so trivial as to be effectively no risk at all. FDA evaluated reports of several toxicity studies of D&C Red No. 19 involving rats, dogs, mice, and rabbits, including acute oral toxicity studies, acute intravenous toxicity studies, subchronic feeding studies, dermal toxicity studies, chronic toxicity studies, metabolism studies, teratogenicity studies, and reproductive toxicity studies (Table II). On the basis of the dermal toxicity studies, FDA found that D&C Red No. 19 was not carcinogenic upon biweekly application to the skin of mice over their lifetimes. However, as discussed above, the newer chronic feeding studies showed that D&C Red No. 19 is an animal carcinogen upon ingestion. From the skin penetration study conducted by CTFA, FDA concluded that radiolabeled material from D&C Red No. 19 passes through the skin in small but measurable amounts, so that some systemic exposure to the color may occur from external uses of D&C Red No. 19 and that ingestion studies are appropriate for evaluating the safety of externally applied uses of the color additive. However, the agency took the position that under the *de minimis* doctrine the Delaney clause does not require it to ban the external uses of D&C Red No. 19, because the risk of carcinogenesis in humans is so low as to be *de minimis* (1 in 9 million or 1.1×10^{-7}). (The *de minimis* doctrine is that the law does not concern itself with trifles (*de minimis non curat lex*).) Thus, FDA found these uses to be safe. The agency clarified its position by saying that D&C Red No. 19 does not induce cancer in man or animal within the meaning of the Act and is safe for external use, because it poses no genuine risk of cancer (45).

The Public Citizen Litigation Group filed objections to this final order, based on their interpretation of the Delaney clause, but FDA denied the objections and confirmed the effective date for the listing of D&C Red No. 19 for use in externally applied drugs and cosmetics as October 6, 1986 (46). However, the Public Citizen Litigation Group filed suit in the U.S. Court of Appeals for the District of Columbia to overturn FDA's decision *(Public Citizen v. Young; No. 86-1548)*. The decision of the appeals court that the agency's *de minimis* interpretation of the Delaney clause is contrary to law and the subsequent refusal of the United States Supreme Court to grant a writ of certiorari (April 18, 1988) forced FDA to revoke the regulation for use of D&C Red No. 19 (47) and to deny the CTFA petition for the permanent listing of the color additive (48). Thus, all uses of D&C Red No. 19 and D&C Red No. 37 were delisted by 1988.

D&C Red Nos. 21 and 22

Supported by the CTFA petition (CAP 6C0043), which was filed by FDA on August 6, 1973 (*33*), D&C Red No. 21 and D&C Red No. 22 remained on the provisional list until they were permanently listed for general use in drugs and cosmetics, effective January 3, 1983 (*49, 50*). The agency previously reviewed reports for a number of other animal toxicity studies on D&C Red No. 21. These included acute, subchronic, and chronic oral toxicity studies in rats and dogs, a dermal study with rabbits, an 18-month skin-painting study with mice, a metabolism and excretion study in rats, a three-generation reproduction study in rats, and teratology studies in rats and rabbits (Table II). These studies did not produce any evidence that the use of this color additive would be unsafe. However. the agency concluded that new chronic toxicity feeding studies would be required to permit a final determination on the listing of D&C Red Nos. 21 and 22 (*15, 16*).

Thus, CTFA sponsored new chronic feeding studies in mice and rats, using D&C Red No. 21. Because D&C Red No. 22 is the disodium salt of D&C Red No. 21, the agency considers the two color additives to be toxicologically equivalent. Thus, any safety conclusions drawn from studies of D&C Red No. 21 apply equally to D&C Red No. 22.

The petitioner reported that the incidence of hepatocellular carcinoma among male mice in the treated groups was higher than in the study controls. FDA conducted its own examination of microslides from male mice in the study and concluded that the color additive did not induce a carcinogenic response in this study because (1) the incidence of hepatocellular carcinoma in the control groups was substantially below the historical incidence of hepatocellular carcinoma in control groups at the testing laboratory in the same strain of mouse in five studies performed within approximately 3 years of the D&C Red No. 21 study, (2) the incidences in all three treated groups were well within the range of historical control incidence, and (3) there was a lack of a dose-response relationship among treated groups. Furthermore, the agency believes that in analyzing the data it is important and proper to combine the incidences of hepatic carcinoma and adenoma for statistical analysis. The FDA interpretation of all the slides yielded values for combined incidences of hepatocellular adenomas and carcinomas that were not considered indicative of a treatment-related effect. Furthermore, the combined incidences of hepatocellular carcinomas and adenomas in each of the treatment groups were well within the range of the historical control incidence of the testing laboratory for the same strain of mouse in five studies performed within approximately 3 years of the D&C Red No. 21 study. These considerations and the agency's review of all the other data in the petition, including the fact that there was no treatment-related increase in incidence of any type of neoplasm in female mice, led FDA to conclude that there is no indication of an association between the occurrence of neoplasms and the administration of D&C Red No. 21 to mice.

In the new rat study the agency found no increases in incidence of tumors in any of the treated groups that could be attributed to exposure to this color additive. However, on the basis of several non-specific, non-neoplastic microscopic changes in the high dose-treated rats, at incidences that were greater than those in the combined control groups, the agency determined that the high dose (2.0%) was an effect level. Therefore, the agency chose the intermediate dose of 1.0% in the diet, corresponding to 500 mg/kg/day, as the no-adverse-effect level and used this rat study and a 100-fold safety factor to calculate an ADI of 5 mg/kg/day or 300 mg/day for a 60-kg person. From its review of the available data on the current uses of D&C Red Nos. 21 and 22, FDA estimated that the upper limit lifetime-averaged internal exposure to these color additives from drugs and cosmetics is 6.3 mg/day (drugs, 5 mg; cosmetics, 1.3 mg), giving a 47-fold margin of safety.

D&C Red Nos. 27 and 28

Supported by the CTFA petition (CAP 6C0044), which was filed by FDA on August 6, 1973 (*33*), D&C Red No. 27 and D&C Red No. 28 remained on the provisional list until they were permanently listed for general use in drugs and cosmetics, effective October 29, 1982 (*51*). As discussed in the final order permanently listing D&C Red Nos. 27 and 28 for general use in drugs and cosmetics (*52*), two new chronic toxicity studies in rats and mice were conducted for the petitioner, using D&C Red No. 27. Because D&C Red No. 28 is the disodium salt of D&C Red No. 27, the agency considers the two color additives to be toxicologically equivalent. Thus, any safety conclusion drawn from studies on D&C Red No. 27 applies equally to D&C Red No. 28.

On the basis of the evaluation of the results of the two new chronic toxicity studies, the agency determined that D&C Red No. 27 and D&C Red No. 28 are not carcinogenic in Charles River Sprague-Dawley CD rats or CD-1 mice after lifetime dietary exposure of 2.0 and 1.0%, respectively. Applying a 100-fold safety factor to the no-effect level in the new chronic feeding study in rats, the agency estimated an ADI for humans of 1.25 mg/kg/day (*52*).

Besides the new chronic studies and its range-finding studies, FDA also reviewed older studies on D&C Red No. 27, including acute oral studies in rats and dogs, subchronic studies in rats and dogs, chronic feeding studies in rats and dogs, dermal studies in rabbits and mice, a lifetime skin-painting study in mice, a three-generation reproductive study in rats, metabolism in rats, and mutagenicity studies (Table II).

Recent Studies by FDA Personnel

FDA scientists have continued to study the photoactivation of xanthene dyes over the past few years. Rosenthal et al. (*53*) tested 43 certified food, drug, and cosmetic dyes allowed by FDA in a screen for chemical photosensitizing activity by using photogeneration of singlet molecular oxygen. Only xanthene dyes substituted with bromine or iodine were photochemically active, with FD&C Red No. 3 and D&C Red Nos. 27 and 28 being the most active, followed by D&C Red Nos. 21 and 22, D&C Orange Nos. 10 and 11, and D&C Orange No. 5; D&C Red Nos. 19 and 37 and D&C Yellow Nos. 7 and 8 were inactive.

Wei et al. (*54*) examined the phototoxicity of fluorescein dyes to human skin fibroblast cells by measuring inhibition of cell growth 3 days after irradiation. D&C

Red No. 28 was the most inhibitory (12.5 μM gave 50% inhibition), and FD&C Red No. 3 was second (50 μM), followed by D&C Red No. 22 (150 μM) and D&C Yellow No. 8 (250 μM). The authors concluded that phototoxicity increases with increasing atomic number of the halogen substituent and that halogen substitution on the benzoic acid moiety results in increased phototoxicity.

Bockstahler et al. (55) found photoinduction of the capacity of human fibroblast cells to support plaque formation by herpes simplex virus in the order FD&C Red No. 3 > D&C Red No. 21 > D&C Red No. 19 > D&C Yellow No. 8, following irradiation in the visible range. FD&C Red No. 3 and D&C Red No. 21 also inhibited plaque formation after irradiation in the near ultraviolet range. The authors hypothesized that FD&C Red No. 3 and D&C Red No. 21 are possibly acting through generation of singlet oxygen, whereas D&C Red No. 19 and D&C Yellow No. 8 may be acting through a direct photochemical mechanism, because the latter dyes have been shown not to produce singlet oxygen upon irradiation in the presence of oxygen.

Conclusion

Table III lists the ADIs calculated by FDA for each of the xanthene dyes that are currently listed for use. Although ADIs are usually based on body weight or on unit weight of the daily diet, it is more valid from a chemist's point of view to compare ADIs on a molar basis. Molar ADIs for the xanthene dyes in Table III may be listed in decreasing order as follows: D&C Red No. 21 > D&C Red No. 27 > D&C Yellow No. 7 > FD&C Red No. 3 $\approx$ D&C Orange No. 5 > D&C Orange No. 10. In this limited sample of xanthene dyes, it appears that as the number of halogens on the fluorescein moiety increases, the ADI increases (see Figure 1 for structures); however, the iodine-substituted fluoresceins are very close in value, as shown by a comparison of D&C Orange No. 10 with FD&C Red No. 3, and D&C Orange No. 5 with D&C Red Nos. 21 and 27. It also appears that substitution on the benzoic acid moiety increases toxicity, e.g., the ADI of D&C Red No. 28 is lower than that of D&C Red No. 21. These results are similar to the trend found by Tonogai et al. (56) in fish. However, comparisons of ADIs calculated from several different types of studies may not be valid. Further work allowing comparison of no-effect levels or lowest-adverse-effect levels in like animal studies might be more valuable.

In examining FDA's decisions on the safety of the xanthene dyes, the author found that a dye's status changed as the perception of its safety changed; however, the criteria used in gaining that perception of safety were determined by the legal framework. Xanthene dyes were delisted either because there were insufficient data to judge their safety under existing criteria or because the available data showed that they were unsafe. On the other hand, dyes were listed for use when the available data established a reasonable certainty of no harm from their intended use.

Literature Cited

1.   Hesse, B.C. U.S. Dep. Agric. Bull. 1912, 147.
2.   Wiley, H.W.; Dunlap, F.L.; McCabe, G.P. Food Inspection Decision 76; U.S. Department of Agriculture: Washington, DC, 1907.

3.   Noonan, J.E.; Meggos, H. In *CRC Handbook of Food Additives*, 2nd Ed.; Furia, T.E., Ed.; CRC Press, Inc.: Boca Raton, FL, 1980, Vol. II; pp 339-383.

4.   Food and Drug Administration. "In the Matter of Public Hearing for Purpose of Receiving Evidence Upon Basis of Which Regulations May Be Promulgated for Listing of Coal-Tar Colors Which Are Harmless and Suitable for Use in Foods, Drugs, and Cosmetics, Drugs and Cosmetics, and Externally Applied Drugs and Cosmetics;...." *Fed. Regist.* **1939**, *4*, 1922-1947.

5.   Food and Drug Administration. "Notice of Proposal to Amend Regulations by Deleting Certain Coal-Tar Colors Subject to Certification." *Fed. Regist.* **1959**, *24*, 2873-2875.

6.   Food and Drug Administration. "Deletion of Certain D&C Coal-Tar Colors From List Subject to Certification; Final Order." *Fed. Regist.* **1959**, *24*, 8065-8067.

7.   Food and Drug Administration. "Findings of Fact and Tentative Order on Proposed Amendment of Color Certification Regulations." *Fed. Regist.* **1960**, *25*, 5582-5588.

8.   U.S. Supreme Court, 358 U.S. 153, Dec. 15, 1958.

9.   Noonan, J.E.; Meggos, H. In *CRC Handbook of Food Additives*, 2nd Ed.; Furia, T.E., Ed.; CRC Press, Inc.: Boca Raton, FL, 1980, Vol. II; p 341.

10.   Public Law 86-618, 86th U.S. Congress, 1960.

11.   Food and Drug Administration. "Transitional Regulations Under Title II of Color Additive Amendments of 1960 to the Federal Food, Drug, and Cosmetic Act." *Fed. Regist.* **1960**, *25*, 9759-9761.

12.   Food and Drug Administration. "Provisional Lists." *Fed. Regist.* **1963**, *28*, 317-318.

13.   Food and Drug Administration. "Temporary Tolerances; D&C Yellow Nos. 7 and 8 and D&C Red No. 37." *Fed. Regist.* **1960**, *25*, 9945.

14.   Food and Drug Administration. "FD&C Red No. 3; for Food." *Fed. Regist.* **1969**, *34*, 7446.

15.   Food and Drug Administration. "Subpart-Provisional Regulations; Postponement of Closing Dates." *Fed. Regist.* **1977**, *42*, 6992-7000.

16.   Food and Drug Administration. "Provisionally Listed Color Additives; Proposed Postponement of Closing Dates." *Fed. Regist.* **1976**, *41*, 41860-41866.

17.   Food and Drug Administration. "Listing of D&C Yellow No. 7 for Use in Externally Applied Drugs and Cosmetics." *Fed. Regist.* **1976**, *41*, 51003.

18.   Food and Drug Administration. "Listing of D&C Yellow No. 8 for Use in Externally Applied Drugs and Cosmetics." *Fed. Regist.* **1976**, *41*, 51004.

19.   H. Rep. 2284, 85th U.S. Congress, 2 1958.

20.   *U.S. Code of Federal Regulations, Title 21*, 70.3(i). U.S. Government Printing Office: Washington, DC, 1994; p 267.

21.   Food and Drug Administration. "Color Additives: Proposed Regulations for Lakes." *Fed. Regist.* **1965**, *30*, 6490.

22.   Food and Drug Administration. "Lakes of Color Additives; Termination of Proposal." *Fed. Regist.* **1979**, *44*, 36411.

23. Food and Drug Administration. "Lakes of Color Additives; Intent to List." *Fed. Regist.* **1979**, *44*, 36411-36415.
24. Food and Drug Administration. "Cosmetic, Toiletry, and Fragrance Association, Inc.; Amendment of Filing of Petitions for Color Additives." *Fed. Regist.* **1976**, *41*, 9584.
25. Food and Drug Administration. "Notice of Filing of Petitions Regarding Color Additives." *Fed. Regist.* **1968**, *33*, 17205.
26. Webb, J. M.; Fonda, H.; Brouwer, E.A. *J. Pharmacol. Exp. Ther.* **1962**, *137*, 141-147.
27. Food and Drug Administration. "Listing of D&C Yellow No. 7 for Use in Externally Applied Drugs and Cosmetics; Confirmation of Effective Date." *Fed. Regist.* **1977**, *42*, 10980.
28. Food and Drug Administration. "Listing of D&C Yellow No. 8 for Use in Externally Applied Drugs and Cosmetics; Confirmation of Effective Date." *Fed. Regist.* **1977**, *42*, 10980.
29. Food and Drug Administration. "Notice of Filing." *Fed. Regist.* **1968**, *33*, 9627.
30. Food and Drug Administration. "FD&C Red No. 3; for Food." *Fed. Regist.* **1969**, *34*, 7446.
31. Food and Drug Administration. "FD&C Red No. 3; Confirmation of Effective Date of Order Listing for Food and Drug Use." *Fed. Regist.* **1969**, *34*, 11542.
32. Joint FAO/WHO Expert Committee on the Food Additives, *Toxicological Evaluation of Certain Food Additives and Contaminants*, World Health Organization, Geneva, 1991, pp 171-181.
33. Food and Drug Administration. "Cosmetic, Toiletry, and Fragrance Association, Inc. Notice of Filing of Petitions Regarding Color Additives." *Fed. Regist.* **1973**, *38*, 21199.
34. Food and Drug Administration. "Termination of the Provisional Listings of FD&C Red No. 3 for Use in Cosmetics and Externally Applied Drugs and of Lakes for FD&C Red No. 3 for All Uses." *Fed. Regist.* **1990**, *55*, 3516-3519.
35. Food and Drug Administration. "Color Additives; Denial of Petition for Listing FD&C Red No. 3 for Use in Cosmetics and Externally Applied Drugs,...." *Fed. Regist.* **1990**, *55*, 3520-3543.
36. Public Health Service, "National Toxicology Program Board of Scientific Counselors Meeting." *Fed. Regist.* **1983**, *48*, 46104.
37. Food and Drug Administration. "D&C Orange No. 5." *Fed. Regist.* **1982**, *47*, 49632-49636.
38. Food and Drug Administration. "General Specifications and General Restrictions for Provisional Color Additives for Use in Foods, Drugs, and Cosmetics; Temporary Tolerances." *Fed. Regist.* **1979**, *44*, 48964-48967.
39. Food and Drug Administration. "D&C Orange No. 5." *Fed. Regist.* **1984**, *49*, 13339-13343.
40. Food and Drug Administration. "D&C Orange No. 10 and D&C Orange No. 11." *Fed. Regist.* **1981**, *46*, 18951-18954.
41. Food and Drug Administration. "D&C Orange No. 10 and D&C Orange No. 11; Confirmation of Effective Date." *Fed. Regist.* **1981**, *46*, 35085-35086.

42. Food and Drug Administration. "Termination of Provisional Listing of D&C Red No. 19 and D&C Red No. 37 for Use in Ingested Drugs and Cosmetics." *Fed. Regist.* **1983**, *48*, 5262-5264.

43. Food and Drug Administration. "Termination of Provisional Listing of D&C Red No. 37 for Use in Externally Applied Drugs and Cosmetics." *Fed. Regist.* **1986**, *51*, 20786-20788.

44. Food and Drug Administration. "Listing of D&C Red No. 19 for Use in Externally Applied Drugs and Cosmetics." *Fed. Regist.* **1986**, *51*, 28346-28363.

45. Food and Drug Administration. "Correction of Listing of D&C Red No. 19 for Use in Externally Applied Drugs and Cosmetics." *Fed. Regist.* **1987**, *52*, 5083-5084.

46. Food and Drug Administration. "D&C Orange No. 17 and D&C Red No. 19." *Fed. Regist.* **1986**, *51*, 35509-35511.

47. Food and Drug Administration. "Revocation of Regulations; D&C Orange No. 17 and D&C Red No. 19." *Fed. Regist.* **1988**, *53*, 26768-26770.

48. Food and Drug Administration. "Color Additives; Denial of Petition for Listing of D&C Red No. 19 for Use in Externally Applied Drugs and Cosmetics." *Fed. Regist.* **1988**, *53*, 26881-26883.

49. Food and Drug Administration. "D&C Red No. 21 and D&C Red No. 22." *Fed. Regist.* **1982**, *47*, 53843-53847.

50. Food and Drug Administration. "D&C Red No. 21 and D&C Red No. 22; Confirmation of Effective Date." *Fed. Regist.* **1983**, *48*, 4463.

51. Food and Drug Administration. "D&C Red No. 27 and D&C Red No. 28; Confirmation of Effective Date." *Fed. Regist.* **1982**, *47*, 53343.

52. Food and Drug Administration. "D&C Red No. 27 and D&C Red No. 28." *Fed. Regist.* **1982**, *47*, 42566-42569.

53. Rosenthal, I.; Yang, G.C.; Bell, S.J.; Scher, A.L. *Food Addit. Contam.* **1988**, *5*, 563-571.

54. Wei, R.R.; Wamer, W.G.; Bell, S.J.; Kornhauser, A. *Photochem. Photobiol.* **1994**, *59*, 31S.

55. Bockstahler, L.E.; Lytle, C.D.; Hitchins, V.M.; Carney, T.G.; Olvey, K.M.; Lamanna, A.; Ellingson, O.L.; Bell, S.J. *In Vitro Toxicol.* **1990**, *3*(2), 139-151.

56. Tonogai, Y.; Iwaida, M.; Tati, M.; Ose, Y.; Sato, T. *J. Toxicol. Sci.* **1978**, 3, 205-214.

RECEIVED August 31, 1995

Chapter 5

Risk Assessment: Phloxine B and Uranine Insecticide Application Trials

D. A. Bergsten

Animal and Plant Health Inspection Service, U.S. Department of Agriculture, 4700 River Road, Riverdale, MD 20737

As part of ongoing efforts to find safe and effective pest control methods, the U.S. Department of Agriculture's (USDA) Animal and Plant Health Inspection Service (APHIS) conducts trial tests with chemicals that show promise for control of pest species while posing a low risk to the human environment. Laboratory testing has shown that Suredye insecticide in bait formulation has good efficacy against Mediterranean and Mexican fruit flies. These nonindigenous fruit flies have the capacity to destroy economically important crops. Field application studies are planned to test further the efficacy of Suredye bait spray. This chapter summarizes analyses of the potential risk of adverse effects to human health, wildlife, and environmental quality from field trials of this formulation.

The National Environmental Policy Act of 1969 requires that Federal agencies prepare environmental documentation for proposed Federal actions that have the potential to pose risk to the human environment. The risk assessment was prepared in anticipation of the future need for compliance with this statute and in determination of the safety of potential use of this insecticide formulation in agency programs.

The risk assessment analyses follow the basic guidelines recommended by the National Research Council (1). The risk assessment is designed to identify potential adverse effects from the insecticide applications in the field trials and to evaluate the risk of such effects occurring. The results of these analyses are presented in sections covering chemical properties and environmental fate of the insecticide, hazard identification, exposure assessment, and risk characterization.

Chemical Properties and Environmental Fate

Formulation of the bait spray involves mixing Suredye with fruit fly attractants (fructose and protein hydrolysate) and diluting the mixture with water. Risk assessment of the attractants has been presented in the *Medfly Cooperative Eradication Program Final Environmental Impact Statement —1993* (2). Both attractants are considered safe to animals, including fish and birds. The proposed rate of application of Suredye for the field trials is 0.25 oz of active ingredient (a.i.) per acre (0.018 kg a.i. per hectare). Suredye is a mixture of two chemicals —69% phloxine B (a red dye) and 31% uranine (a yellow dye) by weight. The risk assessment analyzes potential effects of these chemicals when applied as a mixture at the proposed rate of application.

Phloxine B and uranine are xanthene dyes that are used as color additives in drugs and cosmetics. These dyes occur as powders at room temperature, but they are highly water soluble. The powders decompose or melt at very high temperatures. The water solubility is greater than 120 mg/mL for phloxine B (Riddle, C., Hilton Davis Co., Cincinnati, OH, unpublished data, 1994) and is greater than 100 mg/mL for uranine (3). The molecular weights of these dyes are higher than those of many insecticides —691.91 for phloxine B and 376.27 for uranine. Although most insecticides are lipophilic, these compounds are hydrophilic. Their solubility in lipids is very low compared to the solubility of most other insecticides. The octanol –water partition coefficient is 0.62 for phloxine B at 25 °C (White, E., Hilton Davis Co., Cincinnati, OH, unpublished data, 1994). The octanol-isotonic salt solution partition coefficient is 0.72 for uranine at 25 °C (4).

Neither phloxine B nor uranine are anticipated to persist or remain toxic for very long in the environment. Exposure of these compounds to sunlight is expected to lead to photodegradation (simultaneous photodetoxification) with a half-life of approximately 1 hour (5), so residues will not persist in the atmosphere or on surfaces exposed to direct sunlight, such as soil or plant surfaces. Although some surfaces may be stained by the dyes, their residues do not persist long and readily wash off due to water solubility. Suredye bait spray applications will not be applied directly to water, but drift of some insecticidal particles could occur. Rapid degradation of these dyes in water through photobleaching has been demonstrated (6, 7). Although one might expect some leaching of residues to groundwater due to the high water solubility of phloxine B and uranine, the actual amount of these chemicals reaching groundwater (interstitial soil water) is quite low as indicated in Table I. The limited movement may relate partly to adsorption to soil particles and to the high molecular weights of these compounds. Both compounds have very low affinity for lipids, so bioaccumulation and bioconcentration of residues is not anticipated. Studies of metabolism in rats indicate that these compounds are either unchanged or undergo glucuronide conjugation in the digestive tract (8). These studies found that about 65% of the phloxine B and 44% of the uranine are excreted within the first 2 hours.

Calculations for environmental fate and transport are based on deposition of Suredye at the program application rate (0.018 kg a.i. per hectare). The

Table I. Summary of GLEAMS Modeling Output for Maximum Levels of Insecticide in the Upper 1 cm of Soil ($\mu g/g$) and Interstitial Soil Water ($\mu g/L$)

Chemical/ Media	Brownsville, TX	Gulfport, MS	Los Angeles, CA	Miami, FL	Orlando, FL	Santa Clara, CA
			Site			
Phloxine B						
Soil	0.0180	0.0159	0.0159	0.0167	0.0182	0.0079
Water	2.7	3.4	4.1	3.0	0.9	0.5
Uranine						
Soil	0.0081	0.0072	0.0072	0.0075	0.0082	0.0036
Water	1.2	1.5	1.8	1.4	0.4	0.2

highest concentrations of phloxine B and uranine in surface soil and interstitial soil water (groundwater) as determined using the GLEAMS (Groundwater Loading Effects of Agricultural Management Systems) model (9) are presented in Table I. The model was programmed to project concentrations following a 2-year storm occurring 24–48 hours after insecticide application at six potential program sites. A 2-year storm is an occurrence of precipitation projected to be the highest over a 2-year period. This magnitude of storm was selected to provide representative data for concentrations in groundwater and runoff water for each site. Smaller storms are likely to result in less transport of insecticide from the site of application. Larger storms are likely to result in higher dilution of the insecticide concentrations in runoff and groundwater. Detailed descriptions of the program sites and modeling parameters are presented in the Human Health Risk Assessment for APHIS Fruit Fly Programs (10). This document also provides the equations used in calculations of pesticide concentrations in runoff water from impervious surfaces, surface water, and soil.

Hazard Identification

Humans and Mammals. The active ingredients in Suredye (phloxine B and uranine) have low acute toxicity to mammals by all routes of exposure. The acute dose necessary to kill 50% of the test organisms is referred to as the LD_{50}. The acute oral LD_{50} of phloxine B to male rats is 8,400 milligrams (mg) per kilogram (kg) of body weight (11). No mortality was observed in dogs exposed orally to doses as high as 4,560 mg/kg (12). The acute oral LD_{50} of uranine is 6,700 mg/kg to rats and 4,738 mg/kg to mice (13). The rapid metabolism, low toxicity, and rapid excretion of these compounds in mammals probably account for the low mortality observed (8, 14). The acute intravenous LD_{50} of phloxine B is 310 mg/kg to male mice and 1,000 mg/kg to rats (15, 16). The acute intraperitoneal LD_{50} of uranine is 1,800 mg/kg to mice and 1,700 mg/kg to rats (3).

Phloxine B and uranine can be mild skin and eye irritants. The rate of health-related complaints from the consuming public for cosmetics containing phloxine B (2.6 to 37.1 complaints per million units sold) is considered to be well within reasonable limits of safety, and it is uncertain whether the dye or other agents in the formulated products were the source of the complaints (17). Repeated applications of phloxine B to rabbit skin produced no evidence of immunotoxic responses (18).

Chronic studies of phloxine B and uranine indicate low hazard to mammals. In studies that involved feeding rats and dogs diets that contained up to 1% phloxine B for 2 years, no adverse effects were observed from visible or pathologic examinations (19–22). Lifetime studies of mice found no evidence of tumors when dermal applications of 1% solutions of phloxine B were applied weekly (23, 24). The maximum acceptable daily intake (MADI) of phloxine B by humans as determined by the U.S. Department of Health and Human Services' Food and Drug Administration is 1.25 mg/kg/day (25). To ensure safety, the MADI takes into account differences among responses in study subjects and responses in the general population by applying uncertainty factors.

The lowest effect level (LEL)(the lowest level at which there is an effect on the individuals tested for a series of dose levels) found for humans to chronic exposures of uranine is at 7 mg/kg/day for exposure by intravenous injection (*3*). Neither phloxine B nor uranine are considered to be carcinogenic by either the National Toxicology Program or the International Agency for Research on Cancer (*26, 27*). None of the metabolites or degradation products of these compounds constitute a substantial increase in hazard relative to the parent compound (*5, 8, 14*).

Quantitative toxicological assessments involve the derivation of dose levels associated with an acceptable level of risk. This dose is referred to as the regulatory reference value (RRV). RRVs for noncarcinogenic effects are exposure values designed to estimate the exposure at or below which no adverse effects are expected for a given exposure route and duration. These determinations assume that population thresholds exist for such adverse effects. RRVs are generally derived by taking an experimental threshold dose for the route of exposure and dividing by an uncertainty factor. The uncertainty factor is intended to account for differences between the experimental exposure and the conditions for which the RRV is being derived. Exposure to higher concentrations than the MADI of phloxine B (*25*) is considered to constitute increased hazard for adverse health effects, and this value (1.25 mg/kg/day) is used in this risk assessment as the RRV for phloxine B for both occupational exposures and exposures of the general population. Calculation of the MADI includes uncertainty factors.

The RRV determined for uranine is based on the study of intravenous injections that determined an LEL of 7 mg/kg/day (*16*). Although exposures from pesticide applications of Suredye bait spray will not occur by the intravenous route, the use of this value for uranine assures a conservative approach that makes allowance for the most sensitive route of exposure. An uncertainty (safety) factor of 10 is applied to the LEL of uranine to get an RRV of 0.7 mg/kg/day for the general population. Using this safety factor accounts for the sensitivity of most subgroups within the general population relative to the test study population. The RRV determined for occupational exposure to uranine equals the LEL of 7 mg/kg/day. These values are the same because the potential occupational exposure routes (dermal, ingestion, inhalation) differ from the study (intravenous) and because the actual dose of uranine to the cardiovascular system would be considerably less than the total occupational exposure. This RRV also takes into account the healthy worker effect —the fact that the tolerance of workers to exposure is generally greater than the sensitive subgroups in the general population.

Nontarget Wildlife. The hazards to wildlife from applications of Suredye are primarily to those invertebrates that consume the deposited Suredye bait spray. Dye-sensitized photooxidation reactions and high mortality from toxic effects of phloxine B have been shown for several insect species (*28–33*). Toxicity to terrestrial invertebrates from the dye occurs only through ingestion, and ingestion is likely only for those species attracted to the bait for feeding. Aquatic invertebrates are also affected by exposure to the dye, which is readily soluble in water (*34*). The mode of toxic action of the dye against insects is not completely understood, but toxicity apparently results from released energy from the dye molecules to oxygen molecules (*35*). This "energized" oxygen is an excellent oxidizing agent that can severely damage tissues of the exposed insect. The actual site of toxic action is unknown; mortality likely results from cumulative oxidation at many discrete target organs rather than at a single critical target (*35*).

Unlike their effect on invertebrates, the compounds in Suredye are relatively nontoxic to birds, reptiles, amphibians, and fish (*35–38*). Although phloxine B has been applied as an herbicide, phytotoxicity occurs only at much higher application rates than those used for applying phloxine B as an insecticide (Darlington, L., U.S. Department of Agriculture, Agricultural Research Service, Beltsville, MD, personal communication, 1994).

Some indirect hazards could result from insecticide applications of Suredye. Insectivorous wildlife and insect parasites that depend on certain species of insects for food could be affected if the prey base was significantly reduced due to consumption of the insecticidal dye. A reduced prey base could lead to hunting stress to insectivorous populations or the need to relocate to an area with a more dependable food source. Although indirect effects are likely to be temporary and local for most species, the effects to some sensitive, geographically confined species could be considerable.

Environmental Quality.   The hazards of phloxine B and uranine to environmental quality are minimal, largely as a result of the environmental fate described earlier in this chapter. Neither compound persists for more than a few hours in air or water due to rapid photodegradation and photobleaching. These dyes may stain the surfaces of plants, soil, or other exposed solid materials, but their high water solubility and rapid photodegradation assure that any evidence of their absorption into permeable substrates or adsorption to inert surfaces is not evident shortly after exposure to sunlight, rainfall, or weathering. This rapid breakdown ensures that no permanent effects can be anticipated on the quality of air, soil, and water as a result of insecticide applications of Suredye bait spray.

Exposure Assessment

Exposure Assessment Scenarios. The exposure assessment scenarios analyzed in this risk assessment are classified into three categories (routine, extreme, accidental) based upon the plausible range of exposures. Routine exposure assessments assume that the proposed application rates and recommended safety precautions were followed. Furthermore, routine exposures are based on the most likely estimates of physiological modeling parameters such as water consumption rates and values for skin surface exposure. Extreme exposure assessments assume that recommended procedures and precautions were not followed and use more conservative, but still plausible, modeling parameters that increase the estimate of exposure. Examples of accidental exposure scenarios include some form of equipment failure or gross human error. Although accidental exposure scenarios are *worst case scenarios* within the context of the risk assessment, they are intended, nonetheless, to represent realistic rather than catastrophic events.

The number of scenarios assessed for each program activity was limited because an excessive number of scenarios can obscure rather than clarify the findings of the risk assessment. For example, if analysis of the exposure scenario for toddlers swimming for 4 hours in a pool within the treatment area suggests low potential hazard, then there is no need to calculate exposures for an adult in this scenario or for shorter intervals.

Variability within scenarios for exposure media is considered by estimating exposed or absorbed doses for individuals of different age groups (i.e., adults, young children, toddlers, and infants). Children may behave in ways that increase their exposure to applied pesticides (e.g., long periods of outdoor play, pica, or imprudent consumption of contaminated media). Pica refers to the tendency of individuals to consume unnatural food, such as the toddler who ingests 10 grams of soil per day. In addition, anatomical and physiological factors, such as body surface area, breathing rates, and consumption rates for food and water, are not linearly related to body weight and age (*39–41*). Consequently, the models used to estimate the exposure dose (e.g., mg pesticide per kg body weight per day) based on chemical concentrations in environmental media (e.g., ppm in air, water, soil, vegetation, or food) generally indicate that children, compared to individuals of different age groups, are exposed to the highest doses of chemicals for a given environmental concentration.

Exposure scenarios are designed to consider all likely routes of exposure (dermal, oral, and inhalation). Previous analyses (*10*) have shown that inhalation exposure is minor and inconsequential relative to oral and dermal exposures, so exposure by inhalation was not analyzed. The exposure scenarios analyzed for the general public include consumption of contaminated media (e.g., soil, surface water, runoff water, and vegetation), skin contact with contaminated vegetation, and contact from swimming or bathing. Occupational exposure scenarios were analyzed for pilots, backpack applicators, hydraulic rig applicators, mixers/loaders, and ground personnel.

Dose Estimation. Dose estimates are determined from environmental exposure levels and are converted into common units (e.g., mg pesticide per kg of body weight per day) that can be compared to the RRV. In exposure scenarios where both oral and dermal exposure must be considered, the two exposures can be added to get the total exposure to pesticide residues.

If the amount of chemical consumed is known, the oral dose can be determined directly by dividing the consumed quantity of chemical by the body weight of the person. More often, risk assessors must rely on a concentration estimate for the chemical in a medium. To get the intake of the chemical, risk assessors must estimate the amount consumed and multiply this figure by the concentration estimate for the medium.

Calculation of dermal dose is considerably more complex than calculation of the oral dose. Several methods for calculating dermal dose have been reviewed by the U.S. Environmental Protection Agency (*42*). Information about dermal absorption of many pesticides is limited. Calculation of the dermal absorption depends upon the derivation of the dermal permeability coefficient of the insecticides present to calculate exposure. A rough estimate may be made using an equation to calculate the permeability coefficient from the octanol –water partition coefficient and molecular weight (*42*). Unfortunately, this equation is based on primarily lipophilic compounds, so phloxine B and uranine (hydrophilic compounds) are not likely to be estimated accurately by this method.

A more conservative approach to determine the potential permeability of hydrophilic compounds is to assume a log permeability coefficient of −5 for compounds of high molecular weight and low octanol-water partition coefficient (*43*). This approach assures that the movement of the compound through the skin will not be underestimated. This permeability coefficient may be applied directly to Fick's first law in calculation of exposure of ground personnel and of persons contacting contaminated water or vegetation. Fick's first law of diffusion as applied to skin uses the steady-state assumptions that the concentration gradient of the penetrant across skin provides the driving force for penetration, the rate of penetration at steady state is proportional to concentrations, and the penetrant causes no damage to the skin itself. Applications of this law are reviewed in detail by the U.S. Environmental Protection Agency (*42*).

For other scenarios that are primarily occupational in nature, actual exposure data from similar studies of field applications can be applied. These data generally pertain to different insecticides, different application rates, and different lengths of exposure, so corrections must be made for differences in permeability, application rates, and duration of exposure between the study data and the exposure scenario. Using field data allows risk assessors to project exposure with greater accuracy than direct application of Fick's first law, which tends to overestimate dose, particularly when used to analyze nominal application rates rather than dislodgable residues (*10, 42*).

Human Occupational Exposure. Table II presents exposure doses from applications of Suredye bait spray for pilots, backpack applicators, hydraulic rig applicators, mixers and loaders, and ground personnel (kytoon handlers, flaggers, and quality control crew). The projected daily occupational exposures to phloxine B and uranine are relatively low compared to exposures for similar application rates of more lipophilic insecticides.

The dose calculations for pilots (*44*), backpack sprayers (*45*), and mixers/loaders (*46*) were based upon appropriate field data. Data for the mixers/loaders used nominal application rates, so doses to these workers are likely to be overestimated. Although no field data were available for hydraulic rig applicators, the potential exposure would be similar to hack-and-squirt applicators for which there is good data (*45*). Use of data from these surrogate studies requires adjustments for dermal absorption, application rate, and duration of exposure as discussed in the previous section.

Unlike the other workers, appropriate studies of ground personnel were unavailable. Therefore, determination of the dose required the application of Fick's first law to deposition from nominal application rates. As discussed previously, use of this method tends to overestimate dose. In addition, this scenario assumes that the worker is outside such that all exposed body surfaces get uniform deposition of insecticide and that none of the residue is washed off for 8 hours. These exaggerated conditions actually encompass accidental exposures. The functions of the different ground personnel suggest differences in exposure levels. Flaggers and kytoon handlers guide the flight crew to ensure that the correct areas are treated, but flaggers work from inside vehicles with lighting and signal devices. The quality control crew does not enter the treated site until the application has been completed. Based upon these functions, the exposure levels of ground personnel are expected to be: kytoon handlers > flaggers > quality control crew. The scenarios for ground personnel represent kytoon handlers using the least safety precautions. However, the recent use of electronic navigational techniques, such as Global Positioning Systems, largely restricts the functions of ground personnel to quality control activities.

Table II. Summary of Occupational Exposures to Phloxine B and Uranine From Applications of Suredye Bait Spray

Occupational Group	Exposure Scenario	Dose (mg/kg/day)	
		Phloxine B	*Uranine*
Pilots	Routine	2.7×10^{-6}	1.2×10^{-6}
	Extreme	7.4×10^{-6}	3.3×10^{-6}
Backpack Applicators	Routine	4.3×10^{-5}	1.9×10^{-5}
	Extreme	8.6×10^{-5}	3.9×10^{-5}
Hydraulic Rig Applicators	Routine	2.1×10^{-5}	9.4×10^{-6}
	Extreme	7.8×10^{-5}	3.5×10^{-5}
Mixers/Loaders	Routine	2.0×10^{-5}	1.2×10^{-5}
	Extreme	5.7×10^{-5}	3.2×10^{-5}
Ground	Routine	3.0×10^{-2}	1.3×10^{-2}
	Extreme	7.7×10^{-2}	3.5×10^{-2}

General Public Exposure. Residue exposures of the general population from applications of Suredye bait spray were determined for scenarios of soil consumption, water consumption, vegetation consumption, vegetation contact, and swimming pool exposures (Table III). As with the occupational exposures, the projected daily exposures of the general population to phloxine B and uranine are relatively low compared to exposures for similar application rates of more lipophilic insecticides.

In each of the scenarios, the most sensitive subgroup in the population was examined. The representative body weight selected for adults was 70 kg and for toddlers was 10 kg. The consumption of soil with residual insecticide was calculated for toddlers. Even those toddlers who have the propensity for pica behavior were shown to get low doses of phloxine B and uranine. The dose to a toddler is considerably greater from drinking treated surface water than from drinking runoff water containing residual insecticide, but these doses are still quite low. The swimming pool scenario analyzes oral and dermal exposure to a toddler who spends 4 hours bathing. The scenarios for consumption of vegetation analyze exposure from eating home-grown leafy vegetables based on a USDA food consumption survey (*47*). The scenario for contact with treated vegetation was based on intense exercise on grass for up to 4 hours. The transfer of insecticide in this scenario was based on a study of individuals exercising on a treated carpet (*48*); measurements taken from this study are in close agreement with measurements taken on turf (*49*).

Table III. Summary of General Population Exposures to Phloxine B and Uranine From Applications of Suredye Bait Spray

Exposure Route	Exposure	Dose (mg/kg/day)	
		Phloxine B	Uranine
Soil Consumption	Routine	2.0×10^{-7}	8.8×10^{-8}
	Extreme	4.6×10^{-7}	2.0×10^{-7}
	Pica Behavior	1.8×10^{-5}	8.2×10^{-6}
Consumption of Contaminated Water	Runoff Water	1.8×10^{-8}	8.0×10^{-9}
	Surface Water	1.9×10^{-5}	8.5×10^{-6}
Swimming Pool Exposure	Toddler for 4 hours	4.2×10^{-8}	1.9×10^{-8}
Consumption of Contaminated Vegetation	Routine	4.8×10^{-4}	2.2×10^{-4}
	Extreme	2.5×10^{-3}	1.1×10^{-3}
Contact with Contaminated Vegetation	Routine	4.9×10^{-5}	2.2×10^{-5}
	Extreme	1.2×10^{-4}	5.2×10^{-5}

Wildlife Exposure. Potential exposures of terrestrial wildlife other than insect species to phloxine B and uranine from applications of Suredye bait spray will be very low. Exposures to mammals, birds, reptiles, and terrestrial amphibians are not expected to exceed the doses to humans for the extreme or accidental scenarios. The highest doses determined from those scenarios were in the parts –per –billion range (micrograms chemical per kilogram body weight).

Phytotoxic effects require much higher application rates, so exposure of plants need not be considered.

The only route of exposure to consider for insects is oral because the toxicity of the dyes to terrestrial insects occurs only through ingestion. This exposure may occur through grooming of the body, but doses sufficient to induce toxic responses are limited to feeding on the bait spray. Several insect species may be attracted to and feed on the bait spray. The plant bugs (miridae), ground beetles (carabidae), midges and gnats (nematocerous Diptera), pomace flies, other acalypterate muscoid flies, ants (formicidae), and soil mites (Acari) are attracted to the protein hydrolysate in large numbers (50). These species are most likely to get high exposures. Most terrestrial invertebrates, however, are not attracted to the Nu-Lure bait or fructose in the Suredye bait spray formulation.

Exposures of aquatic species to phloxine B and uranine from Suredye bait spray applications are expected to be very low. The high water solubility of these dyes assures almost instantaneous mixing in the water, but the rapid degradation assures that only short durations of exposure (not expected to exceed several hours) are possible for given treatments. Applying the minimum depth (0.3 meters) considered in analyses of bodies of water in the Nontarget Risk Assessment for the Medfly Cooperative Eradication Program (51) to Suredye bait spray applications, a direct application would result in water concentrations of only 4.67×10^{-3} mg phloxine B per liter and 2.1×10^{-3} mg uranine per liter. The doses of dye taken up by aquatic organisms from this low water concentration will be very low.

Risk Characterization

Risk characterization is the process of comparing the exposure assessment with the toxicological assessment to express the level of concern regarding an exposure scenario or set of scenarios (1). The dose of each chemical insecticide in the formulated material is compared to the RRV for that chemical and exposure group. The overall risk results from exposures to both phloxine B and uranine. Description of the risk from individual chemicals is generally expressed as the hazard quotient (HQ). The HQ is calculated by dividing the exposure to the chemical by the RRV of that chemical and exposure group. A cumulative HQ for exposure to multiple agents with similar mechanisms of toxic action, such as phloxine B and uranine, may be determined by adding the HQ of each agent for a given scenario. The cumulative HQ expresses the likelihood of potential adverse effects for given exposure scenarios. The smaller the HQ, the lower the likelihood of any possible adverse effects. Cumulative HQ values less than 1 are not anticipated to cause any adverse effects; anticipated adverse effects are generally slight when the HQ is at 1 or slightly higher, but individual reactions may vary due to interindividual variability. Although most people are unlikely to respond adversely to conditions where the HQ is 1, individuals who are more sensitive to the chemical exposure could express adverse reactions from even lower exposures.

Human Health. The cumulative HQs for all scenarios indicate that the margin of safety is considerable for both occupational exposures (Table IV) and general population exposures (Table V). The lowest margins of safety for occupational exposures occur in routine and extreme scenarios of ground personnel, and these margins still exceed 10-fold. As stated in the exposure section, this exposure scenario is conservatively biased, and the margins of safety are probably much greater. All other occupational exposures have margins of safety that exceed 1,000-fold. The lowest margin of safety for general population exposures occurs for the individual who consumes leafy vegetables from the garden immediately after treatment with Suredye bait spray. The margin of safety exceeds 1,000-fold for the routine exposure scenario and exceeds 100-fold for the extreme exposure scenario for consumption of contaminated vegetation. The large margins of safety for all scenarios indicate that the risk of adverse effects to humans are very low from applications of Suredye bait spray. No adverse effects to humans are anticipated from these applications, even under accident scenarios.

Wildlife. The margins of safety for most terrestrial wildlife are considerable and of a magnitude similar to those for human health risks from applications of Suredye bait spray. The risks of adverse effects on the survival of mammals, birds, reptiles, and terrestrial amphibians are very low. Although some insectivorous wildlife may have smaller prey bases following applications, the affected prey species would be limited to those insects attracted by the protein hydrolysate bait, and these species are unlikely to be the primary food source for most insectivores. Moreover, these conditions would likely be temporary because movement of these prey species from adjacent areas is expected to readily replace the individuals or populations killed from consuming insecticide bait. Exposures of plants are several orders of magnitude below levels that could cause phytotoxic responses.

The restricted route of toxic action (oral) in terrestrial invertebrates limits the number of species likely to be at risk of adverse effects. Considerable exposure is expected for those invertebrates attracted to the protein hydrolysate. This group includes plant bugs, ground beetles, midges, gnats, acalypterate muscoid flies (such as fruit flies), ants, and soil mites. Populations of these invertebrates are likely to be lowered considerably due to the toxic action of the insecticide. The risk to most other invertebrates is much lower. Species that are not attracted to the protein hydrolysate to feed are at low risk. This group includes honey bees, lacewings, springtails, aphids, whiteflies, tumbling flower beetles, calypterate muscoid flies, and spiders. Many of the species not expected to be affected by Suredye bait spray are currently affected by program applications of malathion bait spray through contact exposure.

Aquatic species are at very low risk of adverse effects. The concentration of these dyes in water is several orders of magnitude less than any concentration

Table IV. Cumulative Hazard Quotients for Occupational Exposures to Insecticide From Applications of Suredye Bait Spray

Occupational Group	Exposure Scenario	Cumulative Hazard Quotient
Pilots	Routine	2.3×10^{-6}
	Extreme	6.4×10^{-6}
Backpack Applicators	Routine	3.7×10^{-5}
	Extreme	7.5×10^{-5}
Hydraulic Rig Applicators	Routine	1.8×10^{-5}
	Extreme	6.7×10^{-5}
Mixers/Loaders	Routine	1.8×10^{-5}
	Extreme	5.1×10^{-5}
Ground Personnel	Routine	2.6×10^{-2}
	Extreme	6.7×10^{-2}

Table V. Cumulative Hazard Quotients for General Population Exposures to Insecticide From Applications of Suredye Bait Spray

Exposure Route	Exposure Scenario	Cumulative Hazard Quotient
Soil Consumption	Routine	2.9×10^{-7}
	Extreme	6.5×10^{-7}
	Pica Behavior	2.7×10^{-5}
Consumption of Contaminated Water	Runoff Water	2.5×10^{-8}
	Surface Water	2.7×10^{-5}
Swimming Pool Exposure	Toddler for 4 hours	6.0×10^{-8}
Consumption of Contaminated Vegetation	Routine	6.9×10^{-4}
	Extreme	3.6×10^{-3}
Contact With Contaminated Vegetation	Routine	7.0×10^{-5}
	Extreme	1.0×10^{-4}

known to affect aquatic organisms adversely. The high water solubility and low octanol –water partition coefficient of these compounds indicate that residues will not bioconcentrate in tissues, so adverse effects are unlikely from residual dye. The short half–life in water assures that any adverse effects from the low concentration would be local and of short duration.

Environmental Quality. The risks from applications of Suredye bait spray to environmental quality are minimal. Neither phloxine B nor uranine persist for more than a few hours in air or water due to rapid photodegradation. These dyes may stain the surfaces of plants, soil, or other exposed solid materials, but their high water solubility and rapid photodegradation ensure that their absorption into permeable substrates or adsorption to inert surfaces is not evident shortly after exposure to sunlight, rainfall, or weathering. This rapid breakdown assures that no permanent effects can be anticipated on the quality of air, soil, and water.

Literature Cited

1.   *Risk Assessment in the Federal Government: Managing the Process.* National Research Council, National Academy Press: Washington, DC, **1983;** 191 pp.
2.   *Medfly Cooperative Eradication Program Final Environmental Impact Statement* . U.S. Department of Agriculture, Animal and Plant Health Inspection Service: Hyattsville, MD, **1993;** 184 pp.
3.   Registry of Toxic Effects of Chemical Substances Database. *Fluorescein, Sodium Salt.* National Institute for Occupational Safety and Health: Cincinnati, OH, **1994;** 2 pp.
4.   Valenzeno, D.P.; Pooler, J.P. *Photochem. Photobiol.* **1982,** 35, 343-350.
5.   Heitz, J. R.; Wilson, W. W. In *Disposal and Decontamination of Pesticides*; Kennedy, M. V., Ed.; ACS Symposium Series 73; American Chemical Society: Washington, DC, **1978;** pp 35-48.
6.   Blum, H. F.; Spealman, C. R. *J. Phys. Chem.* **1933,** 37, 1123-1133.
7.   Imamura, M.; Koizumi, M. *Bull. Chem. Soc. Jpn.* **1955,** 28, 117-124.
8.   Webb, J. M.; Fonda, M.; Brouwer, E. A. *J. Pharmacol. Exp. Therap.* **1962,** 137, 141-147.
9.   Davis, F. M.; Leonard, R. A.; Knisel, W. G. *GLEAMS User Manual* . U.S. Department of Agriculture, Agricultural Research Service: Tifton, GA, **1990;** 60 pp.
10.   *Human Health Risk Assessment, APHIS Fruit Fly Programs.* Syracuse Environmental Research Associates, Inc.: Fayetteville, NY, **1992;** 133 pp.
11.   *Acute Oral Toxicity Studies on Four Materials—Albino Rats.* Industrial Bio–Test Laboratories, Inc.: Northbrook, IL, **1962;** 11 pp.
12.   *Acute Oral Toxicity Studies on Four Materials—Dogs.* Industrial Bio–Test Laboratories, Inc.: Northbrook, IL, **1962;** 11 pp.
13.   Yankell, S.; Loux, J. *J. Peridontol.* **1977,** 48(4), 228-231.

14. Hansen, W. H.; Fitzhugh, O. G.; Williams, M. W. *J. Pharmacol. Exp. Therap.* **1958,** 122, 29A.
15. Lutty, G. A. *Toxicol. Appl. Pharmacol.* **1978,** 44, 225–249.
16. McDonald, T.; Kasten, K.; Hervey, R.; Gregg, S.; Robb, C. A.; and Borgmann, A. R. *Toxicol. Appl. Pharmacol.* **1974,** 29, 97–98.
17. *Report: Safety of Cosmetics in Use.* Toilet Goods Association, Inc.: Washington, DC, **1965;** Entry No. 24, Color Additive Master File No. 9.
18. *Report: Safety Evaluation on Skin of Rabbits, D&C Red No. 27.* Leberco Laboratories: Roselle Park, NJ, **1965;** Entry No. 91, Color Additive Master File No. 9.
19. *Two-Year Chronic Oral Toxicity of D&C Red 27—Beagle Dogs.* Industrial Bio -Test Laboratories, Inc.: Northbrook, IL, **1965;** Entry No. 57, Color Additive Master File No. 9.
20. *Two-Year Chronic Oral Toxicity of D&C Red 27—Albino Rats.* Industrial Bio -Test Laboratories, Inc.: Northbrook, IL, **1965;** 48 pp.
21. *Mouse Long-Term Study.* Litton Bionetics, Inc.: Kensington, MD, **1981;** 1579 pp.
22. *30-Month Chronic Toxicity and Potential Carcinogenicity Study in Rats with In Utero and Lifetime Exposure to D&C Red No. 27 in the Diet.* Litton Bionetics, Inc.: Kensington, MD, **1981;** 3070 pp.
23. *Report: Safety Evaluation on Skin of Rabbits, D&C Red No. 27.* Leberco Laboratories: Roselle Park, NJ, **1964;** 105 pp.
24. *Repeated Dermal Applications —Mice. Final Report.* Hazleton Laboratories, Inc.: Falls Church, VA, **1969;** 16 pp.
25. *Fed. Regist.* **1982,** 47(188), 42566–42569.
26. *Material Safety Data Sheet: Phloxine B.* J. T. Baker, Inc.: Phillipsburg, NJ, **1994;** 7 pp.
27. *Material Safety Data Sheet: Fluorescein, Sodium Derivative, Sodium Salt.* J. T. Baker, Inc.: Phillipsburg, NJ, **1994;** 7 pp.
28. Fondren, J. E., Jr.; Heitz, J. R. *Environ. Entomol.* , **1979,** 8, 432–436.
29. Fondren, J. E., Jr.; Heitz, J. R. *Environ. Entomol.,* **1978,** 7, 843–846.
30. Callaham, M. F.; Broome, J. R.; Lindig, O. H.; Heitz, J. R. *Environ. Entomol.* **1975,** 4(5), 837–841.
31. Clement, S. L.; Schmidt, R. S.; Szatmari-Goodman, G.; Levine, E. *J. Econ. Entomol.* **1980,** 73, 390–392.
32. Callaham, M. F.; Lewis, L. A., Holloman, M. E., Broome, J. R., Heitz, J. R. *Comp. Biochem. Physiol.* **1975,** 51C, 123–128.
33. Broome, J. R.; Callaham, M. F.; Heitz, J. R. *Environ. Entomol.* **1975,** 4(6), 883–886.
34. Schildmacher, H. *Biol. Zentralbl.* **1950,** 69, 468–477.
35. Heitz, J. R. In *Insecticide Mode of Action*; Coats, J. R., Ed.; Academic Press, Inc.: New York, **1982;** pp 429–457.
36. Tonogai, Y.; Ito, Y.; Iwaida, M. *J. Toxicol. Sci.* **1971,** 4, 115–125.
37. Pimprikar, G. D.; Fondren, J. E., Jr.; Greer, D. S.; Heitz, J. R. *Southwest. Entomol.* **1984,** 9(2), 218–222.

38. Marking, L. L. *Prog. Fish-Cult.* **1969,** 31(3), 139–142.
39. *Recommendations for and Documentation of Biological Values for Use in Risk Assessment*. U.S. Environmental Protection Agency, Environmental Criteria and Assessment Office: Cincinnati, OH, **1989;** ECAO-CIN-554.
40. *Interim Methods for Development of Inhalation Reference Concentrations*. U.S. Environmental Protection Agency, Environmental Criteria and Assessment Office: Research Triangle Park, NC, **1990;** EPA/600/8-90/066A.
41. *Reference Physiological Parameters in Pharmacokinetics Modeling*. U.S. Environmental Protection Agency, Office of Health and Environmental Assessment, Office of Research and Development: Washington, DC, **1988;** EPA/600/8-88/004.
42. *Dermal Exposure Assessment: Principles and Applications*. U.S. Environmental Protection Agency, Office of Health and Environmental Assessment, Office of Research and Development: Washington, DC, **1992;** EPA/600/8-91/011B.
43. Flynn, G. L. In *Percutaneous Absorption: Mechanisms –Methodology –Drug Delivery,* 2nd ed.; Bronaugh, R.; Maibach, H. I., Eds.; Marcel Dekker: New York, **1990;** pp 27–51.
44. Nash, R. G.; Kearney, P. C.; Maitlen, J. C.; Sell, C. R.; Fertig, S. N. In *Pesticide Residues and Exposures*; Garner, W. Y.; Harvey, J. Jr., Eds.; ACS Symposium Series 238; American Chemical Society: Washington, DC, **1982;** pp 119–132.
45. Lavy, T. L.; Norris, L. A.; Mattice, J. D.; Marx, D. B. *Environ. Toxicol. Chem.* **1987,** 6, 209–224.
46. Cowell, J. E.; Danhaus, R. G.; Kunstman, J. L.; Hackett, A. G.; Oppenhuizen, M. E.; Steinmetz, J. R. *Arch. Environ. Contam. Toxicol.* **1987,** 16, 327–332.
47. Pao, E. M. *Foods Commonly Eaten by Individuals: Amounts per Day and per Eating Occasion*. Home Economics Research Report No. 44; U.S. Department of Agriculture, Home Nutrition Information Service: Washington, DC, **1982;** 431 pp.
48. Fong, H. R.; Brodberg, R. K.; Formoli, T.; Sanborn, J. R.; Thongsinthusak, T.; Ross, J. *Estimation of Exposure of Persons in California to Pesticide Products that Contain Malathion Used in the Medfly Eradication Project*. California Department of Food and Agriculture: Sacramento, CA, **1990;** 18 pp.
49. Harris, S. A.; Solomon, K. R. *J. Environ. Sci. Hlth.* **1992,** B27(1), 9–22.
50. Troetschler, R. G. *Environ. Entomol.* **1983,** 12(6), 1816–1822.
51. *Nontarget Risk Assessment for the Medfly Cooperative Eradication Program*. U.S. Department of Agriculture, Animal and Plant Health Inspection Service: Hyattsville, MD, **1992;** pp 4–6.

RECEIVED August 18, 1995

Chapter 6

Impact of the Mediterranean Fruit Fly, *Ceratitis capitata* (Wiedemann), on California Agriculture

Ted A. Batkin

California Citrus Research Board, P.O. Box 230, Visalia, CA 93279

The Mediterranean Fruit Fly, *Ceratitis capitata* (Wiedenmann) was first discovered in Southern California in 1975, and again in 1979 in Santa Clara. The 1979 outbreak launched an areawide eradication program which included the use of aerially applied malathion bait. The fly was declared eradicated from the state in 1981, only to reappear in the Los Angeles basin in 1989. Since that time, a continuous program has been in place to eradicate the pest from the region. This paper will focus on the eradication procedures used in the Los Angeles area and the impact of public reaction on the California agricultural industry. The purpose of the presentation is to identify the need to replace malathion with a compound more acceptable for use in an urban environment and how this will impact the short and long term health of the fruit and vegetable production industry.

Officials within California agriculture have learned to react quickly to the threat of the Medfly because the state cannot afford to permit the pest to develop populations that will trigger shipping quarantines or become established. The Medfly attacks and devastates more than 250 species of fruit, berries and vegetables. An established Medfly infestation would imperil the state's $16 billion-a-year commercial agricultural industry as well as wreak havoc on backyard gardens and orchards.

The state first experienced a Medfly outbreak in 1975 in Los Angeles. The infestation was eradicated, but the Medfly reappeared in 1980 in Los Angeles and in the San Jose area. At that time, Governor Jerry Brown and his staff demanded that the treatment be restricted to the use of Sterile Insect Technology (SIT). The Northern California infestation was found to be spread over seven counties during the period from 1980 to 1982. State officials eventually had to aerially treat nearly 1,500 square miles of a heavily populated area with a protein bait spray containing the pesticide

0097–6156/95/0616–0070$12.00/0
© 1995 American Chemical Society

malathion. The eventual use of malathion bait was a result of threats from domestic and foreign trading partners demanding rapid and proven effective eradication methods. The treatment program was effective, and in 1992 the area was declared eradicated and shipping quarantines were lifted.

From 1982 to mid-1987, no Medfly infestations were discovered in California. Discoveries of 44 adult flies and one larval property in 1987 and 54 adult flies and seven larval properties in 1988, both in Los Angeles County, prompted new eradication programs. These sites were considered new introductions because they occurred in areas where very little natural host material existed and the population consisted of ethnic variability considered high risk for importing fruit from known pest population areas of the world. On July 20, 1989, a Medfly was discovered in a trap near Dodger Stadium in Los Angeles. The discovery prompted an eradication program that grew to encompass thirteen treatment locations in the Greater Los Angeles area as new infestations were discovered. From mid-1989 to mid-February of 1990, 272 flies were discovered.

In the fall of 1993, nearly 400 wild Medflies were trapped in the Los Angeles Basin, causing the concern over the existing eradication techniques to reach our foreign trading partners. To address this concern, the combined program of USDA and the California Department of Food and Agriculture (CDFA) formed an international science advisory panel (ISAP) to assist the existing Medfly science advisory panel (MSAP). The ISAP consisted of scientists with proven field experience in eradication. Their recommendation was to expand the Sterile Insect Technique (SIT) program to an area-wide approach rather than treating each new infestation on a localized basis. The result was the formation of the Los Angeles Basin SIT program that releases over 700 million sterile flies per week throughout a 1,500 square mile area (see Figure 1). This is a two- year program started in March of 1994, continuing through February 1996. At the end of the release period, officials will trap through two generations of the fly and determine the effectiveness of the SIT technique and evaluate whether the quarantine can be lifted and the sterile release discontinued.

Public Opinion in the Use of Aerial Application of Malathion Bait

There have been various public reactions to the use of malathion bait in urban populations. These range from mild disgust with the inconvenience of the spray to organized opposition led by environmental and health concern groups. The opposition started in the very early days of the program when the decision was made to put the helicopters in the air in San Jose. Various groups have been organized throughout the years with the majority of activity centered around the 1989 widespread spraying in Los Angeles. These groups included several members of California academia from USC and UCLA along with local and statewide political leaders.

When a single mated female fruitfly was discovered in Corona in December 1993, and the decision to return to aerial applications was made, the public opposition

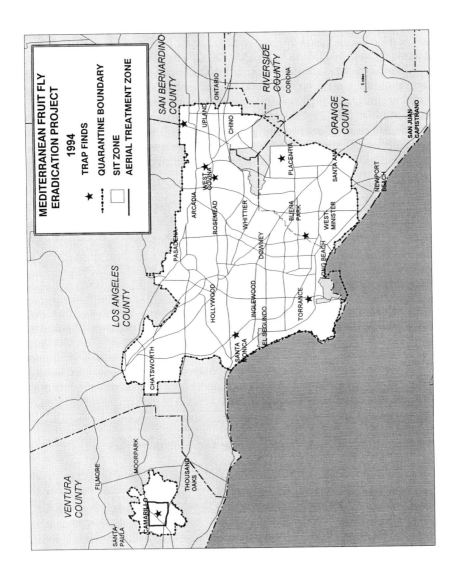

Figure 1. Map of Mediterranean Fruit Fly Eradication Project.

began again with a vengeance. Three different citizen groups formed within the City of Corona. They were each supported by the existing group from the Los Angeles area. The Corona City Council opposed the spraying, even to the extent that one of the city council members agreed to provide financing to one of the citizen groups, C.A.U.S.E. (Citizens Against Urban Aerial Spraying Efforts), to release helium balloons ahead of the helicopters. The media "feeding frenzy" that occurred with this kind of attitude was at an all-time high, and the officials responsible for the eradication program were openly attacked in the press on a daily basis and at public meetings throughout the state. 1994 was a gubernatorial election year which only increased public exposure of the issue.

During the early stages of the Corona eradication program, the agricultural industry was beginning to reorganize a previous public information program to work with the media and show the industry's side of the story. The effort proved to be too little, too late for the Corona eradication program; however, it did encourage the industry to take steps to change the situation. Agricultural communities began to organize information groups to help inform local officials about the economic impact of the Medfly and how an infestation would affect the local economy. This led to an opening of dialog between the opposition groups and the agricultural industry. One result of this ongoing dialog was the eventual adoption of a resolution by the Corona City Council two months after the beginning of the aerial treatments that supported the need for better pest exclusion, alternatives to malathion, and ongoing research into biological techniques. This resolution was presented to the Agriculture Committee of the U.S. House of Representatives in a joint presentation by agricultural leaders and the City of Corona opposition team.

One of the first communities outside of Corona to establish a program was Ventura County. The local economy is highly dependent on agriculture; however, it also has a very large urban population. Due to its proximity to Los Angeles (within a 50-mile radius from the center of the Los Angeles Basin) the industry considered themselves as a high risk area. They organized a group known as the Fruitfly Action Cooperative Taskforce (FACT) that essentially duplicated the industry response program started in Corona. The Ventura FACT group was successful in reaching all the local and county elected officials and informing them of the impact a fly find would have on their community. Later that year, in September of 1994, a major infestation was discovered in the center of Ventura County, in the community of Camarillo. This infestation included the trapping of two wild mated female Medflies followed by the trapping of over 60 wild males within 7 days. This was one of the largest infestations found in the history of the program in a very tightly defined area. All of the flies were trapped within a 200-meter radius on one property. Additionally, over 100 larval sites were identified resulting in the collection of over 2,000 larvae. It was clear that this was a relatively new population, and investigation suggests that the population resulted from fruit being illegally sent to the residents on the property from out of the country.

The public opposition to aerial application of bait in Camarillo was dramatically reduced from the activities in Corona. In fact, the only people showing any organized opposition were from Corona and came to Camarillo to see if they could further their own personal cause. However, as the spraying continued over a 6-month period, even the local residents who supported the eradication effort were beginning to tire of the program and started complaining. It was clear that their patience had run out and the program would be in jeopardy should it need to be continued. Each time government officials are forced to go to the air with malathion, the political pressure gets greater to use some other method of eradication.

Medfly Exclusion Program

Public opposition throughout the state is usually centered around the eradication method of malathion bait. The best defence known to the industry and the regulatory agencies is better exclusion techniques. To help accomplish the task of informing the residents of the state on the dangers of bringing fruit into the country, the FACT concept has been expanded to 9 counties with 5 more in the formation process. These groups are working with local governmental officials, schools and private organizations to insert education programs that will get the message out to as wide a spectrum of the population as possible. This includes working with the public schools through the "Ag In The Classroom" program and through the travel agent industry.

The basic point to remember first and foremost is that Medflies and other exotic fruit flies are brought into the country by the general population, not by the agricultural community. They are brought in by travelers or by people sending fruit into the country through the mail or by parcel services. Agriculture is the victim of these infestations, not the cause of them. The consequences of pest infestations run throughout the society from the inconvenience and perceived risks to the public of eradication treatments to the devastating economic losses to the agricultural industry. Therefore, we are all interested in the same thing: keeping the pests out of the country.

The best method of avoiding future aerial applications of malathion bait is to keep the fly from entering the country. This can be accomplished through diligent attention to the existing laws and regulations against the shipment of host material into the United States. The Medfly IS NOT indigenous to California or the continental United States. It must be brought into the country in host fruits either smuggled or shipped in through illegal means. The sad reality of the problem is that most of the people who bring this material into the country have no understanding of the magnitude of their actions. The majority of the fruit is brought in by people who simply want to have a "taste from home" or to re-live a feeling that they found in some other part of the world.

Once the fruit is in the country, many problems then occur. When the flies are allowed to build up populations, they may be transported throughout a region before the first colony is detected. This creates the possibility of additional colonies becoming developed in other geographical areas of the state. Hence, the fear of the pest becoming "established" becomes greatly accelerated. The consequences of establishment would be devastating to the general public as well as the agricultural industry. One only needs to travel to Central America and bite into a maggot infested mango to recognize the damage that would be caused to the entire population of an area with established populations of the pest.

Several action steps can be initiated to strengthen pest exclusion. They are:

1.   Strengthening our public policy commitment towards the maintenance of a safe, low cost food supply in the United States.

2.   Increasing awareness throughout the public regarding the magnitude of the problem of bringing in host material.

3.   Improving the pest detection programs at points of entry.

4.   Strengthening the laws against the illegal importation of host material into the country.

Response to a Medfly Discovery

The California Department of Food and Agriculture maintains a cooperative statewide system of more that 160,000 traps to detect various agricultural insect pests, including the Mediterranean fruit fly. The detection of one fruit fly in a trap triggers an intensive trapping program to determine whether an infestation exists. Traps are hung in host trees throughout most of California, with county or state employees inspecting the traps on a weekly or bi-weekly basis, depending on the season.

When a single fly is discovered, intensive trapping efforts begin immediately. Trap densities are raised in the square mile around the discovery site within 24 hours and trap densities are increased outward in an 81-square-mile area within 72 hours. If two flies are discovered within a three-mile radius, or if one mated female, or larvae or pupae are found, eradication treatment begins immediately after public notification, within 24 to 72 hours of the find.

Aerial spraying, using a mixture of malathion and protein bait, is conducted over a minimum area of 9 square miles around the fly find. The actual treatment area is determined by the distribution of fly, larval or pupal finds and operational considerations. The spraying, sometimes combined with localized ground pesticide treatment, continues at seven to 10 day intervals until sufficient numbers of sterile flies can be reared and released in the affected area. If the infestation is too widespread for effective use of sterile fly releases, aerial spray treatment will continue until two life cycles of the wild fly have passed with no new discoveries.

Fruit near the property where a fly has been trapped is cut and inspected for larval infestation. If larvae is found, host fruit within a 200-meter radius are stripped and the soil treated with diazinon. In addition, a quarantine is placed on fruit movement unless the fruit is treated in a prescribed manner. In many cases, the quarantine will restrict the movement of fruit to within the United States, thus limiting the export market capability of the industry.

Sterile Medfly Release Program

The sterile insect release method is one technique used to eradicate wild Medfly populations. The program consists of rearing large numbers of flies and then sterilizing then with gamma rays. The sterile flies are released into infested areas to mate with wild flies. The wild population eventually becomes extinct when enough of the wild females mate with sterile males.

To be effective, the sterile fly population should outnumber the wild flies by at least a 100-1 "over-flooding" ratio. When an infestation is identified, the area is first sprayed with malathion bait mixture to knock down the wild fly population. After a one-week wait to permit the malathion to degrade, one million sterile flies are released per square mile per week from the ground, followed up by subsequent aerial releases of 250,000 per square mile per week in a 49-square mile area around the core area. In the Los Angeles Basin program, the SIT program was expanded to include 1,500 square miles due to the wide distribution of the fly finds.

Because of this massive release rate, the sterile fly program is limited by the capacity to produce sterile flies. The State of California has the only state-operated sterile fly rearing facility, operated in conjunction with the U.S. Department of Agriculture. The facility, which can produce 200 million flies per week, is located in Hawaii, where Medflies already are established. (It would be too dangerous to locate in California, where an accidental release of untreated flies could cause a massive infestation.) The USDA also operates a rearing facility in Hawaii that is capable of producing approximately 300 million flies per week. Additionally, some sterile flies are being obtained from a rearing facility in Mexico and from one in Guatemala. The facilities combined can supply about 700 million sterile flies per week. The volume will vary based on quality output from the facilities. During certain times of the year, one or more of the facilities may experience a drop in quality due to weather conditions or dietary problems.

Agricultural Industry Perspective and Economic Impact

This section will discuss the perspective of production agriculture in relation to the priorities being established for exotic pest research activities at USDA-ARS, CDFA, private industry, and the Exotic Pest Research Center within the University of California system. In describing industry needs, it is first important to identify the roles played by the various factions. When the roles are understood, the needs will then make more logical sense. The exotic fruit fly problem in California has been refered to as a three-legged "Milking Stool". We all know that a stool with three legs must have equal strength to each leg if it is to hold the weight placed upon it. In the case of issues such as exotic fruit flies in California, the three legs refer to the (1.) Political or Policy Leg, the (2.) Science or Technological Leg, and the (3.) Economic or Cost Leg. If any one of these structures should fail, the agricultural industry would suffer considerably. Therefore, decisions in research priorities must consider each one carefully.

To begin, let's discuss the political or policy leg. This is the assigned responsibility of the regulatory agencies, CDFA and USDA. Their role in the issue is to protect the industry by administering quarantine procedures driven by the need to protect both our own industry in California, and that of our foreign and domestic trading partners. Therefore it is necessary to provide research programs that will strengthen their role in this protective manner. As an example, they will require trapping and detection methods that will provide early warning for infestations in order to implement eradication protocols as soon as possible. In addition they will establish research priorities based on their best estimate of needs to carry out all of their assigned duties.

The second point of discussion is the importance of the science and technology leg of the issue. It is very easy to believe that this is the only part of the issue to consider when you are totally emersed in the science. However, there are many additional factors to take into account, especially when looking for funding for scientific projects. It is clear that any regulatory effort must be based on solid scientific protocols. Neither the government or the agricultural industry can make adequate decisions without this base of information. Therefore, I see three areas that must be addressed in setting priorities for research:

1. Basic scientific information that studies the biology of the pest, including mating behavior, genetic structure, feeding habits, and any other basic information that will help understand the nature of the pest.

2. Information that will aid in the maintenance of any regulatory activity including, detection, sterile insect technique, chemical controls, exclusion procedures, and any other mechanisms sought by the regulatory community to aid in their assigned responsibilities.

3. Additional mechanisms of eradication or suppression that may look promising to replace current techniques and fit into the future directions of integrated pest management and low risk procedures.

The ways of funding these programs now become a little clearer. As an example, it would not be appropriate to ask an agency charged with the responsibility of maintaining a regulatory program to fund basic informational studies that have no direct application to their efforts. This, however, is often the way we look for help -- then complain when the grants are directed rather than free flowing. The same rules apply to how the agricultural community funds projects. They are looking for some measurable result that will aid in the long term economic health of the industry. This often leads to funding of projects that have a broader application with longer term payouts, however these payouts must be evident at the time of proposals.

In order to achieve the total funding necessary to meet all the objectives of a research program, help must be sought in non-traditional areas. This may include special legislation to provide a base of support that is not directed to any specific area. This will allow the directors of the research operation a certain level of hard money that they can disperse at their sole discretion. Once this base level has been established, additional soft money can be sought to address specific areas of concern, including funding from environmental groups on alternative control and suppression measures that meet their goals and objectives.

The Economic or Cost Leg becomes somewhat more difficult to define. It is broken down into two parts: (1.) The cost of maintaining adequate eradication programs and quarantine procedures, and (2.) The economic loss to the industry should our trading partners choose to embargo our products.

In many respects, the first economic issue relates to the political leg because the cost of maintaining eradication programs becomes more political than economic to the industry. The decisions of how to conduct these programs are often caught up in the cost dilemma. The scientific community and the production agriculture industry must constantly keep an eye on how the regulatory groups fund their programs and make sure that good scientific practices are followed.

The second aspect to the economic leg is the most important to the agricultural community. It truly is the driving force behind all efforts by CDFA and USDA as they carry out their assigned mission of protecting the agricultural interests in the United States. The potential of economic loss to production agriculture is staggering when the implications of worldwide quarantine are considered. As an example, the actual cash loss to the 81-square-mile quarantine area of Ventura County, due to the latest Medfly find in September of 1994, is $52,300,000 as reported by the Ventura County Agricultural Commissioner. The projected loss if the area had to be expanded to the entire county is over $250,000,000. These figures only relate to the actual cash loss from fruit that was destroyed or rerouted to a lower priced market such as processing of avocados and lemons. It is clear to understand that the actual cash losses from multiple quarantines in the state could easily move these figures well in to the billions in a very short time period.

The economic impact that a threatened embargo by Japan would have on California agriculture and the California economy due to an infestation of the Medfly has been summarized by Dr. Jerome Siebert in a preliminary report to the California citrus industry as shown in Table 1. It is expected that other countries would also join in the embargo, primarily Korea, Taiwan, and Hong Kong who normally follow Japan's lead in these matters. The short-run effect (less that two years) would be to significantly decrease the net income, probably to the point of operating at a loss, to growers, packers, and shippers due to denial of lucrative export markets that have been growing over the last five years. This economic impact would come about through a decrease in income from the export sales coupled with a significant domestic price decrease as these quantities are redirected to the domestic market. The long term

Table 1

Economic Impact of Medfly Quarantine To Selected Commodities *(Siebert, 1994)*

Variable	Apples	Apricots	Avocados
Domestic Shipments in lbs.	755,594,444	25,486,000	459,131,825
Export Shipments in lbs.	84,405,556	7,498,000	19,253,225
Total Revenue	347,894,889	10,727,117	177,142,270
Demand Price Elasticity	-1.6	-2.8	-0.54
Revised Total Revenue	332,966,840	10,716,041	171,901,418
Revenue (Lost)	**$1,058,810,775**	**($11,076)**	**$1,068,500,972**

Variable	Grapefruit	Kiwi	Lemon
Domestic Shipments	404,693,000	75,073,180	663,974,000
Export Shipments	207,510,000	14,965,951	240,996,000
Total Revenue	160,211,947	45,019,566	334,374,090
Demand Price Elasticity	-0.45	-2.05	-0.88
Revised Total Revenue	77,749,323	38,597,980	205,807,212
Revenue (Lost)	**$991,279,500**	**$1,067,320,238**	**$945,174,946**

Variable	Valencia Oranges	Navel Oranges	Peach/Nectar
Domestic Shipments	813,450,000	2,186,737,500	842,919,283
Export Shipments	469,800,000	295,650,000	5,435,717
Total Revenue	299,842,600	482,456,850	305,407,800
Demand Price Elasticity	-0.85	-0.85	-0.49
Revised Total Revenue	133,688,374	408,952,892	301,388,459
Revenue (Lost)	**$907,587,598**	**$1,000,237,866**	**$1,069,722,483**

Variable	Plum/Prune	Table Grapes
Domestic Shipments	252,879,552	1,066,204,340
Export Shipments	67,073,488	301,530,741
Total Revenue	118,382,610	578,506,707
Demand Price Elasticity	-0.67	-1.18
Revised Total Revenue	74,771,791	539,181,040
Revenue (Lost)	**$1,030,131,005**	**$1,034,416,157**

TOTAL Revenue (Lost)	**($564,248,075)**

impact would likely be a shrinking of the industries affected to pre-export levels which would mean a significant loss in acres, assets, and jobs, in addition to the loss in sales. The following points are pertinent in the economic evaluation:

1. The products affected are fresh shipments of apples, apricots, avocados, bell peppers, sweet cherries, dates, figs, table grapes, grapefruit, kiwis, lemons, limes, tangerines, oranges, nectarines, peaches, pears, persimmons, plums, and tomatoes.

2. The 1992 farm value of these products was $2,098,529,000; the 1992 farm value of exports was $354,846,000. These products were grown on 655,000 acres in 1992, or 8.5 percent of the total 1992 harvested acreage in California.

3. The 1992 total f.o.b. value of shipments of these products (including both domestic and export shipments), excluding fresh tomatoes for which there was no available data, was $2,859,969,446; the total f.o.b. export value was $605,478,708; the f.o.b. value of shipments to Japan, Korea, Taiwan, and Hong Kong was $376,265,616, or 62.1 percent of total exports for the products involved.

4. The short run economic impact of an embargo would be a decrease of $564,248,075 in net revenue on a f.o.b. basis. When added to previously estimated increases in costs to producers, the total impact on the California agricultural economy would range from a low of $1.057 billion to a high of $1.440 billion.

5. The loss in net revenue to the growers, packers, and shippers from an embargo of the products would have an additional impact on the California economy. It would reduce output in the California economy by $991,778,843, reduce personal income to the California economy by $1,165,003,003, reduce the gross state product by $1,233,051,320, and result in a loss of 14,189 jobs.

6. There have been indications that other states in the U.S. are considering an embargo of shipments of fresh products from California. If such an event were to take place, the economic impact would be much higher than the figures presented in the report. The author is currently estimating the size of this impact.

7. The major impact of an embargo would fall on the citrus industry followed by plums and table grapes. A table is attached which details the changes that would take place in domestic and export shipments, f.o.b. prices, and total revenue.

8. The long-run economic impact of an embargo, assuming it lasts for a period longer than two years, would result in a permanent restructuring of the sectors affected resulting in the acres planted to these crops being shifted to other crops,

a loss of jobs from the packing operations involved (but perhaps partially offset through the shift to other crops), and an uncertain impact on the California economy. The impact of a longer term shift and restructuring of the sectors affected will be the subject of additional studies.

Exotic pests are not just a problem of those in the agricultural community in California, who face the possibility of a $564 million loss within the first year of a Medfly infestation. The problem extends to the entire United States and the world agricultural community. It is truly an international problem that will require international solutions. These solutions will come from coordinated efforts of the scientific, governmental, and private industry sectors. The needs of production agriculture in the area of exotic pest management revolve around the ability of each organization to fulfill their assigned responsibility. The regulatory community must have the tools necessary to carry out their job. This means an on-going research and development effort to find improved methods of detection at points of entry and eradication techniques once populations of pests have reached the areas of concern. The growers must have the tools necessary to deal with long term introductions of existing pests and any new pests that may be determined of economic importance.

All of this involves a major effort to provide the research resources necessary to maintain a steady improvement in detection and eradication techniques. Improvement in the pest exclusion and eradication arena must be coupled with a strengthening of the national food policy to provide a safe, low cost supply of food to the population of the United States and the world. This policy has eroded severely over the past 20 years due to attacks from many directions including environmental organizations that try to take levels of food safety to extremes. It appears that some form of balance is being introduced to the debate at the legislative level. This effort must continue if our food supply is to be protected from new challenges by exotic pests that are not part of the existing eco-system. One major bright spot on the horizon is the development of photo-activated dyes as replacements for many of the more toxic compounds currently in use. This effort must continue and, if possible, be accelerated if the American agricultural industry is to continue to provide food to the tables of the nation and the world.

RECEIVED September 15, 1995

Chapter 7

Light-Activated Toxicity of Phloxine B and Uranine to Mediterranean Fruit Fly, *Ceratitis capitata* (Wiedemann) (Diptera: Tephritidae), Adults

Nicanor J. Liquido, Grant T. McQuate, and Roy T. Cunningham

Biology, Ecology, Semiochemicals, and Field Operations Research Unit, Tropical Fruit and Vegetable Research Laboratory, Agricultural Research Service, U.S. Department of Agriculture, P.O. Box 4459, Hilo, HI 96720

Experiments were conducted to determine the light-activated toxicity of a 1:1 molar mixture of phloxine B and uranine or phloxine B alone to Mediterranean fruit fly, *Ceratitis capitata* (Wiedemann), adults. Fully-fed or slightly starved test adults were fed either for two or four hours with dye concentrations ranging from $2.5 \times 10^{-5}\ M$ to $1.28 \times 10^{-2}\ M$ in a stock of various types of baits. After feeding, adults were kept 10 cm below four fluorescent lights yielding 200 $\mu E/sec/m^2$ and mortality observed for 48 hours. LC_{50} and LT_{50} values were calculated. At a concentration of $1.28 \times 10^{-2}\ M$ of the 1:1 molar mixture of phloxine B and uranine mixed with 20% autolyzed yeast hydrolysate and 20% fructose, 100% mortality occurred within 12 hours of exposure to light. At the same concentration of phloxine B alone mixed in a sugar-free formulation of autolyzed yeast hydrolysate, over 99% mortality occurred within 12 hours of exposure to light. Phloxine B alone, or in combination with uranine, should be considered as a possible replacement for malathion in area-wide bait spraying programs.

Light-activated toxicity of several substituted xanthene dyes to many species of insects has been documented in the laboratory and field (*1*, *2*). Among these is the dye-sensitized photoactive toxicity of phloxine B to adult housefly, *Musca domestica* L. (*3*), face fly, *M. autumnalis* De Geer (*4*), boll weevil, *Anthonomus grandis* Boheman (*5*), black cutworm, *Agrotis ipsilon* (Hufnagel) (*6*), and imported fire ant, *Solenopsis richteri* (Forel) (*7*, *8*). Light-activated toxicity has also been reported in uranine-fed adult house fly (*3*), face fly (*4*), and imported fire ant (*7*). Although dye-sensitized photooxidative toxicity from uranine is considerably less than phloxine B, uranine is known to cause synergistic effects when combined with other xanthene dyes (*1*, *2*). Phloxine B and uranine are color additives in drugs and cosmetics.

The insecticidal property of phloxine B and uranine is of importance in finding suitable alternatives to toxicants that are mixed with food baits in sprays (9) that are used in eradicating or suppressing many tephritid fruit fly populations. For instance, one of the methods that has been successfully used in eradicating introduced populations of Mediterranean fruit flies, *Ceratitis capitata* (Wiedemann), in California is the spraying, either by air or ground, of hydrolyzed protein bait mixed with malathion (10). In addition, a mixture of malathion and food bait spray is also used as one of the components of integrated management measures aimed at suppressing established populations of an assemblage of fruit fly species infesting fruits and vegetables (11). While it is a proven efficacious approach, spraying food bait mixed with malathion in urban areas has received very strong public opposition because of perceived public health and environmental concerns. Thus, in an effort to find an acceptable alternative to malathion, we conducted experiments to determine the toxicity of a mixture of phloxine B and uranine, as well as phloxine B alone, to Mediterranean fruit fly adults.

Materials and Methods

Insects. Mediterranean fruit fly pupae were obtained from the mass rearing facility of the USDA-ARS, Honolulu. Pupae were kept in an insectary at 24-27 °C, 65-70% RH, and a 12:12 hour (L:D) photoperiod. Emergent adults were held under the same temperature, relative humidity, and photoperiod as the pupae, and were fed with water (in gel, consisting of 9.975 parts water and 0.025 part Gelcarin [FMC Corp., Rockland, ME]) and a diet consisting of 3 parts sucrose, 1 part protein yeast hydrolysate (Enzymatic, United States Biochemical Corporation, Cleveland, OH) and 0.5 part Torula yeast (Lake States Division, Rhinelander Paper Co., Rhinelander, WI). Five-day old adults were used in the succeeding experiments.

Dyes. Phloxine B (94% purity) and uranine (77.1% purity) were obtained from Hilton-Davis (Cincinnati, OH). Phloxine B is chemically known as 2',4',5',7'-tetrabromo-4,5,6,7-tetrachlorofluorescein, disodium salt, and registered as D&C (Drug and Cosmetic) Red Dye #28. Uranine is chemically known as 3',6'-dihydroxyspiro (isobenzofuran-1(3H),9'(9H)-xanthen)-3-one, disodium salt, and registered as D&C Yellow Dye #8.

Relationship Between Concentration of 1:1 Molar Mixture of Phloxine B and Uranine and Mortality of Adults.

Test 1. Dyes Mixed in 10% Molasses. Ten molarity (M) concentrations consisting of 2.5×10^{-5}, 5.0×10^{-5}, 1.0×10^{-4}, 2.0×10^{-4}, 4.0×10^{-4}, 8.0×10^{-4}, 1.6×10^{-3}, 3.2×10^{-3}, 6.4×10^{-3}, and 1.28×10^{-2} of 1:1 molar mixture of phloxine B (829 g per mole) and uranine (376 g per mole) in a stock of 10% molasses were prepared. Test adults confined in feeding chambers (2 liter transparent plastic cups [Sweetheart Cup Co., Chicago, IL]) were provided with 12 ml dye-molasses preparations saturated in two cotton wicks (2.0 cm diam. x 2.5 cm long), one inserted through the bottom and the other through the side of the chamber. Control adults were provided with wicks containing only 10% molasses. Each feeding chamber had either 20 males or 20 females. Adult exposure to dye-molasses food commenced at 0600 hour and terminated at 0800 hour, providing a 2-hour feeding duration. From the onset of feeding, each feeding chamber was kept 10 cm below four high intensity, cool white fluorescent lights (Slimline, General Electric, Cleveland, OH) yielding surface intensity of 200 μE/sec/m^2; the same lighting condition was maintained throughout the 12-hour light phase (ending at 1800 hour daily) of the experiment duration. Following removal of wicks containing the dyes, flies were provided with the standard adult rearing diet consisting of sucrose, protein hydrolysate, torula yeast, and water. Mortality counts were made 1, 2, 10, 22, and 46 hours after the end of the feeding period. Twenty groups of 20 flies were bioassayed for each dye concentration, with male and female flies tested separately. Adults used in the test were not starved; i.e., 24 hours before their use in the experiment, adults had sugar-protein-water diet during the 12-hour light phase and were provided with water during the 12-hour dark phase (the non-feeding period). For each sex, the bioassay followed a randomized complete block design with 20 replications of 10 dye concentrations and a control.

Test 2. Dyes Mixed in 1% NuLure. Five molarity (*M*) concentrations consisting of 2.5 x 10^{-5}, 1.0 x 10^{-4}, 4.0 x 10^{-4}, 8.0 x 10^{-4}, and 1.6 x 10^{-3} of 1:1 molar mixture of phloxine B and uranine were prepared in a stock of 1% NuLure (Miller Chemical and Fertilizer, Hanover, PA), an acid hydrolysate of corn proteins. Control flies were fed with 1% NuLure. The bioassay procedure was similar to that in Test 1. Test adults were not starved. The bioassay followed a randomized complete block design with 14 replications of 5 dye concentrations and a control (20 females per replication).

Test 3. Dyes Mixed in 20% Autolyzed Yeast Hydrolysate. Females were fed with dyes at 8.0 x 10^{-4}, 3.2 x 10^{-3}, and 1.28 x 10^{-2} molarity concentrations that were prepared in a stock of 20% autolyzed yeast hydrolysate (ICN Biomedicals, Inc., Aurora, OH) and presented on two saturated cotton wicks per chamber, each wick holding 6 ml. Control females were fed with two wicks holding 20% autolyzed yeast hydrolysate solution. Each feeding chamber had 20 females. Female exposure to dye-containing food commenced at 0800 hour and terminated at 1200 hour, providing a 4-hour feeding duration. During the feeding period, the light intensity was 40 μE/sec/m^2. After 4-hour feeding, the light intensity inside the feeding chamber was increased to 200 μE/sec/m^2 and maintained throughout the light-phase of the bioassay. Following removal of wicks containing the dyes, females were provided with water and the standard rearing diet and mortality counts were made every 2 hour up to 1800 hour, when lights were automatically turned off, and at the 24th and 48th hours after wick removal. Adults used in the test were not given food during the 8-hour light phase before the bioassay. The experiment followed a randomized complete block design with 6 replications of 3 dye concentrations and a control (20 females per replication).

Test 4. Dyes Mixed in 20% Autolyzed Yeast Hydrolysate and 20% Fructose Aqueous Solution. The bioassay procedure was as described for Test 3 except that the three molarity (M) concentrations were prepared in an aqueous stock of 20% autolyzed yeast protein hydrolysate and 20% fructose and two groups of females were used in the tests: one group was fully fed; the other, starved for 8 hours. The experiment followed a randomized complete block design with 11 replications, for each prior feeding condition, of 3 dye concentrations and a control (20 females per replication).

Test 5.   Dyes mixed in 20% Mazoferm.   The bioassay procedure was as described for Test 3 except that the three molarity (M) concentrations were prepared in an aqueous stock of 20% Mazoferm E802 (Corn Products, Argo, IL), an enzymatic hydrolysate of corn proteins.   Control females were fed with 20% Mazoferm. The experiment followed a randomized complete block design with 7 replications of 3 dye concentrations and a control (20 females per replication).

Test 6.   Dyes mixed in 20% Mazoferm and 20% Fructose Aqueous Solution.   The bioassay procedure was as described for Test 3 except that the three molarity (M) concentrations were prepared in an aqueous stock of 20% Mazoferm E802 and 20% Fructose.   Control females were fed with 20% Mazoferm and 20% fructose solution.   The experiment followed a randomized complete block design with 7 replications of 3 dye concentrations and a control (20 females per replication).

Relationship Between Concentrations of Phloxine B Alone and Mortality of Adults.

Phloxine B Alone Mixed in 20% Autolyzed Yeast Hydrolysate.   The bioassay procedure was as described for Test 3 except that the three molarity (M) concentrations were prepared with phloxine B alone (no uranine added), females used in the test were not starved, and additional mortality counts were taken at the 18th, 20th, and 22nd hour after feeding. The additional mortality counts were taken to test for dark phase mortality, as the 18th hour after feeding was the end of the first 12 hour dark period following the feeding period.   Control females were fed with 20% autolyzed yeast hydrolysate. The experiment followed a randomized complete block design with 12 replications of 3 dye concentrations and a control (20 females per replication).

Feeding Behavior Studies

Effect of the Concentration of the 1:1 Molar Mixture of Phloxine B and Uranine Mixed in 10% Molasses on the Ingestion Rate by Females. Twenty 5 to 6 day old females were placed in feeding chambers between 1500 and 1600 hour. No sucrose, protein, yeast, or water was provided to test females. Between 0830 and 1100 hour the following morning, a 38 x 22 mm plastic feeding board was introduced into each chamber. Each board held ten 20 µl drops, of one of the following three treatments: 10% molasses (control); 8.0 x 10^{-4} M of a 1:1 molar mixture of phloxine B and uranine prepared in a stock of 10% molasses; and 1.28 x $10^{-2} M$ of a 1:1 molar mixture of phloxine B and uranine prepared in a stock of 10% molasses. The weight of the drops added to the boards was determined both before and after a one-half hour feeding period. Flies were observed during this feeding period and the number of times the flies fed and the total time spent feeding was recorded using a behavioral observation program developed in Toolbook (ver. 3.0a; Asymetrix Corporation, Bellevue, WA). For each feeding board introduced to a feeding chamber, there was a comparable board ('evaporation board') prepared which was left open to ambient conditions, but not subject to feeding by flies. Weights of the evaporation board before and after the one-half hour feeding behavior observation were used to estimate evaporative loss of the solution during the feeding period. Total food consumption was calculated using the formula

$$W_c = W_i - EW_i - W_f$$

where W_c is the calculated total food consumption, E is the evaporation rate estimated from the evaporation board ([Initial weight - Final weight]/Initial Weight), W_i is the initial food weight, and W_f is the food weight after feeding. Calculation of food consumption per fly was determined by dividing the total food consumption by the number of flies in the chamber. A total of nine observations of each treatment was made for a 30 min duration per observation. Data gathered were food consumption (mg) per female, mean duration of probing, and number of adults with red abdomens.

Statistical Analyses

Calculations of LC$_{50}$ and LT$_{50}$. Probit analysis (*12*) was used to determine the LC$_{50}$. The LT$_{50}$ was determined using the "pseudo first-order rate constant" proposed by Callaham *et al.* (*5*) for insects fed with photoactive dyes. A linear relationship between the natural logarithm of survival and time was assumed:

$$ln(S(t)) = -k_1 t$$

where S is percentage survival and t is time after termination of feeding. The LT$_{50}$ was estimated using $-ln(0.5)/k_1$. The goodness of fit of regressions for LC$_{50}$ and LT$_{50}$ was assessed by X^2 method, with $n - p$ (n is the number of concentrations [or time periods] for which survival was determined and p is the number of parameters in the model) degrees of freedom (*13*). POLO-PC (Le Ora Software, Berkeley, CA) was used for probit analysis and TableCurve 2D (Jandel Scientific, San Rafael, CA) was used to determine a regression line of the form

$$ln(S(t)) = -k_1 t.$$

Analysis of Feeding Behavior Observations. Food consumption per fly and duration of feeding per fly were $log_{10}(x + 0.5)$ transformed, while percentage with reddened abdomens was arcsin transformed before analysis by ANOVA (analysis of variance) using PROC GLM, with Waller-Duncan K-ratio T tests for separation of treatment means (Release 6.08, SAS Institute, Cary, NC).

Results

Behavior of Dying Adults Fed With Either 1:1 Molar Mixture of Phloxine B and Uranine or Phloxine B Alone

Adults that have fed on food containing either a 1:1 molar mixture of phloxine B and uranine or phloxine B alone exhibited the same behavioral and morphological manifestations. Adults developed reddish abdomens; the intensity of redness increased with the increase in the concentration of dyes in the bait. Adults fed phloxine B alone had vivid, purplish red abdomens when observed under direct light; those that were given a mixture of phloxine B and uranine had dark, maroon red abdomens. With increased dye concentration in the food (especially at dye concentrations $\geq 4.0 \times 10^{-4}$ M), adults increasingly regurgitated the reddish content of their crop and defecated red-colored excreta. It was common to observe adults feeding on the reddish regurgitate and excreta of their cohorts. Following one hour exposure to high light intensity, dye-fed adults (especially at dye concentrations $\geq 4.0 \times 10^{-4} M$) progressively became disoriented and uncoordinated, often falling onto their dorsa and making slow to vigorous rotations. Partially immobilized flies exhibited leg and antennal twitchings, as well as inward and outward movements of the labella. Completely immobilized adults with darkened, or black, rather than their natural crimson, compound eyes were considered dead. Dead flies had legs folded onto their thorax.

Relationship Between Concentration of 1:1 Molar Mixture of Phloxine B and Uranine and Mortality of Adults.

Test 1. Dyes Mixed in 10% Molasses.

LC_{50}. Mortality of males and females increased with the increase in concentration of dyes from $2.5 \times 10^{-5} M$ to $1.6 \times 10^{-3} M$. Forty six hours after feeding, mortality was highest for females at $1.6 \times 10^{-3} M$ (85%) and for males at $1.6 \times 10^{-3} M$ - $3.2 \times 10^{-3} M$ (88%), and then progressively decreased with increasing concentration to $1.28 \times 10^{-2} M$. At $1.28 \times 10^{-2} M$, 61% of males and 45% of females were dead at the end of the 46-h observation period (Figures 1, 2). The LC_{50}, 46 hours after wick removal, was $1.7 \times 10^{-4} M$ for males, $1.2 \times 10^{-4} M$ for females, and $1.4 \times 10^{-4} M$ for pooled male and female data (Table I).

LT_{50}. Table II shows the calculated LT_{50} at $4.0 \times 10^{-4} M$, $8.0 \times 10^{-4} M$, $1.6 \times 10^{-3} M$, and $3.2 \times 10^{-3} M$ for males, females, and pooled male and female data. The LT_{50} at $1.6 \times 10^{-3} M$ was 6.6 hours for males, 5.6 hours for females, and 6.1 hours for pooled data.

Test 2. Dyes Mixed in 1% NuLure

LC_{50}. When the food carrier for the two xanthene dyes was 1% NuLure, there was no significant increase in mortality from $2.5 \times 10^{-5} M$ to $4.0 \times 10^4 M$; an increase in mortality was observed from $4.0 \times 10^{-4} M$ to $1.60 \times 10^{-3} M$ (Figures 2, 3). The LC_{50} for females fed with the dyes mixed in NuLure, 46 hours after wicks were removed, was $1.64 \times 10^{-3} M$ (Table I).

LT_{50}. The LT_{50} at $1.6 \times 10^{-3} M$ was 35.5 hours (Table II).

Table I. Mortality of Mediterranean fruit fly adults 46 hours after fed with 1:1 molar mixture of phloxine B and uranine diluted in either 10% molasses or 1% NuLure.[1,2]

| | 10% Molasses | | | 1% NuLure |
	Males	Females	Pooled	Females
Slope ± SEM	1.442 ± 0.076	$1.334 \pm 0.846 \times 10^{-1}$	$1.416 \pm 0.652 \times 10^{-1}$	2.054 ± 0.274
LC_{50} (M)	1.70×10^{-4}	1.20×10^{-4}	1.40×10^{-4}	1.64×10^{-3}
95% CL(M) Lower	1.50×10^{-4}	0.80×10^{-4}	1.10×10^{-4}	1.40×10^{-3}
Upper	1.90×10^{-4}	1.60×10^{-4}	1.70×10^{-4}	2.05×10^{-3}
χ^2	2.756	14.429	6.354	0.023
d.f.	5	4	4	2

[1] Refer to Fig. 2 for the concentrations used in LC_{50} calculations.

[2] $\chi^2_{(0.05)(2)} = 5.992$; $\chi^2_{(0.05)(4)} = 9.488$; $\chi^2_{(0.05)(5)} = 11.071$.

Table II. LT_{50} of Mediterranean fruit fly adults fed with different concentrations of 1:1 molar mixture of phloxine B and uranine diluted in either 10% molasses or 1% Nulure.[1]

	slope ± SEM	LT_{50} (h) (95% CL)	χ^2	df
Males (10% molasses)				
$4.0 \times 10^{-4} M$	0.070 ± 0.005	9.637 (7.407-13.033)	0.135	3
$8.0 \times 10^{-4} M$	0.091 ± 0.015	7.060 (4.271-13.073)	0.728	3
$1.6 \times 10^{-3} M$	0.101 ± 0.010	6.65- (4.795-9.822)	0.344	3
$3.2 \times 10^{-3} M$	0.112 ± 0.022	5.648 (3.140-11.549)	1.209	3
Females (10% molasses)				
$4.0 \times 10^{-4} M$	0.077 ± 0.005	8.855 (7.045-11.463)	0.118	3
$8.0 \times 10^{-4} M$	0.105 ± 0.010	6.378 (4.702-9.077)	0.295	3
$1.6 \times 10^{-3} M$	0.117 ± 0.015	5.601 (3.682-9.242)	0.632	3
$3.2 \times 10^{-3} M$	0.099 ± 0.007	6.746 (5.298-8.877)	0.158	3
Pooled data (10% molasses)				
$4.0 \times 10^{-4} M$	0.074 ± 0.005	9.158 (7.264-11.906)	0.115	3
$8.0 \times 10^{-4} M$	0.098 ± 0.012	6.691 (4.517-10.666)	0.074	3
$1.6 \times 10^{-3} M$	0.108 ± 0.013	6.110 (4.207-9.477)	0.462	3
$3.2 \times 10^{-3} M$	0.105 ± 0.014	6.206 (4.101-10.174)	0.531	3
Females (1% Nulure)				
$1.6 \times 10^{-3} M$	0.018 ± 0.002	35.534 (26.577-49.763)	1.472	5

[1] $\chi^2_{(0.05)(3)} = 7.815$; $\chi^2_{(0.05)(5)} = 11.071$.

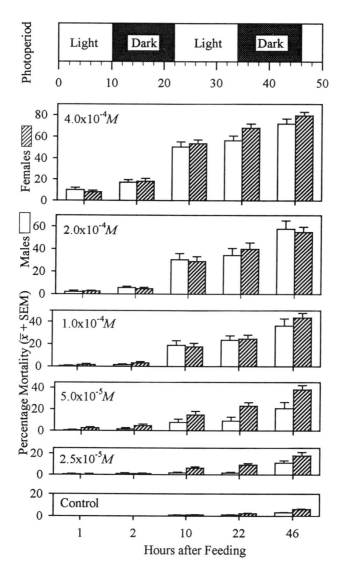

Figure 1. Mortality of adult male and female Mediterranean fruit flies 1, 2, 10, 22, and 46 hours after feeding on $2.5 \times 10^{-5} M$ to $1.28 \times 10^{-2} M$ concentrations of a 1:1 molar mixture of phloxine B and uranine prepared in a stock of 10% molasses and exposure to high intensity fluorescent lights.

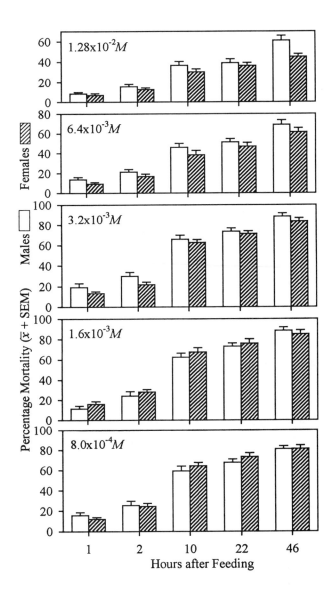

Figure 1. *Continued*

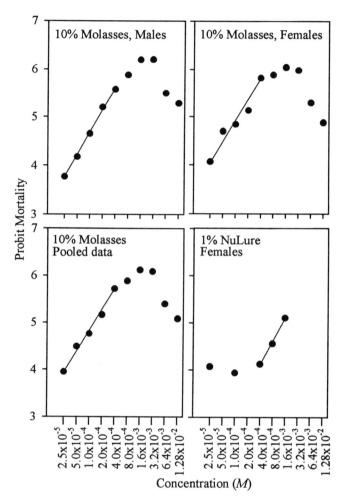

Figure 2. Probit mortality of Mediterranean fruit fly adults 46 hours after feeding on 2.5 x $10^{-5}M$ to 1.28 x $10^{-2}M$ concentrations of a 1:1 molar mixture of phloxine B and uranine prepared in stocks of 10% molasses or 1% NuLure and exposure to high intensity fluorescent lights. The straight line designates the concentrations used in the probit analyses.

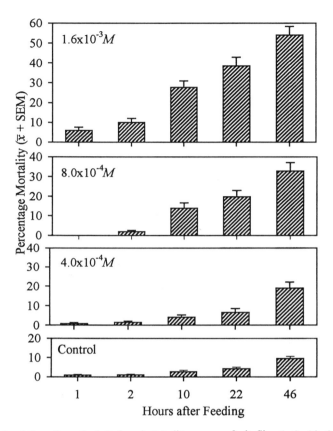

Figure 3. Mortality of adult female Mediterranean fruit flies 1, 2, 10, 22, and 46 hours after feeding on $4.0 \times 10^{-4} M$, $8.0 \times 10^{-4} M$, and $1.6 \times 10^{-3} M$ concentrations of a 1:1 molar mixture of phloxine B and uranine prepared in a stock of 1% NuLure and exposure to high intensity fluorescent lights.

Test 3. Dyes Mixed in 20% Autolyzed Yeast Hydrolysate. When the bait used was 20% autolyzed yeast hydrolysate, no decrease in mortality was observed at dye concentrations greater than $1.6 \times 10^{-3} M$. Six hours after feeding on autolyzed yeast hydrolysate bait containing $8.0 \times 10^{-4} M$, $3.2 \times 10^{-3} M$, and $1.28 \times 10^{-2} M$ dyes, 56%, 91%, and 97% of females were dead (Figure 4). The LT_{50} at $8.0 \times 10^{-4} M$, $3.2 \times 10^{-3} M$, and $1.28 \times 10^{-2} M$ concentrations were 5.8 hours, 1.5 hours, and 1.2 hours, respectively (Table III).

Test 4. Dyes Mixed in 20% Autolyzed Yeast and 20% Fructose Aqueous Solution. Females, irrespective of their feeding status at the start of the bioassay, exhibited similar rates of mortality. After 12 hours of exposure to light, 100% mortality was observed at $3.2 \times 10^{-3} M$ and $1.28 \times 10^{-2} M$ concentrations for both starved and nonstarved flies (Figures 5 and 6, respectively). The LT_{50} at $8.0 \times 10^{-4} M$, $3.2 \times 10^{-3} M$, and $1.28 \times 10^{-2} M$ concentrations were 7.2 hours, 1.5 hours, and 0.6 hours,

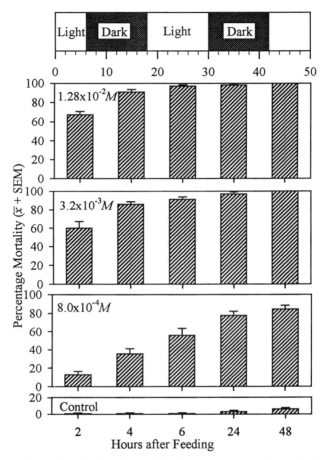

Figure 4. Mortality of adult female Mediterranean fruit flies 2, 4, 6, 24, and 48 hours after feeding on $8.0 \times 10^{-4} M$, $3.2 \times 10^{-3} M$, and $1.28 \times 10^{-2} M$ concentrations of a 1:1 molar mixture of phloxine B and uranine prepared in a stock of 20% autolyzed yeast hydrolysate and exposure to high intensity fluorescent lights.

respectively, for adults which were starved for 8 hours. For fully fed adults, the LT_{50} was 5.9 hours, 1.7 hours, and 0.9 hours, respectively, at $8 \times 10^{-4} M$, $3.2 \times 10^{-3} M$, and $1.28 \times 10^{-2} M$ concentrations (Table III).

Test 5. Dyes Mixed in 20% Mazoferm. Two hours after feeding and initial exposure to full light, 40% and 86% mortality were observed at $3.2 \times 10^{-3} M$ and $1.28 \times 10^{-2} M$, respectively. Four hours after light exposure, 86% and 99% mortality were observed at $3.2 \times 10^{-3} M$ and $1.28 \times 10^{-2} M$, respectively (Figure 7). The LT_{50} at $3.2 \times 10^{-3} M$ and $1.28 \times 10^{-2} M$ concentrations were 1.9 hours, and 0.7 hours, respectively (Table III).

Table III. LT$_{50}$ of Mediterranean fruit fly adults fed with different concentrations of 1:1 molar mixture of phloxine B and uranine diluted in either 20% autolyzed yeast hydrolysate or 20% Mazoferm with or without 20% fructose.[1,2]

	slope ± SEM	LT$_{50}$ (h) (95% CL)	χ^2	df
20% Autolyzed yeast hydrolysate				
$8.0 \times 10^{-4}M$	0.125 ± 0.020	5.823 (3.703-10.191)	1.161	3
$3.2 \times 10^{-3}M$	0.461 ± 0.021	1.502 (1.246-1.803)	1.146	3
$1.28 \times 10^{-2}M$	0.572 ± 0.014	1.214 (1.098-1.340)	0.176	3
20% Autolyzed yeast hydrolysate and 20% fructose				
$8.0 \times 10^{-4}M$	0.106 ± 0.024	7.203 (4.458-12.926)	3.649	4
$3.2 \times 10^{-3}M$	0.477 ± 0.032	1.465 (1.106-1.608)	1.994	3
$1.28 \times 10^{-2}M$	1.091 ± 0.060	0.635 (0.506-0.802)	12.650	3
20% Autolyzed yeast hydrolysate and 20% fructose (not starved)				
$8.0 \times 10^{-4}M$	0.130 ± 0.031	5.906 (3.519-10.917)	6.911	4
$3.2 \times 10^{-3}M$	0.427 ± 0.072	1.669 (0.771-3.124)	6.089	3
$1.28 \times 10^{-2}M$	0.755 ± 0.025	0.918 (0.796-1.058)	2.613	3
20% Mazoferm				
$3.2 \times 10^{-3}M$	0.373 ± 0.082	1.938 (0.602-4.146)	8.006	3
$1.28 \times 10^{-2}M$	0.984 ± 0.030	0.704 (0.619-0.803)	2.211	3
20% Mazoferm + 20% fructose				
$3.2 \times 10^{-3}M$	0.070 ± 0.015	10.421 (6.605-18.209)	2.399	3
$1.28 \times 10^{-2}M$	0.396 ± 0.109	1.839 (0.0-4.506)	14.772	3

[1] $\chi^2_{(0.05)(3)} = 7.815$; $\chi^2_{(0.05)(4)} = 9.488$.
[2] Unless otherwise stated, test adults were starved for 8 hours prior to the start of the feeding period.

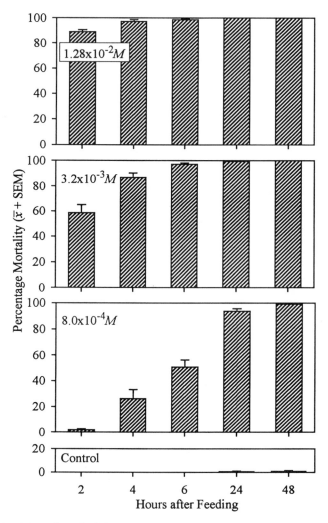

Figure 5. Mortality of adult female Mediterranean fruit flies, starved for 8 hours prior to the onset of the feeding period, 2, 4, 6, 24, and 48 hours after feeding on $8.0 \times 10^{-4}M$, $3.2 \times 10^{-3}M$, and $1.28 \times 10^{-2}M$ concentrations of a 1:1 molar mixture of phloxine B and uranine prepared in a stock of 20% autolyzed yeast hydrolysate and 20% fructose aqueous solution and exposure to high intensity fluorescent lights.

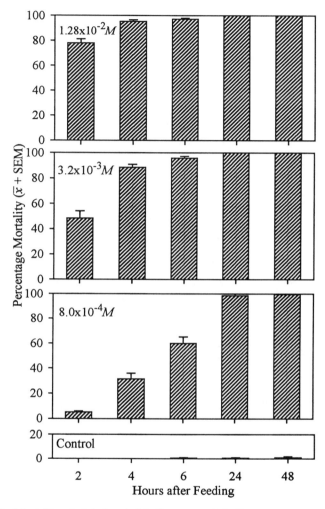

Figure 6. Mortality of adult female Mediterranean fruit flies, not starved prior to the onset of the feeding period, 2, 4, 6, 24, and 48 hours after feeding on $8.0 \times 10^{-4} M$, $3.2 \times 10^{-3} M$, and $1.28 \times 10^{-2} M$ concentrations of a 1:1 molar mixture of phloxine B and uranine prepared in a stock of 20% autolyzed yeast hydrolysate and 20% fructose aqueous solution and exposure to high intensity fluorescent lights.

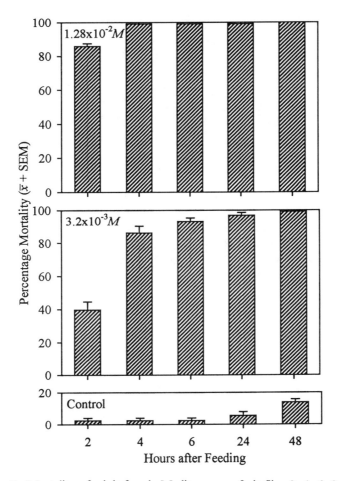

Figure 7. Mortality of adult female Mediterranean fruit flies 2, 4, 6, 24, and 48 hours after feeding on $3.2 \times 10^{-3} M$ and $1.28 \times 10^{-2} M$ concentrations of a 1:1 molar mixture of phloxine B and uranine prepared in a stock of 20% Mazoferm and exposure to high intensity fluorescent lights.

Test 6. Dyes Mixed in 20% Mazoferm and 20% Fructose Aqueous Solution. Addition of fructose in Mazoferm decreased the rate of mortality of females. Two hours after feeding, only 2% and 39% mortality were observed at $3.2 \times 10^{-3} M$ and $1.28 \times 10^{-2} M$, respectively. This increased to 41% and 97%, respectively, six hours after initial exposure to full light; 100% mortality was observed at $1.28 \times 10^{-2} M$ after 24 hours exposure to light (Figure 8). The LT_{50} at $3.2 \times 10^{-3} M$ and $1.28 \times 10^{-2} M$ concentrations were 10.4 hours, and 1.8 hours, respectively (Table III).

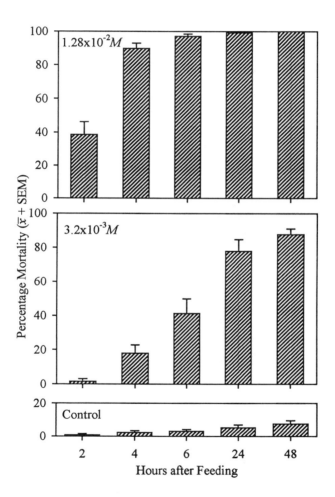

Figure 8. Mortality of adult female Mediterranean fruit flies 2, 4, 6, 24, and 48 hours after feeding on $3.2 \times 10^{-3} M$ and $1.28 \times 10^{-2} M$ concentrations of a 1:1 molar mixture of phloxine B and uranine prepared in a stock of 20% Mazoferm and 20% fructose aqueous solution and exposure to high intensity fluorescent lights.

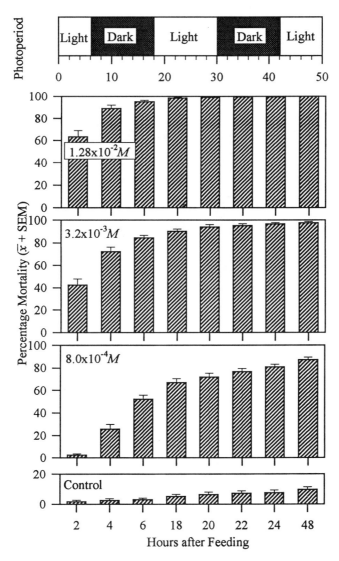

Figure 9. Mortality of adult female Mediterranean fruit flies 2, 4, 6, 18, 20, 22, 24, and 48 hours after feeding on $8.0 \times 10^{-4}\,M$, $3.2 \times 10^{-3}\,M$, and $1.28 \times 10^{-2}\,M$ concentrations of phloxine B alone prepared in a stock of 20% Autolyzed yeast hydrolysate and exposure to high intensity fluorescent lights.

Relationship Between Concentrations of Phloxine B Alone and Mortality of Adults.

Phloxine B Alone Mixed in 20% Autolyzed Yeast Hydrolysate. Females fed with phloxine B alone mixed in 20% autolyzed yeast hydrolysate showed a level of mortality comparable to the tests in which uranine was added. Four hours after initial exposure to full light, 26%, 72%, and 89% mortality were observed for females fed with phloxine B at $8.0 \times 10^{-4} M$, $3.2 \times 10^{-3} M$, and $1.28 \times 10^{-2} M$ concentrations, respectively. Six hours after light exposure, mortality increased to 52%, 85%, and 95%, respectively. Eighteen hours after feeding (6 hours of light exposure and 12 hours in the dark), mortality increased to 67%, 90%, and 98%, respectively, showing some increase in mortality during the dark period. After 12 hours of light exposure, 97% and over 99% of females fed with $3.2 \times 10^{-3} M$ and $1.28 \times 10^{-2} M$ dye concentrations were dead, respectively; after 24 hours of exposure to light (i.e., 48 hours after feeding), 100% mortality was observed at $1.28 \times 10^{-2} M$ (Figure 9). The LT_{50} at $8.0 \times 10^{-4} M$, $3.2 \times 10^{-3} M$, and $1.28 \times 10^{-2} M$ concentrations were 9.6 hours, 2.3 hours, and 1.4 hours, respectively (Table IV).

Effect of the Concentration of the 1:1 Molar Mixture of Phloxine B and Uranine Mixed in 10% Molasses on the Ingestion Rate by Females. There were significant differences among treatments in consumption per fly ($F = 3.43$; df $= 2, 8$; $P = 0.0139$; control $= 8.0 \times 10^{-4} M > 1.28 \times 10^{-2} M$), mean duration of feeding per fly ($F = 7.70$; df $= 2, 8$; $P = 0.0002$; control $> 8.0 \times 10^{-4} M > 1.28 \times 10^{-2} M$), and percentage with reddened abdomens ($F = 6.50$; df $= 2, 8$; $P = 0.0005$; control $> 8.0 \times 10^{-4} M > 1.28 \times 10^{-2} M$) (Figure 10). Females fed with 10% molasses containing $1.28 \times 10^{-2} M$ dyes had significantly shorter probing duration, significantly less ingestion of bait, and a significantly smaller proportion of flies with reddened abdomens than females given molasses with $8.0 \times 10^{-4} M$ dyes. Females that were given only molasses (control) and those which were given molasses containing $8.0 \times 10^{-4} M$ dyes had similar amount of ingested food, but controls had significantly longer probing duration and no occurrence of reddened abdomens.

Discussion

Although the exact mechanism of the toxicity of xanthene dyes to insects is not known, a proposed mechanism that is commonly accepted is as follows: the dyes collect visible light energy and convert it into a form that allows energy transfer to oxygen, thereby forming the reactive and toxic singlet oxygen (*14, 15*). The resulting high energy, singlet oxygen molecules probably oxidize a multitude of cellular components that result in the death of the insect. Similarly, the site of phototoxic dye reaction has not fully been elucidated. Light and electron microscope data gathered by Yoho (*16*) indicate that a site of action in houseflies may include midgut epithelial cells. In cockroaches, *Periplaneta americana* L., abnormal changes in hemolymph volume (*17*), body weight, and levels of total protein, total lipids, certain enzymes, and free amino acids (*5, 18, 19, 20*) have been observed. Although the toxicity of xanthene dyes is associated with the absorbance of visible light energy, it has been reported that

Table IV. LT$_{50}$ of Mediterranean fruit fly females fed with different concentrations of phloxine B diluted in 20% autolyzed yeast hydrolysate.[1]

	slope ±SEM	LT$_{50}$ (h) (95% CL)	χ^2	df
8.0x10^{-4}M	0.072 ± 0.016	9.632 (6.154-16.092)	7.959	5
3.2x10^{-3}M	0.306 ± 0.016	2.294 (1.897-2.768)	0.313	3
1.28x10^{-2}M	0.512 ± 0.019	1.356 (1.164-1.576)	0.176	3

[1] $\chi^2_{(0.05\chi(3))} = 7.815$; $\chi^2_{(0.05\chi(5))} = 11.071$.

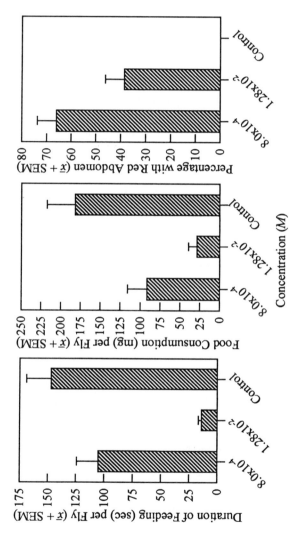

Figure 10. Duration (sec) of feeding per fly and consumption (mg) per fly during 30 min feeding observation of female Mediterranean fruit flies feeding on $8.0 \times 10^{-4} M$, $1.28 \times 10^{-2} M$ concentrations of a 1:1 molar mixture of phloxine B and uranine prepared in a stock of 10% molasses or 10% molasses alone (control) and percentage of flies with reddened abdomen 2 hours after the start of feeding.

mortality occurs even when the test insects are kept in the dark, suggesting an additional mode of toxic action (2).

Because the transfer of light energy to ground state oxygen is facilitated by the halogens, a direct relationship between the degree of halogenation and the observed toxicity is expected. In the present study, phloxine B was selected over other highly halogenated xanthene dyes (e.g., rose bengal) because of its safety record as a drug and cosmetic additive. Uranine is an established synergist that enhances the toxicity of halogenated xanthene dyes; however, the mechanism of action of synergism observed between uranine and other xanthene dyes has not been explained (2). As found in previous studies (1, 2), results presented here and preliminary results of our continuing studies have indicated that the toxicity of phloxine B and uranine to Mediterranean fruit flies is affected by the duration and intensity of illumination and the amount of ingested dye. The amount of ingested dyes may be affected by two variables: one, by the degree of the orientational and feeding stimulation that the bait exerts on the insects; and two, by the insects' innate feeding propensity and gut capacity.

Here, we clearly show that phloxine B and uranine, or phloxine B alone, are toxic to adults of Mediterranean fruit fly. In addition, we demonstrated that the food or bait that serve as the dye carrier significantly affects the level of mortality. As our data show, the LC_{50} of the dyes was significantly lower and the LT_{50} significantly shorter when dyes were mixed in 10% molasses than in 1% NuLure. In addition, the LT_{50} values were significanlty shorter when dyes were mixed in either 20% autolyzed yeast hydrolysate or 20% Mazoferm, with or without the addition of 20% fructose. More importantly, our data show that adult feeding deterrence that occurs when given high dye concentrations mixed in 10% molasses or 1% NuLure can be prevented by using more phagostimulatory baits; 100% mortality of females was achieved when high dye concentrations were mixed in the more phagostimulatory 20% autolyzed yeast hydrolysate or 20% Mazoferm, with or without 20% fructose. Our data support our contention of the importance of adult feeding behavior and the bait phagostimulation to achieve the desired level of mortality. In our continuing studies, we focus on formulating food bait with the following attributes: 1) volatiles from the bait attract adults; 2) the bait stimulates adult feeding; and 3) the dye can be mixed in the bait at high concentrations, to maximize lethality, without feeding deterrence problems.

The usual bioassay procedure in determining the toxicity of halogenated xanthene dyes is to severely starve (24 hours or more without food) the test insects and then feed them with known concentration of dyes in the dark for an extended number of hours or days; test insects are then subjected to lights of known intensity for a specific period of hours. It is indicated in previous studies that feeding in the dark and keeping the insects in the dark phase for a period of time before subjecting them to light causes a faster and more intense mortality effect (2, 21). Our assay procedure deviates from this norm. Test adults were either well fed or slightly starved and were subjected to either subdued or intense, artificial light (approximately 1/10th the intensity of the visible spectrum of sunlight) while feeding on dye-containing food bait and were exposed to intense artificial light throughout the light phase of the observation period (48 hours at 12:12 L:D photophase). Our data show that mortality of adults fed with autolyzed yeast or Mazoferm containing dyes at high concentrations is not affected by up to 8 hours of starvation period.

Our results have direct importance to area-wide control programs that aim to suppress and eradicate Mediterranean fruit flies without, or with minimal use of, organophosphate insecticides, such as malathion. As we have mentioned earlier, spraying of hydrolyzed protein bait mixed with malathion, either by ground or air, is one of the proven methods used in suppressing or eradicating incipient or established populations of Mediterranean fruit flies; however, the application of this eradication method has been a public relations nightmare. Unlike the case with malathion, the Food and Drug Administration has established a daily intake allowance for the xanthene dyes when used as coloring agents in drugs and cosmetics. These xanthene dyes also do not act as contact or fumigant insecticides, as does malathion, but must be consumed by the insect in order to cause mortality. Because the mixture of phloxine B and uranine is toxic only when ingested, selectivity may be conferred by using food bait that exerts strong attraction and phagostimulation to fruit fly adults but not to non-target organisms such as honeybees and parasitic wasps. This results in anticipated lessened effects of these sprays on non-target insects compared to the case when malathion is used as the insecticide. Additional advantages of the use of phloxine B and uranine include: that only a small quantity need be consumed for a toxic effect because the toxicity is light-activated; and that the dye does not kill the insect immediately, making it possible for other individuals in the population to consume some quantity of the pesticide secondarily, through feeding on food regurgitated by an impacted adult; this regurgitation behavior being common following consumption of higher dye concentrations.

Our study presents the first information on the light-activated toxicity of phloxine B or a mixture of phloxine B and uranine to Mediterranean fruit fly adults. In another chapter we present toxicity of these two dyes upon ingestion by adults of oriental fruit fly, *Bactrocera dorsalis* (Hendel). Krasnoff *et al.* (*21*) earlier reported that erythrosin B is highly toxic to adults of apple maggot, *Rhagoletis pomonella* (Walsh). The toxicity of phloxine B and uranine to Mexican fruit fly, *Anastrepha ludens* (Loew.), is reported in this symposium by the USDA-ARS research group at Weslaco, Texas.

We contend that phloxine B and uranine are potential, environmentally safe, replacements for malathion. As such, we feel that they may fill some of our research objectives in trying to develop more benign and efficient eradication technologies for the Mediterranean fruit fly and related pests.

Acknowledgments

We thank Napoleon Annanyartey, Paul Barr, Peter Bianchi, Taina Burke, Michelle Fernandez, Enrique Francia, Robert Gibbons, Harvey Llantero, Sarah Marshall, Excelsis Matias, Connie Molarius, Gandharva Ross, John Ross, Charmaine Sylva, and Judy Yoshimoto for reliably and consistently meeting the many logistical requirements of these studies. We appreciate the cooperation of James R. Heitz and Fred Putsche, Jr. of PhotoDye International, Inc. We additionally thank Paul Barr for invaluable assistance in data management.

Literature Cited

1. Heitz, J. R. In *Insecticide Mode of Action*; Coats, J. R., Ed.; Academic: New York, 1982; pp 429-457.
2. Heitz, J. R. In *Light-Activated Pesticides*; Heitz, J. R. and Downum, K. R., Eds.,; ACS Symposium Series 339; American Chemical Society: Washington, D.C., 1987; pp 1-21.
3. Fondren, J. E., Jr.; Heitz, J. R. *Environ. Entomol.* **1979**, 8, 432-436.
4. Fondren, J. E., Jr.; Heitz, J. R. *Environ. Entomol.* **1978**, 7, 843-846.
5. Callaham, M. F.; Broome, J. R.; Lindig, O. H.; Heitz, J. R. *Environ. Entomol.* **1975**, 4, 837-841.
6. Clement, S. L.; Schmidt, R. S.; Szatmari-Goodman, G.; Levine, E. *J. Econ. Entomol.* **1980**, 73, 390-392.
7. Callaham, M. F.; Lewis, L. A.; Holloman, M. E.; Broome, J. R.; Heitz, J. R. *Comp. Biochem. Physiol.* **1975**, 51, 123-128.
8. Broome, J. R.; Callaham, M. F.; Heitz, J. R. *Environ. Entomol.* **1975**, 4, 883-886.
9. Steiner, L. F. *J. Econ. Entomol.* **1952**, 45, 838-843.
10. Jackson, P.S.; Lee, B. G. *Bull. Entomol. Soc. Amer.* **1985**, 31, 29-37.
11. Roessler, Y. In *Fruit Flies: Their Biology, Natural Enemies and Control*; Robinson, A. S.; Hooper, G., Eds.; Elsevier: Amsterdam, 1989, Vol. 3B; pp 329-336.
12. Finney, D. J. *Probit Analysis*; Cambridge University Press: Cambridge, England, 1971.
13. Hosmer, D.W., Jr.; Lemeshow, S. *Applied Logistic Regression*; Wiley: New York, 1989.
14. Spikes, J. D.; Livingston, R. *Adv. Radiat. Biol.* **1969**, 3, 29-121.
15. Wilson, T.; Hastings, J. W. *Photophysiology* **1970**, 5, 49-95.
16. Yoho, T. P. Ph.D. Dissertation, West Virginia University, Morgantown, 1972.
17. Weaver, J. E.; Butler, L.; Yoho, T. P. *Environ. Entomol.* **1976**, 5, 840-844.
18. Broome, J. R.; Callaham, M. F.; Poe,W. E.; Heitz, J. R. *Chem. Biol. Interact.* **1976**, 14, 203-206.
19. Callaham, M. F.; Palmertree, C. O.; Broome, J. R.; Heitz, J. R. *Pestic. Biochem. Physiol.* **1977**, 7, 21-27.
20. Callaham, M. F.; Broome, J. R.; Poe, W. E.; Heitz, J. R. *Environ. Entomol.* **1977**, 6, 669-673.
21. Krasnoff, S. B.; Sawyer, A. J.; Chapple, M.; Chock, S.; Reissig, W. H. *Environ. Entomol.* **1994**, 23, 738-743.

RECEIVED September 11, 1995

Chapter 8

Light-Activated Toxicity of Phloxine B and Fluorescein in Methyleugenol to Oriental Fruit Fly, *Bactrocera dorsalis* (Hendel) (Diptera: Tephritidae), Males

Nicanor J. Liquido, Grant T. McQuate, and Roy T. Cunningham

Biology, Ecology, Semiochemicals, and Field Operations Research Unit, Tropical Fruit and Vegetable Research Laboratory, Agricultural Research Service, U.S. Department of Agriculture, P.O. Box 4459, Hilo, HI 96720

Experiments were conducted to determine the light-activated toxicity of a 1:1 molar mixture of phloxine B and fluorescein or phloxine B alone in methyl eugenol to oriental fruit fly, *Bactrocera dorsalis* (Hendel), males. Containers with 4 entry holes and wicks holding dye concentrations ranging from $1.28 \times 10^{-2}\ M$ to $8.58 \times 10^{-2}\ M$ were set out in an orchard, in a location shaded from direct sunlight. After either 1.5 or 2 hours, the entry holes were covered with screen, entrapping any flies which had been attracted into the container, and the traps were moved to full daylight conditions. Mortality counts were then taken periodically over the next 2 - 3 hours. At a concentration of $4.29 \times 10^{-2}\ M$ of the 1:1 molar mixture of phloxine B and fluorescein and at concentrations of $4.29 \times 10^{-2}\ M$ and $8.58 \times 10^{-2}\ M$ of phloxine B alone, 100% mortality occurred within two hours of exposure to full daylight. For times less than 2 hours, for both phloxine B alone and the phloxine B and fluorescein mixture, higher dye concentrations resulted in higher percentage mortalities. Phloxine B alone, or in combination with fluorescein, warrants further testing as a toxicant for use with methyl eugenol in oriental fruit fly male annihilation programs.

The oriental fruit fly, *Bactrocera dorsalis* (Hendel) is widely distributed in Oriental Asia and the Pacific (*1*). Its larvae feed on a broad range of tropical, subtropical, and temperate species of suitable host plants; thus, it is considered one of the world's most economically important tephritid pest (*2*). Males of oriental fruit fly are strongly attracted to methyl eugenol (1,2-dimethoxy-4-[2-propenyl]benzene) (*3*); for many years, methyl eugenol has been used to detect, delimit infestations, and eradicate accidentally-introduced, incipient populations of oriental fruit fly in the mainland United States, particularly California (*4*). The combination of methyl eugenol with a toxicant (usually naled [dimethyl-1,2-dibromo-2,2-dichloroethyl phosphate] or malathion [O,O-dimethyl-S-(1,2-dicarbethoxyethyl)dithiophosphate]) (*5, 6, 7, 8, 9, 10*) has been used successfully to eradicate endemic oriental fruit fly populations in Rota (*6*) and Okinawa (*11*). The present emphasis on the development and use of biorational, environmentally-acceptable pesticides has led us to investigate the light-activated toxicity of phloxine B and fluorescein, as well as phloxine B alone, mixed in methyl eugenol to oriental fruit flies.

Materials and Methods

Insects. Oriental fruit fly pupae were obtained from the mass rearing facility of the USDA-ARS, Honolulu. Pupae were kept in an insectary at 24-27 °C, 65-70% RH, and a 12:12 h (L:D) photoperiod. Emergent adults were held under the same temperature, relative humidity, and photoperiod as the pupae, and were fed with water (in gel, consisting of 9.975 parts water and 0.025 part Gelcarin [FMC Corp., Rockland, ME]) and a diet consisting of 3 parts sucrose, 1 part protein yeast hydrolysate (Enzymatic, United States Biochemical Corporation, Cleveland, OH) and 0.5 part Torula yeast (Lake States Division, Rhinelander Paper Co., Rhinelander, WI). Ten-d old adults were used in the succeeding experiments.

Dyes. The water insoluble form of phloxine B (99.0% purity) and fluorescein (95.9% purity) were obtained from Hilton-Davis (Cincinnati, OH). Phloxine B (water-insoluble form) is chemically known as 2',4',5',7'-tetrabromo-4,5,6,7-tetrachlorofluorescein, and registered as D&C (Drug and Cosmetic) Red Dye #27. Fluorescein is chemically known as 3',6'-dihydrospiro[isobensofuran-1(3H,9'-[9H]xanthen]-3-one, and registered as D&C Yellow Dye #7.

Relationship Between Concentration of 1:1 Molar Mixture of Phloxine B and Fluorescein in Methyl Eugenol and Mortality of Adults.

Test 1. Mortality in Methyl Eugenol + Dyes compared to Treatments with Methyl Eugenol alone or Methyl Eugenol + dog collar (15% Naled). Containers (0.95 liter [Highland Plastics, Pasadena, CA]) with four 2.5 cm diameter openings on the sides and holding a 2.5 cm long x 1.0 cm diameter cotton wick containing 1.0 ml of either methyl eugenol alone (Treatments ME I and ME II) or a 1.28×10^{-2} M concentration of phloxine B + fluorescein in methyl eugenol were emplaced in a macadamia nut (*Macadamia integrifolia* Maiden and Betche) orchard at 1200 hour, with supplemental release of irradiated oriental fruit fly adults. Treatment ME II, additionally, had a 1-inch strip of dog flea collar (containing 0.19 g Naled; Bansect flea and tick collar for dogs, ConAgra Pet Products, Richmond, VA) added to the interior of the container as a knock-down toxicant. One and one-half hours after the fly release (1330 hour), containers were collected with screen added to the openings to trap all contained flies. Containers were then exposed to full daylight and mortality counts were taken every half hour for the next two hours, after which the total number of flies was counted in each chamber. The experiment followed a randomized complete block design with 5 replicates per treatment.

Test 2. Mortality of Oriental Fruit Fly Males when Exposed to Dye Concentrations of $1.28 \times 10^{-2} M$ or $4.29 \times 10^{-2} M$ Mixed with Methyl Eugenol. Containers (2 liter [Sweetheart Cup Co., Chicago, IL]) with four 2.5 cm diameter openings on the sides and holding a 1.8 cm long x 1.0 cm diameter cotton wick containing 1.0 ml of either methyl eugenol alone (Treatments ME I and ME II) or a $1.28 \times 10^{-2} M$ or $4.29 \times 10^{-2} M$ concentration of phloxine B + fluorescein in methyl eugenol were emplaced in a citrus (mixture of *Citrus reticulata* Blanco and *Citrus sinensis* (L.) Osbeck) orchard known to have a low density wild population of oriental fruit flies. After two hours, containers were recovered with screens added to the holes to entrap any contained flies. Wicks in all treatments except ME I were replaced with water saturated wicks. In Treatment ME I, wicks saturated with methyl eugenol were not exchanged or removed. Containers were moved to full daylight and mortality counts were taken every 15 min for the first hour, then every 30 min for the next 2 hours, after which the total number of flies was counted in each container, and the number of live flies with reddened abdomens in the containers which had included dye in their initial wicks. The experiment followed a randomized complete block design with 3 replicates per treatment.

Relationship Between Concentration of Phloxine B Alone in Methyl Eugenol and Mortality of Adults.

Test 3. Mortality of Oriental Fruit Fly Males when Exposed to Phloxine B Concentrations of $4.29 \times 10^{-2} M$ or $8.58 \times 10^{-2} M$ Mixed with Methyl Eugenol. Containers (2 liter [Sweetheart Cup Co., Chicago, IL]) with four 2.5 cm diameter openings on the sides and holding a 1.8 cm long x 1.0 cm diameter cotton wick containing 1.0 ml of either methyl eugenol alone (Treatments ME I and ME II) or a $4.29 \times 10^{-2} M$ or $8.58 \times 10^{-2} M$ concentration of phloxine B (without the addition of fluorescein) in methyl eugenol were emplaced in a citrus orchard known to have a wild population of oriental fruit flies. After two hours, containers were recovered with screens added to the holes to entrap any contained flies. Wicks in all treatments except ME I were replaced with water saturated wicks. In treatment ME I, wicks saturated with methyl eugenol were not exchanged or removed. Containers were moved to full daylight and mortality counts were taken every 15 min for the first hour, then every 30 min for the next hour, after which the total number of flies was counted in each container. The experiment followed a randomized complete block design with 6 replicates per treatment. The solar light intensity during the entire test was measured using a Li-Cor LI190SB quantum sensor (Li-Cor, Lincoln, Nebraska).

Results and Discussion

Test 1. Mortality of Oriental Fruit Fly Males in Methyl Eugenol Mixed with Phloxine B and Fluorescein Compared to Treatments with Methyl Eugenol Alone or Methyl Eugenol + dog collar (0.19 g Naled). Figure 1 shows the average mortality in the three treatments over time. There was a high initial rate of mortality in

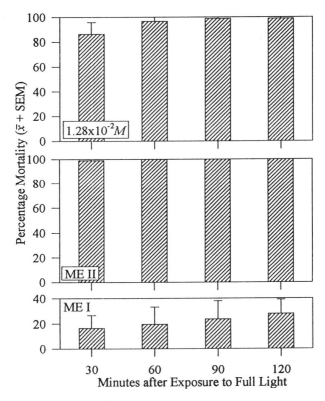

Figure 1.   Percentage mortality of wild and released irradiated oriental fruit fly males over 2 hour period of exposure to full daylight following 1.5 hour exposure (in a shaded macadamia nut orchard environment) to cotton wicks holding either methyl eugenol (without [ME I] or with [ME II] attached dog collar strip providing naled insecticide as a knock-down toxicant) or methyl eugenol mixed with 1.28 x 10^{-2} M concentration of a 1:1 molar mixture of phloxine B and fluorescein.

the treatment including the dog collar strip (ME II).   One and a half hours after exposure to full sunlight, there was greater than 98% mortality of oriental fruit fly males in containers to which the 1.28 x 10^{-2} M dye concentration had been added to the methyl eugenol wicks, compared to 100% mortality in containers which had naled insecticide in a strip of dog collar (Treatment ME II) and less than 25% mortality in chambers which lacked dog collar strips (Treatment ME I) and had no dye added to the methyl eugenol contained in the wicks .

Test 2.   Mortality of Oriental Fruit Fly Males when Exposed to Phloxine B + Fluorescein Concentrations of 1.28 x 10^{-2} M or 4.29 x 10^{-2} M Mixed with Methyl Eugenol. Figure 2 shows the increase in mortality over time in the 4 treatments. By 30 min after the exposure of oriental fruit fly males to full daylight, 67% of the males in the 4.29 x 10^{-2} M dye treatment containers had died compared to 14% of those in the 1.28

x 10^{-2} M dye treatment containers and 8% and 5% in ME I and ME II, respectively. After 2 hours, these percentages had increased to 100% for males in the 4.29 x 10^{-2} M dye treatment containers, 66% for those in the 1.28 x 10^{-2} M dye treatment containers, and 18% and 14% in ME I and ME II, respectively. None of the males in the 1.28 x 10^{-2} M dye treatment containers which continued to survive at the end of the experiment had reddened abdomens.

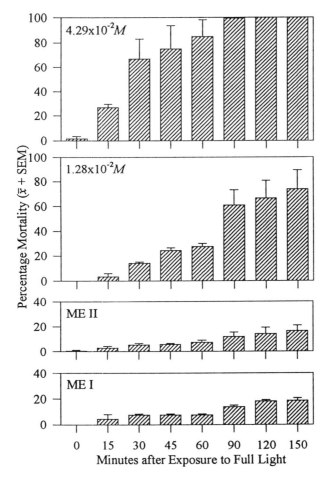

Figure 2.   Percentage mortality of wild oriental fruit fly males over 3 hour period of exposure to full daylight following 2 hour exposure (in a   shaded citrus orchard environment) to cotton wick holding methyl eugenol alone (ME I and ME II) or methyl eugenol mixed with 1.28 x 10^{-2} M or 4.29 x 10^{-2} M concentration of a 1:1 molar mix of phloxine B and fluorescein. Prior to the daylight exposure period, all wicks were replaced by water saturated cotton wicks except in Treatment ME I, where trapped oriental fruit fly males were continually exposed to methyl eugenol-containing wicks.

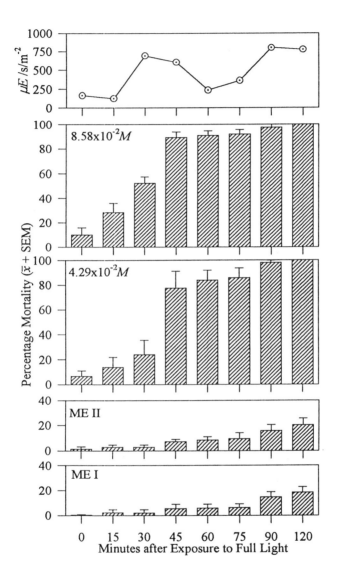

Figure 3. Percentage mortality of wild oriental fruit fly males over 2 hour period of exposure to full daylight following 2 hour exposure (in a shaded citrus orchard environment) to cotton wick holding methyl eugenol alone (ME I and ME II) or methyl eugenol mixed with 4.29×10^{-2} M or $8.58 \times 10^{-2}M$ concentration of phloxine B. Prior to the daylight exposure period, all wicks were replaced by water saturated cotton wicks except in Treatment ME I, where the methyl eugenol-containing wick was not removed. Light intensity, measured in μE, during the full daylight exposure period is also presented.

Test 3. Mortality of Oriental Fruit Fly Males when Exposed to Phloxine B Concentrations of 4.29 x 10^{-2} M or 8.58 x 10^{-2} M Mixed with Methyl Eugenol. Figure 3 shows the increase in mortality over time in the 4 treatments. After 30 min of exposure to full daylight conditions, 52% of the flies had died in the 8.58 x 10^{-2} M dye treatment compared to 24% in the 4.29 x 10^{-2} M dye treatment and 3% and 2% in ME I and ME II, respectively. The difference in total mortality between the two dye treatments, however, became less over time, both reaching 100% after 2 hours of exposure; at which point, mortality in ME I and ME II was 20% and 18%, respectively.

Conclusion

Our data show that a 1:1 molar mixture of phloxine B and fluorescein, or phloxine B alone, warrants further tests as a light-activated biorational toxicant for use with methyl eugenol in the male annihilation of oriental fruit fly populations. The use of these xanthene dyes as toxicants for oriental fruit fly male annihilation programs may be tested using the following formulations:
(1) Methyl eugenol (70 to 89.5% by weight) + 1:1 molar mixture of phloxine B and fluorescein or phloxine B alone (0.5 to 5.0%) + Min-U-Gel 400 (a thickener; a fine grade attapulgite clay, anhydrous magnesium aluminum silicate; Floridin Co., Pittsburgh, PA) (10 to 25%) . This formulation could be applied every two weeks to tree trunks, telephone poles, or other surfaces in 10 to 15 g globs at a rate of 230 globs per km^2. Alternatively, it could be applied as a thickened spray by aircraft every 1 to 2 weeks at a rate of 3.5 kg per km^2. (2) Methyl eugenol (90.0 to 99.5% by weight) + 1:1 molar mixture of phloxine B and fluorescein or phloxine B alone (0.5 to10.0%). This formulation could be absorbed into various solid carriers such as cotton strings or wicks, cigarette filter tips, or cellulosic fiber wallboard cut into pieces of various sizes. These saturated solid carrier pieces could be distributed every two to eight weeks either by aircraft or hung on foliage at a rate providing 3.5 to 17.5 kg total liquid formulation per km^2. Alternatively, this liquid formulation could be sprayed weekly without any thickener or solid carrier at a rate of 3.5 kg per km^2. If these formulations are found efficacious, they may be used not only in area-wide suppression and eradication of oriental fruit fly populations but also in small-scale integrated population management (IPM) programs.

Acknowledgments

We thank Paul Barr, Peter Bianchi, Sarah Marshall, Excelsis Matias, Gandharva Ross, John Ross, Charmaine Sylva, and Judy Yoshimoto for reliably and consistently meeting the many logistical requirements of these studies. We appreciate the cooperation of James R. Heitz and Fred Putsche, Jr. of PhotoDye International, Inc. We additionally thank Paul Barr for invaluable assistance in data management.

Literature Cited

1. White, I. M.; Elson-Harris, M. M. *Fruit Flies of Economic Significance: Their Identification and Bionomics*; CAB International: Wallingford, U.K., **1992**.

2. United States Department of Agriculture. *Host List: Oriental Fruit Fly;* Biological Assessment Support Staff, Plant Protection and Quarantine, USDA-APHIS: Hyattsville, MD, **1983**.
3. Steiner, L. F. *J. Econ. Entomol.* **1952,** 45, 241-248.
4. United States Department of Agriculture. *Action Plan: Oriental Fruit Fly;* Animal and Plant Health Inspection Service, USDA: Hyattsville, MD, **1989**.
5. Steiner, L. F.; Lee, R. K. S. *J. Econ. Entomol.* **1955,** 48, 311-317.
6. Steiner, L. F.; Mitchell, W. C.; Harris, E. J.; Kozuma, T. T.; Fujimoto, M. S. *J. Econ. Entomol.* **1965,** 58, 961-964.
7. Steiner, L. F.; Hart, W. G.; Harris, E. J.; Cunningham, R. T.; Ohinata, K.; Kamakahi, D. C. *J. Econ. Entomol.* **1970,** 63, 131-135.
8. Cunningham, R. T.; Chambers, D. L.; Forbes, A. G. *J. Econ. Entomol.* **1975,** 68, 861-863.
9. Cunningham, R. T.; Chambers, D. L.; Steiner, L. F.; Ohinata, K. *J. Econ. Entomol.* **1975,** 68, 857-860.
10. Cunningham, R. T.; Suda, D. Y. *J. Econ. Entomol.* **1986,** 79, 1580-1582.
11. Koyama, J.; Teruya, T.; Tanaka, K. *J. Econ. Entomol.* **1984,** 77, 468-472.

RECEIVED September 11, 1995

Chapter 9

Development of Phloxine B and Uranine Bait for Control of Mexican Fruit Fly

Robert L. Mangan and Daniel S. Moreno

Crop Quality and Fruit Insects Research, Subtropical Agriculture Research Laboratory, Agricultural Research Service, U.S. Department of Agriculture, 2301 South International Boulevard, Weslaco, TX 98596

The Mexican fruit fly (*Anastrepha ludens* Loew) has been recognized as a threat to citrus production and marketing in the southwestern United States since 1900. We proposed to develop a bait system for this species, and 5-8 other economically damaging *Anastrepha* species. The bait system should be toxic to insects only when ingested, is non-toxic to vertebrates, is relatively stable, and has delayed toxicity. Phloxin B + Uranine has these properties. We have developed a bait using dye, protein hydrolysate, and simple sugars to attract, stimulate feeding, and kill adult fruit flies. Field caged tree trials with these ingredients indicated that adults of both sexes are killed during sexual development and 60-90% of populations can be killed as they mature in 5 days. Under field cage conditions in south Texas concentrations of dye from 0.25% to 2.0% were equally effective. Addition of surfactants (2.0%) increased fly mortality by 250%.

Pest control systems require efficacy of the pesticidal agent to control the target pest and specificity to avoid injury to non-target organisms. The combination of bait and insecticide in sprays to control fruit flies has been used since the beginning of the 20th century (1). At that time inorganic insecticides were recommended in combination with carbohydrates and fermenting substances. McPhail (9) was probably the first to carry out studies on the attraction of fruit flies to protein hydrolysates in Mexico. He compared the attractiveness of fermenting brown sugar with decomposing proteins on the Mexican fruit fly and the guava fly (*A. striata*) and found that the hydrolyzed proteins were most attractive to the guava fly. Protein hydrolysate baits were first used in Hawaii to control the oriental fruit fly, *Bactrocera dorsalis* Hendel, (15,6). In the successful campaign against the Mediterranean fruit fly in 1956 in Florida, the combination of malathion and protein hydrolysate bait spray were used. This combination was probably instrumental in the successful outcome of the campaign (16).

Experience over the last 30 years shows that most organophosphates can efficiently control fruit flies. Thus malathion and hydrolyzed protein baits became the standard to control fruit flies. The Mexican fruit fly was eradicated from southern California, El Cajon area in 1991 and the Mediterranean fruit fly has been eradicated a number of times from California by the California Department of Agriculture and the Animal Plant Health Inspection Service by using a malathion-NuLure bait. The Mediterranean fruit fly was also successfully eradicated in 1962-1963 from Florida and 1966 from Texas using protein-insecticide bait sprays (2). The main reason for the wide use of malathion is its low mammalian toxicity as well as its relatively low price (13).

A number of protein hydrolysates could be used, but NuLure was selected because it is relatively inexpensive. The standard treatment by air in the U.S. is NuLure (previously PIB-7 or Staley's Protein Insecticide Bait No. 7) in combination with malathion (4 : 1 respectively) at ultra low volume sprays of 0.9 liters per hectare. The recommendation in Israel is the use of 100 ml poisoned bait mixture (75 ml protein hydrolysate and 25 ml malathion 95% ULVC) per 1000m^2 (13).

Malathion is a stomach and contact poison insecticide and even at low dosages can cause considerable disruption to parasite or predator based pest biocontrol programs. Malathion has acquired an undesirable reputation by the general public. Among the undesirable attributes of malathion is damage to certain painted surfaces on cars. This damage was verified by Ohinata and Steiner (11) when they found that PIB-7 by itself caused no paint damage, and malathion alone caused no less damage than when it was used with the attractant. They also found that formulations of malathion, fenthion, or ethion containing sugar, molasses, or glycerine decreased but did not prevent damage to car paints. An alternative insecticide that becomes toxic to flies only when consumed, has practically no mammalian toxicity, is relatively stable, and is effective at low dosages may be more acceptable to a greater portion of the public for treatments in urban habitats than the current malathion formulation.

In addition to environmental and public perception considerations, any candidate bait spray for the fruit flies must meet certain efficacy criteria. The toxicant must be capable of killing adult flies in a reasonable time and the toxicant must have sufficient residual toxicity so that applications can be made at reasonable intervals. Other characteristics are insect dependent . There must be an effective attractant to lure the flies to surfaces with the toxicant, and there must be a feeding stimulant (which may be the attractant itself) to insure that sufficient toxicant is ingested.

Mexican Fruit Fly Research. USDA entomology programs have continuously performed research on the Mexican fruit fly since 1928 (14), in Mexico City (closed 1968), in Monterey, Mexico (closed 1981), Brownsville, Texas (intermittent 1950's and 1960's) and in Weslaco and Mission Texas (1982 to present). The USDA-ARS research effort at the Subtropical Research Laboratory in Weslaco focused on Mexican fruit fly eradication after the sterile fly mass production facility was transferred from Monterey to Mission. Following the withdrawal of ethylene dibromide as a fumigant for tropical fruit in 1987, the Crop Quality and Fruit Insects Research Unit (formerly the Fruit Insects Research Unit) has had responsibility for development of post harvest disinfestation technology and eradication programs to maintain fruit fly free zones against the Mexican fruit fly and other *Anastrepha* species.

Preliminary Research with Phototoxic Dyes. We encountered reports of the toxicity of certain xanthene related dyes during the mid 1980's when we were performing literature searches and preliminary screening of dyes for internal markers of screwworm (*Cochliomyia hominivorax* (Coq.)) flies (RLM) and Mexican fruit flies (DSM). At that time we were searching for better methods to mark the mass produced, sterile flies that are released in sterile insect programs. At that time we were involved in our respective insect rearing research and the fact that certain chemicals are toxic to our flies was hardly impressive. The toxicity of these compounds was a disappointment because they would have provided cheap, easy to handle, and safe markers for sterile flies.

We were contacted in spring 1993 by officers of the newly formed PhotoDye International and asked to test Phloxin B, and the Phloxin B + Uranine combination on the Mexican fruit fly. Our preliminary approach was to determine: 1. Was the dye toxic to fruit flies? 2. Would the flies readily eat the dye? 3. Could the dye be used under tropical and subtropical conditions? 4. Could a practical, economical and effective application system be designed?

Preliminary tests. We began testing the Phloxin B (alone) toxicity using dilute (4%) sugar water on dental wicks as the feeding medium with various concentrations of dye. Our first effort was to test a series of concentrations of dye to develop a dose response curve. These tests were performed by Nada Tomic Carruthers in our laboratory. There were two components to "dose" in these tests, the concentration of dye in the sugar water and the amount of sugar water the flies consumed. Dr. J. Heitz had suggested that in other Diptera concentrations of dye above 1% were likely to inhibit feeding, so we used ranges of dye concentrations well below 1% (=10g/L). Flies were allowed to feed for 20 hours in laboratory room with fluorescent lights placed below the cages then placed outdoors in South Texas sunlight (July - August) for 4 hours. In south Texas, the temperatures and wind are stressful to exposed flies. We tested the flies in wood frame, sleeved screen cages with shaded areas available on the sleeves and edges. Very few flies were exposed to direct sunlight for more than a few minutes in any test.
Data in Table 1 show that there is a dose response curve relating Phloxin B concentration to mortality in light. These data show only that there is a relationship between mortality and concentration of dye. The flies in these tests had only the dye-sugar water as a food source and had the flies not fed on this material they would have died from starvation or desiccation. We had no reason to believe that sugar water would attract flies in the field. Essentially these tests provided answered the first condition for effectiveness, the dye appeared to be toxic to Mexican fruit flies at reasonable concentrations.

Preliminary NuLure Tests. Hydrolyzed protein or yeast culture based attractants have long been used as attractants for fruit fly traps or toxic bait systems. The Nu-Lure bait system currently used for fruit fly control programs in California has been tested extensively under laboratory and field conditions for delivery of other insecticides to target populations. Martinez and Moreno (8) and Moreno et al. (10) showed that insecticides that must be ingested in order to be effective are not readily ingested by the Mexican fruit fly when delivered using the currently used delivery system of NuLure mixed with insecticide. Evidence indicated that, while NuLure is a strong attractant, pure NuLure has a feeding inhibition effect on the flies

Table 1. Mortality effects of feeding Phloxin B to Mexican fruit fly at various concentrations in 4% sugar water for 20-24 hr. then placing cage in sunlight for 4 hr.

Test Group	Concentration Phloxin B (g/L)	Number Replications	Total Flies	Pre-Light Mortality (Total)	% Mortality in Light Mean	SD
1	0.3	4	119	14	100	-
	0.1	4	118	8	98.3	3.3
	0.03	4	122	9	71.9	32.8
	0.01	4	118	4	28.2	26.6
2	0.1	9	271	5	84.9	23.6
	0.05	9	271	6	84.9	33.3
	0.01	9	270	3	59.8	39.1
	5×10^{-3}	9	271	2	3.3	5.5

Table 2. Mortality Rates for the Mexican fruit fly fed mixtures of NuLure (50%), water (30%), and fructose (30%) baits with various concentrations of Phloxin B + Uranine, and Malathion.

Trial	Insecticide Treatment	Concentration (g/L)	Total Flies (7-8 Females)	% Mortality
1	Phloxin B + Uranine	0.5	14	100
	Malathion	0.5	15	100
	Control	0	15	6.6
2	Phloxin B + Uranine	0.25	15	26.6
	Malathion	0.25	16	100
	Control	0	15	0
3	Phloxin B + Uranine	0.25	15	100
	Malathion	0.25	14	100
	Control	0	15	0
4	Phloxin B + Uranine	0.1	15	86.6
	Malathion	0.1	15	100
	Control	0	16	37.5
5	Phloxin B + Uranine	0.1	14	100
	Malathion	0.1	15	100
	Control	0	16	56.3

NuLure based media with various fructose mixtures were tested with Phloxin B + Uranine (the Uranine was expected to enhance toxicity) during the summer and fall of 1993. Tests were designed so that flies were fed for 12 hours in a dark room (overnight), then exposed to daylight for 4 hours the following morning in the same type of cages used in the sugar - water tests. We confirmed that females offered only NuLure droplets (on waxed paper) and water died within 12 hours. When we dissected these flies, we found the crops to be either empty or containing only water, indicating that the flies did not feed on the NuLure. Flies fed on a mixture of NuLure mixed with 10-20% fructose had approximately the same survival as 4% fructose in water controls.

The preliminary evidence for effectiveness of Phloxin B + Uranine in a NuLure bait for Mexican fruit fly control is given in Table 2. Our primary goal in these tests were to show that the photodynamic action we found for the Phloxin B + Uranine + 4% sugar water was not inhibited by the NuLure. The data in Table 2 show that, although the Phloxin B + Uranine system frequently killed 100% of the flies, it was never better than the same malathion concentration. Much of the apparent superiority of malathion was related to the design problems, e.g. the flies had to be kept in the dark during the feeding phase in order to estimate the kill rate in 4 hr. of daylight. The malathion bait was toxic during the 12 hr feeding period (whether the flies fed or not) and we observed that all flies had died in every malathion test before the flies were placed in the light. The addition of fructose to the bait has an additional benefit, especially for droplets that land on plant surfaces. Fructose has strong hydroscopic activity so droplets should remain moist in a moderate humidity environment and be more attractive for feeding than pure Nulure-insecticide baits.

Efficacy of Phloxin B + Uranine in Field Cage Tests. The tests with NuLure suggestedthat sugar stimulates feeding and there was a lack of stimulation to feed in the NuLure without sugar, or that NuLure contained some feeding inhibitory factors that were overcome by the sugar. The accompanying paper (Moreno and Mangan) has presented the experiments that clarify feeding stimulants and inhibitory factors in the protein hydrolysates. All tests in the field cages were carried out following his investigation of feeding rates related to sugars, salts, and amino acids in various protein baits. The bait used in our field cage tests was composed of an enzymatic hydrolysate by-product from corn syrup extraction marketed as "Argo-Steepwater" or "MazoFerm ", sugar, water, certain surfactants and either Phloxin B (alone) or Phloxin B + Uranine.

Efficacy of the Phloxin B + Uranine bait formulation was tested under field conditions on caged, mature, fruit-bearing grapefruit trees. The field cage consisted of four individually caged "Ruby Red" grapefruit trees that were about 5 years old. Adult Mexican fruit flies were released into the cage either as they emerged from pupae (replicates 1, 2) or as 3-4 day old adults (replicates 3-7). Trees were heavily infested with various homopteran pests (scale insects, mealybugs) and mites during the first two replicates, we attempted to control these pests at about the level found in a commercial orchard for the later tests. In each test, two sections were treated with a dye-bait solution (dye treatment) and two sections with the bait alone (control). Water was provided by wicks protruding from plastic containers in all tests. In sequential tests the two treatments were rotated with identical treatments located diagonally from each other in the

cage. In all tests the dye was the mixture of Phloxin B and Uranine. In the original efficacy trials, bait consisted of a formulation of 70% mazoferm, 20% fructose, 0.1% surfactant (in this case a material marketed as "SM-9") and 2.0% Phloxin B + Uranine with
the remainder water.

Baits were not applied to the trees in any tests. In the first test a cup with ca.
100 ml of bait was suspended below a cap (for shade and rain protection) as the feeding
site. In all other tests, baits (with or without dye) were soaked into sponge sheets, the
sponges were stapled to rigid plastic (yellow) backing and folded to make a roof structure with the sponge on the underside. The stations were hung in the trees. Flies were
observed to feed on the bait, no flies were ever observed to get stuck on the surface of
the sponges. We tested various numbers and shapes of feeding stations keeping the total
surface area and amount of bait constant. We could not discover any effect of these factors (thus far) and those results are not discussed here.

The procedure was to release flies on day one, allow them to feed on the bait
stations and move freely about the cage for 4.5 days, then remove the stations and water. On the afternoon of the 5th day, 2 McPhail type traps with ammonium acetate or
other attractants were placed in each cage (identical traps and attractants were used in
each cage). Traps were left in the cage for 3.5 days then removed and numbers of males
and females in each trap were recorded. During tests 1 and 2 we allowed 2,000 flies to
emerge in each cage, for tests 3-7 (and following tests on various alternative formulations) we released 3000 4 -6 day old flies. The sex ratio in all was about 1:1. Flies were
provided only sugar and water prior to release. Data collected included collections of
dead flies on the bottom soil and leaf surfaces every 2 hours from 7:30 am to 3:30 PM
during the 4.5 day "feeding" period, counts of numbers of flies on the bait stations taken
at the same times, and the numbers of flies recovered during the 3.5 day "trap back"
period.

Results of the first field cage series are given in Table 3 and Figures 1 and 2.
These include total flies trapped during the final 3 days of the tests. In this table each
cell total for bait or bait plus dye represents the sum of two cells. In the early tests (1
and 2) the survival rate for flies was very low due to ant and spider predation on the
emerging adults and feeding rate at the bait stations may have been low because the flies

Table 3. Summary of recapture (survival) rates of Phloxin B + Uranine and Control
treatments of fruit flies in field cages.

Test	Recaptured flies Bait Only		Recaptured flies Bait + Dye		Total Flies Recaptured
	Number	%	Number	%	
1	163	69.3	72	30.7	235
2	174	80.5	42	19.5	216
3	609	70.1	260	29.9	869
4	2,064	84.8	280	15.2	2,344
5	879	77.9	230	22.1	1,127
6	335	56.7	256	43.3	591
7	1,184	77.4	317	22.7	1,530

had many alternative food sources from the pest infestations on the trees. Survival rates were much higher in tests 3-7 but cool weather and rain storms slowed the feeding and trapping rate during some of the tests. Statistics were calculated for the effects of the dye as an Analysis of Variance with the percentage of flies recaptured as dependent variable (Table 4). The effect of the dye was highly significant and the cage effect was also significant .

Table 4. Analysis of Variance results for effects of cage, dye treatment, on percentage of flies recaptured in 7 tests.

Source	DF	Mean Square	F-Ratio	Prob.
Cage	3	269.52	3.37	0.04
Treatment	1	4,043.55	50.59	<.0001
Error	23	79.92		

Time Relationships in Feeding Behavior and Mortality. Numbers of males and females feeding and their mortality rates are shown in Figures 1 and 2. Feeding rate in the Phloxin B + Uranine cages reached a slight peak at about 09:30 and was slightly higher in the morning than in the afternoon. In the control cages (bait but no Phloxin B + Uranine) feeding was greatest at 07:30 and declined during the day. Feeding was also much higher (ca. 27 flies) in the control cages than in the Phloxin B + Uranine cages (ca. 7.5). Control mortality remained uniformly low in the control cages but increased in the afternoon in the Phloxin B + Uranine cages reaching a peak at 13:30. The lag between peak feeding and peak mortality in the Phloxin B + Uranine cage suggests that there is a minimum 4 hour delay between feeding and mortality. Average daily feeding rate is plotted in Figure 3. The number of flies observed feeding increased during the period of the test in the control cages but remained remarkably constant in the Phloxin B + Uranine cages. The data suggest that as flies mature they begin feeding and continue feeding for some time. In the Phloxin B + Uranine cage, flies died as they began feeding so there was no accumulation of feeding flies.

The mode of action of the Phloxin B + Uranine bait system requires feeding and exposure to light. The design of this series of 7 tests allowed us to independently measure comparative feeding rates and mortality rates in the treatment (dye + bait) cages and the control (bait only) cages. The Mexican fruit fly adult normally requires 10-15 days to mature and adults feed continuously on protein sources from day 5 through the rest of their lives. We were able to cause highly significant reduction in fly survival (based on trap sampling) averaging over 70% in 5 days.

Dose - Response in field cages. A series of trials were run in the screen cages to determine the effects of Phloxin B + Uranine concentration on mortality and survival. In these test each of the four cages received a different concentration and the test was repeated 4 times so that each dose was run in each of the 4 cages. Concentrations were chosen after we considered the mortality rates observed in laboratory cages. The data in Table 5 summarizes the survival rate (recaptures) and the mortality rates.

The recapture and mortality data were analyzed by analysis of covariance with cage as a categorical independent variable and concentration as a continuous variable.

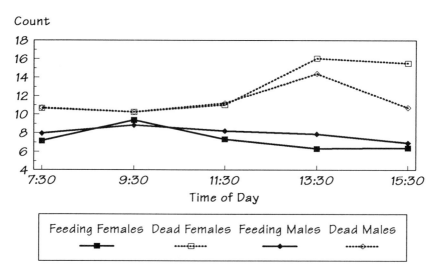

Figure 1. Feeding rate and mortality rate related to time of day for Mexican fruit flies in field cage with Phloxin B + Uranine bait stations.

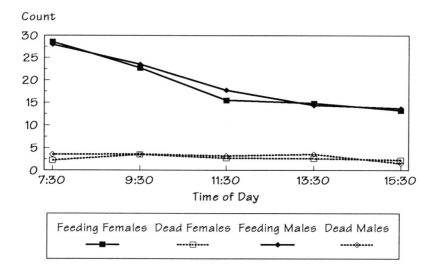

Figure 2. Feeding rate and mortality rate related to time of day for Mexican fruit flies in field cage with control bait stations.

Average Feeding Rate

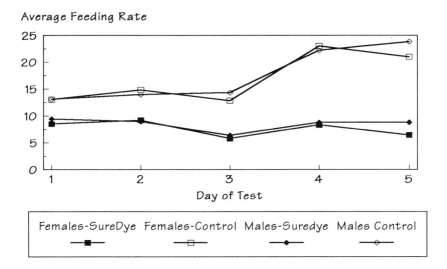

Females-SureDye Females-Control Males-Suredye Males Control

Figure 3. Average numbers of flies recorded per day feeding on Phloxin B + Uranine or control bait stations.

These analyses indicated that there was no significant effect for either cage or dye concentration effects when the number of flies was used but cage was significant for percentages of total dead flies were used as dependent variable. Concentration of dye was not significant in any of the tests. We interpret these results to suggest that under the field conditions over a five day period, even flies that feed on 0.25% bait will die. Other factors that may be ecologically dependent (weather, alternative protein sources, fly maturation rate) differed among the test dates.

Table 5. Mortality rates based on number of dead flies collected during peak sunlight periods and survival rates based on flies trapped in 48h in cages with bait stations containing various concentrations of Phloxin B + Uranine.

	Mortality				Survival				
	Concentration				Concentration				
Trials	2%	1%	0.5%	0.25%	2%	1%	0.5%	0.25%	Date
1	149	152	206	221	233	163	178	144	11-28-94
2	286	205	139	138	373	190	429	361	12-05-94
3	606	427	374	504	86	174	143	53	12-12-94
4	532	526	333	539	70	224	96	182	12-19-94
Mean	393.25	327.5	263	350.5	190.5	187.7	211.5	185	

Use of Surfactants. Phloxin B + Uranine bait system toxicity can be greatly increased by addition of certain surfactants that enhance the solubility of the dye in the bait solution and, possibly, the absorption of the phototoxic agents though membranes. A number of these agents are registered for use in both herbicide and insecticide delivery systems. We tested several surfactants and found that one (SM-9®), a secondary alcohol ethoxylate was particularly effective in enhancing the activity of Phloxin B + Uranine. Following the laboratory cage tests described by Moreno and Mangan (this volume), we performed a series of field cage tests using 2% SM-9 (the apparent optimum dose of the surfactant) vs. no surfactant (which was statistically as effective as 0.1% used in the first efficacy tests).

As is shown in Table 6 the addition of surfactant reduced survival in the cages by >60% over the total trials and similarly mortality rates were about >60% higher. These results held even though the overall tests were significantly different. Further laboratory tests have shown that the dye concentration can be reduced to below 0.5% with no detectable change in mortality rate when proper surfactants are used. Other work has shown that there is are a number of surfactants that are superior to SM-9.

Table 6. Mortality rate and survival rate of flies in cages with and without addition of a surfactant (SM-9) to bait formulation.

Trial	Survival Rate		Mortality Rate		Date
	Concentration of Surfactant		Concentration of Surfactant		
	0%	2%	0%	2%	
1	2,107	659	692	625	12-29-94
2	723	106	439	2,259	1-17-95
Total	2,830	765	1,131	2,884	

Fly hunger and bait system effectiveness. All tests carried out up to this point had used 5-8 day old flies that had been maintained on sugar and water only. This was based on the observations that the Mexican fruit fly develops specific hunger at about day 8. In these tests we tested a group of flies maintained with sugar and protein available from time of emergence until flies were released in the cage at 5-8 day adult age against a group maintained with only sugar and water. We were particularly concerned that older flies or flies that had already had a protein meal before a bait treatment is applied may not be susceptible to the treatment. Our daily counts of numbers of flies feeding at bait stations (at 2 h intervals) indicated that in control cages the numbers of flies feeding increased during the 5 day feeding period. But we could not determine whether this was due to accumulation of feeding flies over time or due to increasing numbers of flies reaching the protein hunger age (and feeding once) during the test period.

In these tests we observed that the treatment was highly effective in the February trials for both feeding treatments with very low recapture rates and very large numbers of dead flies recovered (Table 7). In the February trials, both survival and mortality data indicate that protein + sugar flies were more susceptible to the bait than flies fed only sugar. Activity in January was much lower due to cold wet weather but under these conditions protein + sugar flies also averaged greater susceptibility to the Phloxin B + Uranine bait treatment. These results were surprising to us because we were expecting, at best, that the protein fed flies would be only slightly less susceptible to the bait. We prefer to believe that some metabolic pathway is opened by protein feeding and the flies continue to feed at high rates after the first meal. Other possibilities such as some sort of learning process in the fly population cannot be eliminated with the knowledge we have at present.

Table 7. Survival and mortality rates for sugar-only fed and sugar- protein fed adult flies in field cage with optimal bait formulation.

Trial	Survival (Flies trapped)		Mortality (Dead flies collected)		
	Sugar	Sugar+Protein	Sugar	Sugar+Protein	Date
1	202	300	783	1,019	1-23-95
2	277	235	1,614	1,544	1-30-95
3	1,158	118	2,160	3,810	2-6-95
4	355	11	2,390	4,236	2-13-95
Average	1,992	664	887.7	1,506	

Conclusions. The development of the Phloxin B + Uranine bait system for the Mexican fruit fly was largely dependent on the 65 years (1928-1965) of applied and basic research carried out before initiation of this project. The most important aspect of this system that contrasts with the malathion bait system is that the dye must be ingested in order to have any pesticide effect. This provides an opportunity to formulate a bait system that is preferred by the target organism and ignored by beneficial or non-target organisms. After formulating a bait that is effective in killing adult Mexican fruit flies, we placed emphasis on improving the activity of the bait by examining the effects of concentration, surfactants, and feeding status of the target insects. We showed that increasing the concentration of dye in bait stations has little effect on pesticide activity. Addition of surfactants, however, increased toxicity 3-5 fold. We also showed that protein fed flies are more, not less, susceptible to the bait than protein starved flies.

There are a number of requirements of the Phloxin B + Uranine bait system that we have not investigated. A major area for investigation is the use of sprays rather than bait stations, Characteristics such as optimum formulation and spray rates for ultra low volume aerial sprays, stability of the Phloxin B + Uranine in sunlight or rainy weather (especially when applied as small droplets), and the attractance and palatability of the bait to Mexican fruit flies over time can only be tested under field conditions that are imperfectly mimicked in field cages. Other required information includes tests of attractance, feeding, and toxicity on non-target or beneficial insect populations (parasitic wasps, lady birds, lacewings, honeybees), and effects of bait on sterile flies.

Acknowledgements
Nada Tomic Carruthers performed the lab. series of sugar-water and NuLure tests. Andres Morales performed the field cage tests. Ismael Saenz, Robert Rivas and Guadalupe Gonzales helped with various technical aspects of the tests. Extensive support for this program was provided by the California Department of Food and Agriculture.

Literature Cited

1. Back, E. A.; Pemberton,C. E. . USDA Bull. **1918**. 538, 118 pp.
2. Chambers, D. L.; In Chemical Control of Insect Behavior, Theory and Application; H. H. Shorey & J. J. McKelvey, Jr. Ed; John Wiley & Sons, New York, **1977**; pp. 327-344.
3. Crounse, N. N.; Heitz.,J. R. United States Patent 4,320,140, March 16, **1982**.
4. Duggan, K. E.; Corneliussen, P. E. Pesticide Monitoring Journal. **1972**. 5, 331-341.
5. Federal Register, Tuesday, Sept. 28, **1982**. 47, #188, pg 42566,
6. Gow, P. A.; J. Econ. Entomol. **1954**. 47:153-160.
7. Heitz, J. R. In: Insecticide Mode of Action.Coats, J. R. Ed , ACS Press **1982**, pp. 429-457.
8. Martinez, A. J.; Moreno, D. S. ; J. Econ. Entomol. **1991**, 84:1540-1543.
9. McPhail, M.; J. Econ Entomol. **1939**, 32;758-761.
10. Moreno, D. S.; Martinez, A. J.; Sanchez-Riviello, M. J. Econ. Entomol. **1994**, 87; 202-211.
11. Ohinata, K.; Steiner, L. F. J. Econ. Entomol. **1967**, 60; 704-707.
12. Robacker, D.C.; Mangan, R.L.; Moreno D.S.; Tarshis Moreno, A.M., J. Ins. Beh. **1992**, 4; 471-487.
13. Roessler, Y.; In *Fruit Flies: Their biology, natural enemies and control, World Crop Pests, 3(B)*; Robinson, A. S. ; Hooper, G. Eds, Elsevier, Amsterdam. **1989**, pp. 329-336,
14. Shaw, J.G.; Lopez,-D, F. ; Chambers, D.L. Bull. Entomol. Soc. Amer. **1970**, 16; 186-193.
15. Steiner, L. F. J. Econ. Entomol. **1952**, 45; 838-843.
16. Steiner, L. F.; Rohwer, G. G.; Ayers E. L.; Christenson, L. D. J. Econ. Entomol. **1961**, 54; 30-35.
17. White, I. M.; Elson-Harris, M. M . *Fruit flies of economic significance: Their identification and bionomics.* CAB International, U.K., **1992**.

RECEIVED August 30, 1995

Chapter 10

Potential Markets for the Photoactivated Pesticides

G. D. Pimprikar

Kalyani Agro Corporation Limited, Kalyani Steels Limited Building, First Floor, Mundhwa, Pune: 411 036, Maharashtra, India

Since the pesticidal activity of photoactivated molecules has been success-fully proven in various insect species, fungal organisms and vegetation, these molecules have great potential in different market segments including the public health segment (mosquitoes and other disease vectors), poultry (house flies), farm animals (face fly), horticultural & agricultural sector (various insects and fungal species), and public nuisance segments (fire ants, cockroaches, etc.). These pesticides are quite effective, environmentally friendly and economical with less residual effect. Global markets could be developed for the photoactivated pesticides with focus in Africa (disease-vector segment), South America (horticultural and agricultural segment), and Asia (poultry & farm animal segment). Work on formulation and application technology needs to be done on a priority basis for commercial exploitation of these molecules.

Photoactivated molecules have been proven to be effective for several insect species and other organisms (*1*). There has been more emphasis on agricultural and horticultural pests since the previous symposium on photoactivated molecules held in 1987 and more numbers of insect species have been added to this category for photodynamic action. Work has also been done on the practical field application of these molecules. There have also been attempts in registering some of these chemicals for insecticide use in the U.S.A. In this paper, attempts are made to identify and describe the market segments and the scope in the international markets for this new class of pesticide.

   The following unique selling features of the photoactivated insecticides make them potential and successful candidates for global marketing.

*   They are effective at considerably low dosage rates for various insects species belonging to different groups.
*   They are very selective and specific to the targeted organisms (e.g., since these molecules have to be ingested into the target insects, they have no contact or systemic action).

0097–6156/95/0616–0127$12.00/0
© 1995 American Chemical Society

- They have been shown to be safe toward non-target parasites and predators.
- They have low residual activity.
- They are ecologically harmless and have no adverse impact on human health, wildlife, or environmental components.
- They are economical compared to currently favored pesticides.

Since various photoactivated pesticides effectively control insect species at various stages of development, they are good prospects to become an effective tool in integrated pest management programs for several species of insects.

Several researchers have done exploratory research on a spectrum of activity for photoactivated molecules and have reported insecticidal, bactericidal, fungicidal, antiviral and herbicidal activities (2). Except for insect studies, most research is in initial stages. Detailed studies of selectivity (especially for herbicidal activity), field trials, effective stages and dosages, etc. remain to be conducted. Hence, the discussions in this paper are limited to the insecticidal activity of photoactivated molecules.

In regard to the insecticidal activity, most of the molecules are gastric toxins and must be ingested into the digestive system of the insects (1). There is no contact or systemic activity as in many of the presently marketed insecticides such as the organophosphorus, carbamate or synthetic pyrethroid groups. The insecticidal activity of these chemicals is more prominent in the presence of light, but at higher dosages, insecticidal action has been reported even in the absence of light (3-4). Photoactivated insecticides are effective at various stages of insect development ranging from egg and larval stages through adult (5). Biotic effects also have been reported in subsequent generations (5). In addition, morphological and physiological abnormalities ultimately resulting in abortive molting have been reported at lower dosages (5-9). These effects are similar to IGR activity and probably reflect the effect of the chemicals on hormonal imbalances during development (5,8). Symptoms are also observed in the absence of light. Other effects observed in insects due to photoactivated molecules are changes in behavior and activity in presence of light (10). In social insects like imported fire ants, workers generally carry the bait to the mound and the queen is fed with the chemical impregnated bait. The effects of these chemicals can be amplified under such situations which can lead to destruction of the colony or mound.

The above mentioned multifaceted activity patterns of photoactivated molecules on insect development have a cumulative effect and could result in effective control of selectively targeted insects (5). These multiple patterns of activity can be a major tool in positioning the photoactivated molecules in various market segments.

Market Segmentation for Light Activated Molecules

Markets for the photoactivated molecules can be segmented as follows depending on the effective target group:

Public Health   Horticultural Crops
Poultry   Agricultural Crops
Animal Health   Nuisance Pests

Public Health.

Several insect vectors are responsible for spreading deadly diseases in various countries. The critical success factor for positioning of any chemical for this market segment is its selectivity to the targeted insect vector and safety to human, wildlife and other environmental components.

A variety of Dipteran insect species are world-wide vectors of such serious diseases as St. Louis encephalitis, dengue fever, malaria, leishmaniasis, cholera, dysentery, etc. Important insect vectors, like mosquitoes and house flies, could be effectively controlled by the application of photoactive insecticides during larval and adult stages (7-8, 11-16). These chemicals could be successfully used in integrated pest management programs for control of these disease vectors. It is necessary to develop and implement protocols for the use of light-activated insecticides in integrated vector management programs through either WHO or private institutions and organizations. This market segment could be "The LEADING" area for photoactivated molecules and could get the maximum market share in short period.

Poultry.

In layer poultry houses, houseflies are a serious problem that cause unhygienic conditions and effect the quality of poultry products. Several organophosphates and synthetic pyrethroids are currently used to suppress fly populations in poultry houses. However, due to their continuous use, resistance to both organophosphate and pyrethroid insecticides has developed. Some of the photoactive chemicals have been successfully used to reduce fly populations in chicken houses (17). These chemicals are sprayed directly on the chicken manure. There is a rapid mortality of adult flies after they feed and are exposed to visible light. Housefly eggs and larvae are also susceptible to photoactivated insecticides. There also is mortality in immature stages due to ingestion or absorption of the chemicals followed by light exposure. In addition to the direct mortality in eggs, larvae and adults, physiological and morphological abnormalities in pupal and adult developmental stages have been observed after treatment with photoactive chemicals (5). Apparently, this results from improper formation of chitin during ecdysis or from chemically induced energy stress on insects, which diminishes their ability to undergo metamorphosis (7). In addition to direct larval mortality, these initial sublethal effects indicate that larval treatment with photoactive chemicals may also cause latent mortality of the subsequent larval instars, pupae, or emerging adults.

The cumulative effect of exposure to photoactivated insecticides for all developmental stages (due to multiple mechanisms) results in control of the fly population in poultry houses. There is a good prospect for photoactive molecules in integrated fly control programs for poultry houses. This market segment is ideal for photoactive chemicals and has tremendous potential.

Animal Health.

Targeted insect species in this market segment include the housefly, facefly and stable fly. Photoactive chemicals have shown to be effective in controlling these insect species (6,9,14-16,18). In this particular segment, the chemicals may be fed to the insects through gelatin capsules applied to manure for maximal insecticidal activity. In order, to succeed in this market segment, more developmental work in terms of effective formulation is needed.

Horticultural and Agricultural Crops.
In the recent years, considering the favorable properties of the photoactive molecules, more attention is being given to the horticultural and agricultural market segment. One of the major limitations in entering this segment is the color of many photoactive molecules. However, with the advent of newer bait technology, this problem can be overcome. Some of the important target insect species in this segment include several species of fruit fly (*19-20*), apple maggot flies (*21*), bollweevils (*4,10*) cut worms (*22*), cabbage butterfly (*23*), codling moth (*24*), cabbage looper (*25*) and corn earworm (*25*).

The critical success factors in this market segment are high safety margins to the end user and low residual effects. Photoactive molecules meet both of these criteria and should have a good potential in this market segment - especially for use in the fruit and vegetable industry. Photoactive insecticides may also be useful additions to integrated pest management programs for specific crops.

Nuisance Pests.
The two insect species in this segment which have exhibited a good susceptibility to light-activated insecticides are imported fire ants (*3*) and cockroaches (*26*). In both cases bait formulations are effective. In the case of fire ants, the workers of the colony carry the bait impregnated with the insecticide to the mound where the queen ingests the bait. Eggs laid by this queen show morphological and physiological abnormalities during developmental stages and also at adult emergence. Deformed adults, as well as changes in sex-ratio, have been observed (Pimprikar, G.D., Mississippi State University, unpublished data, 1986). The ability of photoactive chemicals to cause toxicity from the egg to the adult stage through multiple mechanisms results in effective control of household pests like cockroaches and social insects like imported fire ants.

Apart from these major potential market segments, there is a good prospect for the photoactive molecules as herbicides, bactericides and fungicides. However, research in these areas is in the preliminary stage. Before commercialization, additional laboratory and field performance data will be needed as well as specific information regarding selectivity, especially in the case of herbicides.

GLOBAL MARKET FOR PHOTOACTIVE MOLECULES

Geographically the markets for photoactive molecules can be divided into the following regions.

Africa
North/South America
Europe
Asia
Australia

Africa.
In many of the African countries (*e.g.*, Niger, Chad, Zaire, Ethiopia, Nigeria, Angola etc.) there are serious disease problems transmitted by mosquitoes, and flies. The predominant diseases from these insect vectors include malaria, dengue, leishmaniasis, encephalitis and dysentery. WHO has taken up this region on a priority basis for protecting human health.

Several broad-spectrum organophosphates and carbamates are presently being used in these countries. Keeping in mind the desired properties of photoactivated molecules, they can be positioned very successfully in African markets for controlling the vectors. Regional integrated vector management programs may be developed in which photoactivated chemicals play a dominant role. This market segment is mainly controlled by governmental agencies and institutions.

North/South America.

Work on practical applications of photoactivated chemicals was initiated in this region. During the last decade there has been considerable emphasis placed on applying the photodynamic concept in the control of insect pests that affect horticultural and agricultural crop plants. Fruit flies are of particular interest to the USDA (*19-20*). Considering the strict safety standards and restrictions on the use of pesticides monitored by the regulatory agencies in this market segment, the photoactive chemicals could play an increasingly important role in programs aimed at controlling insect pests of various horticultural and agricultural crops.

Other areas which are a good prospect in this market segment are nuisance pests and, the animal and poultry segments. Imported fire ants are a great nuisance in many of the southern states in the U.S. Photoactive molecules can be effective in controlling these nuisance pests. House flies impose serious problems in poultry houses and cattle ranches all over the U.S. Pressures from governmental regulatory agencies and nearby housing developments make house fly control mandatory in most poultry houses and cattle ranches in the U.S.

Keeping in mind the total pesticide market of this geographic segment, the photoactive molecules should gain a substantial share from this region. In addition to the U.S., the other major countries from this geographical segment are Canada, Mexico, Brazil, Chile and Argentina.

Europe.

This geographical market segment is relatively small in size. However, this segment is localized and intensive in application. The areas in this segment are mainly horticultural crops.

Asia.

There is a vast potential for this geographic segment consisting of India, Bangladesh, Pakistan, Thailand and other eastern countries. Major application areas are public health and poultry. Governmental institutions play a dominant role in monitoring the vector control. However, in poultry, the private organizations have a major market share. The economic status of this region does not permit the use of expensive pesticides for controlling the vectors. Photoactive chemicals can be a great boon to this region and this may be a potential market.

Australia.

This market segment is quite large in size consisting of Australia and New Zealand. The market is limited to the animal health area.

Marketing Strategy for Photoactivated Molecules

The functional or geographical market segmentation highlights the potential areas for exploitation of global markets for the photoactivated molecules. While developing strategic marketing policies for these markets it is necessary to consider and focus on the following points;

- The insect species in each case is varied in nature, habitat, habits and also the nature of damage.
- The ecosystems and the environments are different.
- There is a geographical isolation of markets which have different characteristics, requirements and restrictions.
- In order to fulfill the requirements of the market, there is a need to have appropriate formulation and application techniques as well as equipment needed to selectively reach the target organisms.
- The purchase behavior and buying authorities could be different in each of the segments (e.g., private companies, farmer cooperatives, governmental agencies and institutions).

Different marketing strategies need to be implemented in each segment (geographical or functional). Universal marketing plans or differentiated marketing plans may not be appropriate in these cases. The best marketing approach may be a "**Selective and geographical niche marketing strategy.**" Selective positioning of photoactive molecules into regional integrated pest management programs could yield a better success in capturing targeted markets. In order to successfully implement a selective niche marketing strategy, the following areas need considerable adjustments (*27*).

- Product adjustments (intrinsic quality, packing, formulation, dose etc.)
- Distribution adjustments (channels, institutions, government agencies,)
- Promotional adjustments (discounts, media, publicity, etc,)
- Pricing adjustments (price threshold, price quality etc.)

While developing a global marketing strategy, it is necessary to concentrate on the following market globalization drivers (*28*):

- Market requirements
- Costs and economies of scale
- Government policies and restrictions
- Competitive market conditions.

For the photoactive molecules, the strategy of geographical niche marketing is ideal at targeting specific and localized needs. In this particular case, it is ideal to "*think globally but act locally*".

Areas of Concern Before Market Exploitation

The following four areas are of major concern for the photoactivated molecules in order to successfully enter potential markets.

*   Formulation technology
*   Application technology
*   Government regulation and restriction on usages
*   Worldwide registration policies

Formulation technology.

Photoactive molecules do not have contact or systemic activity. They are gustatory poisons and, hence, it is necessary to ingest the chemical into the targeted system. Commercially available formulations for organophosphate, carbamates or synthetic pyrethroids are not suitable or effective for photoactivated chemicals. There is a need for developing entirely new formulations for photoactive chemicals to selectively kill targeted organisms. Different types of baits for fire ants (3), fruit flies (21,22) and houseflies, have been developed and are quite effective for each respective species. In controlling the face fly and house fly, in cattle ranches, "gelatin capsules" are effective (6,20). In poultry houses, microencapsulated formulations may yield better effects. In the case of mosquitoes, a slow release formulation with a floating bait is probably needed. Innovation in developing and selecting the formulation is required. The success of the photoactivated molecules rests on innovative approaches in formulation so that the active component reaches the target site selectively.

Application technology.

There is a need to develop innovative, practical, and economical application technologies for the photoactivated molecules. The technology has to be different for various target species. In the case of fire ants, aerial broadcasting of the bait in the affected area could be effective and economical in certain markets (*e.g.*, U.S.A.). Application technology requirements are totally different with household pests like cockroaches. Research on application technology needs to be conducted on a priority basis before entering the different market segments.

Government regulations, restrictions on usages and global registration.

Some of the photoactivated molecules are registered by FDA and are found to be quite safe. However, there is a need to register these chemicals for the specified usage with EPA in the U.S. and with other regulatory agencies worldwide.

Conclusion

The potential geographic and functional market segments have been identified and described. There is a need to identify additional market segments globally. The selective and geographical niche marketing strategy will be ideal for photoactive pesticides. Work on formulation and application technology as well as global registration needs to be done on priority basis before commercial exploitation of photoactive molecules for the above market segments can begin.

Literature Cited

1.   Heitz, J.R. In *Insecticide Mode of Action*; Coats, J.P., Ed.; Academic Press: New York, 1982, pp. 429-457.
2.   Ellis, S.M.; Balza, F.; Constabel, P.; Hudson, J.B.; Towers, G.H.N. In *Light-Activated Pest Control*; Heitz, J.R.; Downum, K.R., Eds.; ACS Symposium Series No. 616;   American Chemical Society,: Washington, D.C., 1995, pp. 164-178.
3.   Broome, J.R.; Callaham, M.F.; Lewis, L.A.; Ladner, C.M.; Heitz, J.R. *Comp. Biochem. P.* **1975**, 51C, 117-121.
4.   Broome, J.R.; Callaham, M. F.; Poe, W.E.; Heitz, J. R. *Chem. Biol. Interact.* **1976**, 14, 203-206.
5.   Pimprikar, G.D.; Noe, B.L.; Norment, B.R.; Heitz, J. R. *Environ. Entomol.* **1980**, 9, 785-788.
6.   Fairbrother, T.E.; Essig, H.W.; Combs, R.L.; Heitz, J. R. *Environ. Entomol.* **1981**, 10, 506-510.
7.   Carpenter, T.L.; Heitz, J. R. *Environ. Entomol.* **1980**, 9, 533-537.
8.   Pimprikar, G. D.; Norment, B.R.; Heitz, J.R. *Environ. Entomol.* **1979**, 9, 856-859.
9.   Fondren, Jr., J.E.; Heitz, J.R. *Environ. Entomol.* **1978**, 7, 843-846.
10.   Callaham, M.R.; Broome, J.R., Lindig, O.H.; Heitz, J.R. *Environ. Entomol.* **1975**, 4, 837-841.
11.   Pimprikar, G.D.; Fondren, J.E.; Greer, D.S.; Heitz, J.R. *Southwestern Entomologist* **1984**, 9, 218-222.
12.   Barbosa, P.; Peters, T.M. *Med. Entomol.* **1970**, 7, 693-696.
13.   Barbieri, A. *Riv. Malariol* **1928**, 7, 456-463.
14.   Yoho, T.P.; Weaver, J.E.; Butler, L. *Environ. Entomol.* **1973**, 2, 1092-1096.
15.   Yoho, T.P.; Butler, L.; Weaver, J.E. *J. Econ. Entomol.* **1971**, 64, 972-973.
16.   Fondren, Jr., J.E.; Norman, B.R.; Heitz, J.R. *Environ. Entomol.* **1978**, 7, 205-208.
17.   Pimprikar, G.D.; Norman, B.R.; Heitz, J.R. *Environ. Entomol.* **1979**, 8, 856-859.
18.   Fairbrother, T.E. 1978. Ph.D. Diss. Mississippi State Univ., 131p.
19.   Liquido, N.J.; McQuate, G.T.; Cunningham, R.T. In *Light Activated Pest Control*; Heitz, J.R.; Downum, K.R. Eds.; ACS Symposium Series;   Am. Chem. Soc., Washington, D.C. pp. 82-114.
20.   Mangan, R.L.; Moreno, D.S. In *Light Activated Pest Control*; Heitz, J.R.; Downum, K.R. Eds.; ACS Symposium Series;   Am. Chem. Soc., Washington, D.C. pp. 115-126.
21.   Krasnoff, S.B.; Sawyer, A.J.; Chapple, M.; Chock, S.; Reissig, W.H. *Environ. Entomol.* **1994**, 23, 738-743.
22.   Clement, S.L.; Schmidt, R.S.; Szatmari-Goodman, G.; Levine, E. *J. Econ. Entomol.* **1980**, 73, 390-392.
23.   Lavialle, M.; Dumortier, B. *C.R. Hebd. Seances Acad. Sci. 1978*, 287, 875-878.
24.   Hayes, D.K.; Schecter, M.S. *J. Econ. Entomol.* **1970**, 63, 997.
25.   Creighton, C. S.; McFadolen, T.L.; Schalk, J.M. *J.Georgia Ent. Soc.* **1980**, 15, 66-68.
26.   Weaver, J.E.; Butter, L.; Yoho, T.P. *Environ.Entomol.* **1976**, 5, 840.
27.   Sheth, J.N. *European Res.* **1977**, 6, 3-12.
28.   Yip, G.S. *Sloan Management Rev.* **1989**, 31, 29-41.

RECEIVED September 7, 1995

Chapter 11

The Occurrence of Photosensitizers Among Higher Plants

Kelsey R. Downum and Jinghai Wen

Department of Biological Sciences, Florida International University, Miami, FL 33199 and Fairchild Tropical Garden, 10901 Old Cutler Road, Miami, FL 33156

More than 100 photosensitizers or phototoxins have been identified from higher plant tissues. These light-activated metabolites are broad-spectrum biocidal agents which have been implicated in a variety of plant defensive responses. The phototoxic metabolites isolated to date belong to 15 different phytochemical classes and are products of at least four biosynthetic pathways - fatty acid, polyketide, shikimate and terpenoid. A review of the phytochemical literature confirms the presence of phototoxic components in 35 families. Using disk-diffusion antimicrobial bioassays to survey higher plants for phototoxic activity, we have found light-activated bioactivity in extracts of taxa belonging to a variety of taxonomically disparate families including the: Acanthaceae, Campanulaceae, Gesnariaceae, Loganiaceae, Malpigiaceae, Papaveraceae, Phytolaccaceae, Piperaceae and Sapotaceae. This brings the total number of families that have been found to contain photosensitizers or exhibit phototoxic activity to 44.

Botanical phototoxins or photosensitizers are a structurally-diverse assemblage of plant metabolites that are grouped together because they mediate similar biological actions; namely, they catalyze toxic reactions following the absorption of light energy (*1*). These reactions are often lethal toward organisms that compete with or are otherwise harmful to plants, including pathogens, parasites, herbivores and other plants (*1*). Since the mechanisms of action, cellular targets and organismal effects of a variety of plant photosensitizers are reviewed elsewhere (*1-3*), these issues will not be dealt with in the present paper. Instead, an overview of the current state of knowledge regarding the phytochemical diversity, biosynthetic origins and taxonomic occurrence of photosensitizers among flowering plant families will be presented.

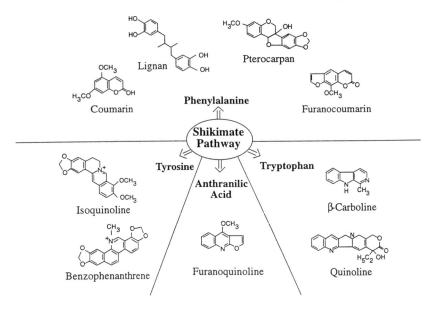

Fig. 1: Phototoxins derived from products of the Shikimate Pathway

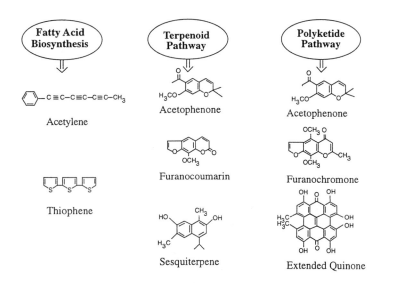

Fig. 2: Phototoxins derived from products of other biochemical pathways

Biosynthetic Origins

More than 100 phototoxic natural products have been identified since their discovery in higher plants slightly more than 20 years ago. Compounds from at least 15 different phytochemical classes are known to enter a "phototoxic" state following the absorption of light energy. Molecules that share this type of biological activity have diverse biochemical origins and are derived from four distinct metabolic pathways - the shikimate, fatty acid, terpenoid and polyketide pathways.

The biosynthetic origins of phototoxins identified to date are summarized in Figures 1 & 2. Photosensitizers derived from the shikimic acid pathway, the major biochemical pathway that gives rise to most of the phototoxic metabolites, are shown in Figure 1. Aromatic amino acids produced by this pathway are precursors for many of the polycyclic, aromatic photosensitizers. Phenylalanine serves as the starting molecule from which various coumarin, furanocoumarin, lignan and pterocarpan phototoxins are built. Isoquinoline and benzophenanthrene phototoxins are derived from tyrosine, while β-carboline and quinoline photosensitizers result from structural modifications to tryptophan. Anthranilic acid, another product of the shikimate pathway, gives rise to furanoquinoline-type photosensitizers.

The fatty acid, terpenoid, and polyketide pathways give rise to a variety of essential higher plant metabolites as well as to other photosensitizer types (Fig. 2). Linear polyacetylenes, thiophenes and various other ring-stabilized polyynes are formed from C_{18} fatty acid precursors following shortening and desaturation of the hydrocarbon chain (*4*). Terpenoids, the phytochemical class most recently shown to mediate phototoxic reactions (*5-8*), are formed by the condensation and modification of isoprene (C_5) units. Enzymes of the polyketide pathway are responsible for the conversion of 2-carbon acetate units into dictamnine (a furanochromone), extended quinones like hypericin and the benzyl functions of acetophenones.

Furanocoumarins and acetophenones are somewhat unique among plant photosensitizers in that they require the coordinated action of two biosynthetic pathways for their final structures, *i.e.*, they have mixed biosynthetic origins. The furan ring of furanocoumarins and the pyran ring of acetophenones are formed by the addition of isoprene units to the coumaryl portion of furanocumarins (formed via shikimate pathway) and the benzyl moiety of acetophenone derivatives (formed via the polyketide pathway).

Taxonomic Occurrence

A survey of the literature reveals that a substantial number of flowering-plant families contain phototoxic constituents. Phytochemical studies have confirmed the presence of derivatives from one or more classes of photosensitizers discussed above in at least 35 families (Tables I & II); bioassay studies suggest their presence in an even greater number of families. Twelve plant families have taxa that make multiple types of photosensitizers (Table I). Members of the Rutaceae (citrus family) are of particular note because of their ability to synthesize the broadest array of photosensitizers with specific examples reported in the literature from the β-carbolines, coumarins, furanocoumarins, furanoquinolines, isoquinolines and sesquiterpenes. Acetophenone, acetylene, thiophene and furanocoumarin

Table I. Flowering plant families with two or more classes of photosensitizers.

FAMILY	PHYTOCHEMICAL CLASS	REFERENCE
Apiaceae	Coumarins, Furanochromones, Furanocoumarins	9
Asteraceae	Acetophenones, Acetylenes, Furanocoumarins, Thiophenes	4,10,11
Dipsacaceae	Benzophenanthrenes, Furanocoumarins	9,12
Fabaceae	β-Carbolines, Furanocoumarins, Pterocarpans	9,13, 18,19
Fagaceae	β-Carbolines, Extended Quinones, Furanocoumarins	9,13,17
Moraceae	Coumarins, Furanocoumarins	9,14
Papaveraceae	Benzophenanthrenes, Isoquinolines	12
Polygonaceae	β-Carbolines, Coumarins	9,13
Rubiaceae	β-Carbolines, Isoquinolines, Quinolines	12,13
Rutaceae	β-Carbolines, Coumarins, Furanocoumarins, Furanoquinolines, Isoquinolines, Sesquiterpenes	5,6,9,12,13
Solanaceae	β-Carbolines, Furanocoumarins	9,13
Zygophyllaceae	β-Carbolines, Lignans	13,16

Table II. Flowering plant families from which one class of photosensitizer has been reported.

PHYTOCHEMICAL CLASS	FAMILIES	REFERENCE
β-Carbolines	Apocyanaceae, Bignoniaceae, Calycanthaceae, Chenopodiaceae, Combretaceae, Cyperaceae, Malpigiaceae, Passifloraceae	9
Benzophenanthrenes	Sapindaceae	12
Isoquinolines	Annonaceae, Berberidaceae, Euphorbiaceae, Juglandaceae, Magnoliaceae, Menispermaceae, Ranunculaceae	12
Furanocoumarins	Amaranthaceae, Orchidaceae, Pittosporaceae	9
Extended Quinones	Hypericaceae	17
Sesquiterpenes	Malvaceae	7,8
Quinolines	Nyssaceae	12

photosensitizers occur in various tribes of the Asteraceae (sunflower family). β-Carbolines and furanocoumarins have the widest occurrence among other families that synthesize multiple types of photosensitizers.

Taxa belonging to 23 other families contain metabolites from at least one of the phototoxic phytochemical classes (Table II). β-Carboline and isoquinoline derivatives occur in the majority of these families with benzophenanthrenes, furanocoumarins, extended quinones, sesquiterpenes and quinolines occurring in various other families. Little work has been done to document the structural variation and biological activity of phototoxins from these plants. Further study is necessary to identify the specific phytochemicals responsible for phototoxicity and to determine the scope of organisms susceptible to their detrimental effects.

The phytochemical diversity, albeit patchy occurrence, of phototoxic phyto-chemicals among higher plants, makes chromatographic methods of screening botanical tissues for specific types of photosensitizers of limited use. Instead, a rapid and sensitive antimicrobial bioassay can be used to survey large numbers of plants for phototoxic activity (*14*). Since a range of organisms (*e.g.*, pathogens, insect herbivores, etc.) are sensitive to the toxic action of most plant photosensitizers (*20*), antimicrobial bioassays offer an effective method of screening plants for phototoxic metabolites that can potentially affect a wide variety of organisms. For the past 10 years, my laboratory has employed disk-diffusion antimicrobial bioassays to survey neotropical plants for phototoxic activity. We have bioassayed nearly 2,000 plant species from 250 genera and *ca.* 80 families. Methanolic extracts of fresh plant tissue either field collected in Mexico, Costa Rica or Brazil, or obtained from the living collections at Fairchild Tropical Garden or the USDA Plant Induction Station in Miami, FL were bioassayed using established methodology (20).

Extracts from approximately 5% of the species bioassayed tested positive for phototoxins (Table III). These plants belong to 16 families - four of which had previously been shown to have photosensitizer-containing species (Asteraceae, Hypericaceae, Moraceae and Rutaceae). HPLC analysis of extracts from photosensitizer-containing plants revealed the presence of compounds with UV absorption spectra characteristic of acetylenic and thiophenic components (Asteraceae), extended quinones (Hypericaceae), and coumarins/furanocoumarins (Moraceae and Rutaceae) (*14-15, 28* and unpublished results). Nine of the families listed in Table III were not previously known to produce phototoxic phytochemicals. Indication of phototoxic components in extracts from plants belonging to the Acanthaceae, Campanulaceae, Gesnariaceae, Loganiaceae, Malpigiaceae, Papaveraceae, Phytolaccaceae, Piperaceae and Sapotaceae represent important new findings. Preliminary examination of the extracts by reverse-phase HPLC suggest that the phototoxic agents in these plants do not absorb in spectral regions characteristic of any of the known phototoxin classes.

We have become particularly interested in identifying the phototoxic constituents from the genus *Piper* (Piperaceae) because of the prevalence of phototoxins in leaf extracts (30 of 33 species were phototoxic against *Bacillus cereus*) and the ecological importance of the genus in the New World tropics. Efforts to isolate the UV-activated constituent(s) from various *Piper* extracts using bioassay-directed fractionation of extracted leaf material have been unsuccessful as phototoxicity disappears during the fractionation procedure. It is not yet clear whether the loss of bioactivity is due to chemical degradation or the result

Table III. Tropical/subtropical plants that have tested positive for phototoxic components using the disk difussion bioassay for antimicrobial activity (20). Numbers in parentheses represent the number of species that were phototoxic.

FAMILIES	PHOTOTOXIC GENERA
Asteraceae	Ambrosia (1), Coreopsis (1), Dyssodia (2), Eclipta (1), Erigeron (1), Flaveria (4), Gaillardia (1), Rudbeckia (1), Tithonia (1)
Acanthaceae*	Herpetacanthus (1)
Campanulaceae*	Hippobroma (1)
Gesnariaceae*	Bresera (1), Palicoria (1)
Hypericaceae	Hypericum (2)
Loganiaceae*	Spigelia (1)
Malpigaceae*	Banisteriopsis (2), Marscagnia (1)
Moraceae	Clarisia, Dorstenia (5), Fatoua (1), Ficus (2), Tropis ()
Papaveraceae*	Argemone (1)
Phytolaccaceae*	Phytolacca (1)
Piperaceae*	Piper (33), Pothomorphe (1)
Rubiaceae	Psycotria (2), Cephalis (4)
Rutaceae	Afraegle (1), Atalantia (1), Citrus (10), Microcitrus (1), Severinia (1), Swinglea (1)
Sapotaceae*	Pouteria (1)
Solanaceae	Solanum (1), Witheringia (2)

*Represents a new report of photosensitizers/phototoxic activity in this plant family.

of physical separation of extract components during the isolation procedure. We suspect the latter, since HPLC analysis of fraction components does not suggest chemical degradation.

A range of phytochemicals have been reported from *Piper* that could be responsible for the phototoxic action of extracts - these include various acetophenones (21), alkaloids (22-24), lignans (25-26) and terpenoids (27). In an effort to test some of these compounds, samples of dillapiole (a methylene dioxyphenylpropene) and three lignans isolated from *P. decurrans* were obtained from Dr. Thor Arnason's laboratory (Univ. of Ottawa) for antimicrobial bioassay. Although these compounds mediated varying degrees of antimicrobial activity toward *B. cereus*, they were not responsible for the phototoxic action observed with crude plant extracts (Table IV). Efforts to isolate the phototoxic compounds from *Piper* extracts are ongoing.

Survey studies to bioassay new plant extracts for phototoxic antimicrobial activity continue in an effort to establish the prevalence of phototoxic phytochemicals among higher plants, as do phytochemical studies to isolate and identify new phototoxic components that are found in our surveys. Ninety-five percent of the plants examined thus far have tested negative for phototoxic activity. It is not practical to list all of the genera that lacked phototoxic activity here. Plant families lacking phototoxicity have been identified in previous publications (1, 28). A listing of the genera and species that have been tested is maintained and can be provided upon request.

Table IV. Antimicrobial activity of lignans from *Piper decurrens* against *Bacillus cereus*.

NAME	STRUCTURES	INHIBITION ZONE * Dark	UVA
Dillapiole		7.4 ± 0.06	7.8 ± 0.47 **
HG1 (Conocarpan)		9.3 ± 0.46	9.7 ± 0.85
HG2		—	—
HG2'		7.4 ± 0.06	7.7 ± 0.64

* Inhibition zones expressed in mm ± standard deviation; (-) indicates the absence of an inhibition zone. All lignans were bioassayed with doses of 10 μg/disk; dillapiole was bioassayed at a dose of 0.8 mg/disk.
** The data were analysed using ANOVA and no significant differences were detected between dark and UVA treatment at $p=0.05$.

Conclusion

Phytochemical and bioassay studies have identified a large number of photosensitizers and phototoxin-containing plant species since their initial discovery more than two decades ago. Studies of the potential applications of several types of plant-derived photosensitizers as biocontrol agents are ongoing in a number of laboratories in Canada and the U.S. (see following chapters). To date, fewer than 100 higher plant families (*ca.* 30% of the extant families) have been analyzed for phototoxins or phototoxic activity. With the exception of a few families (*e.g.*, Apiaceae, Asteraceae, Moraceae, and Rutaceae), only a small fraction of the species in each family has been examined. Since phototoxins or phototoxic activity seem to occur in taxa belonging to at least 10-15% of the higher plant families, it is reasonable to suspect that many more phototoxic phytochemicals remain to be discovered, and that additional families of plants will be found to contain light-activated chemicals.

Continued exploratory research will help to: i) clarify the taxonomic relationships between plants that contain phototoxic phytochemicals; ii) assess the prevalence of light-activated defenses among higher plants; and iii) provide new photosensitizers with potential as biocontrol agents.

Acknowledgments. The authors would like to thank the following individuals for providing lignan standards - Denise Schauaret, Thor Arnason and Tony Durst (University of Ottawa) and identification of various plant materials - Luis Poveda and Pablo Sanchez (Universidad Nacional de Costa Rica).

Literature Cited

1.   Downum, K.R. *New Phytol.* **1992**, 122, 401-420.
2.   Towers, G.H.N. *Can. J. Bot.* **1984**, 62, 900-2911.
3.   Downum, K.R.; Rodriguez, E. *J. Chem. Ecol.* **1986**, 823-834.
4.   Bohlmann, F.; Burkhardt, T.; Zdero, C. *Naturally Occurring Acetylenes.* Academic Press, London, 1973.
5.   Asthana, A.; Larson, R.A.; Marley, K.A.; Tuveson, R.W. *Photochem. Photobiol.* **1992**, 56, 211-222.
6.   Green, E.S.; Berenbaum, M.R. *Photochem. Photobiol.* **1994**, 60, 459.
7.   Sun, T.J.; Melcher, U; Essenberg, M. *Physiol. Mol. Plant Path.* **1988**, 33, 115-136.
8.   Sun, T.J.; Essenberg, M; Melcher, U. *Mol. Plant-Microbe Interact.*, **1989**, 2, 139-147.
9.   Murray, R.D.H.; Mendez, J.; Brown, S.A. *The Natural Coumarins: Occurrence, Chemistry and Biochemistry.* J. Wiley & Sons Ltd., Chichester, 1982.
10.   Miyakado, M.; Ohno, N.; Yoshiokka, H.; Mabry, T.J. *Phytochem.* **1978**, 17, 143-144.
11.   Proksch, P.; Rodriguez, E. *Phytochem.* **1983**, 22, 2335-2348.
12.   Raffauf, R.F. *A Handbook of Alkaloids and Alkaloid-Containing Plants.* J. Wiley & Sons, New York, NY, 1970.
13.   Allen, J.R.F.; Holmstedt, B.R. *Phytochem.* **1980**, 19, 1573-1582.
14.   Swain, L.A.; Downum, K.R. *Biochem. Syst. Ecol.* **1990**, 18, 153-156.
15.   Downum, K.R.; Provost, D.; Swain, L. In *Bioactive Molecules: Chemistry and Biology of Naturally-Occurring Acetylenes and Related Compounds (NOARC)*, Lam, J; Breteler, H.; Arnason, T.; Hansen, L., Eds.; Elsevier: Amsterdam, 1988, Vol. 7; pp. 151-158.
16.   MacRae, W.D.; Towers, G.H.N. *Phytochem.* **1984**, 23,1207-1221.
17.   Thompson, R.H. *Naturally Occurring Quinones*; Academic Press, London, 1971.
18.   Robeson, D.J. ; Harborne, J.B. *Phytochem.* **1980**, 2359-2365.
19.   Smith, D.A.; Banks, S.W. *Phytochem.* **1986**, 25, 979-995.
20.   Downum, K.R.; Swain, L.A.; Faleiro, L.J. *Arch. Insect Biochem. Physiol.* **1991**, 17, 201-211.
21.   Diaz D., P.P.; Arias C., T; Joseph-Nathan, P.J. *Phytochem.* **1987,** 26, 809-811.
22.   Shah, S.; Kalla, A.K.; Dhar, K.L. *Phytochem.* **1986**, 25 1997-1998.
23.   Duh, C-Y.; Wu, Y-C; Wank, S-K. *Phytochem.* **1990**, 29, 2689-2691.

24.   Dominguez, X.A.; Verde S., J.; Sucar, S.; Trevino, R. *Phytochem.* **1986**, 25, 239-240.
25.   Badheka, L.P.; Prabhu, B.R.; Mulchandani, N.B. *Phytochem.* **1987**, 26, 2033-2036.
26.   Koul, S.K.; Taneja, S.C.; Pushpangadan, P.; Dhar, K.L. *Phytochem.* **1988**, 27, 1479-1482.
27.   Russell, G.R.; Jennings, W.G. *J. Agr. Food Chem.* **1969**, 17, 1107-1112.
28.   Swain, L.A.; Downum, K.R. In *Naturally Occurring Pest Bioregulators*; Hedin, P.A., Ed.; ACS Symp. Ser. No. 449; ACS, Washington, D.C., 1991; pp. 361-370.

RECEIVED August 10, 1995

Chapter 12

Fate of Phototoxic Terthiophene Insecticides in Organisms and the Environment

J. T. Arnason[1], T. Durst[1], M. Kobaisy[1], R. J. Marles[3], E. Szenasy[1], G. Guillet[1], S. Kacew[2], B. Hasspieler[4], and A. E. R. Downe[4]

[1]Ottawa–Carleton Institutes of Biology and Chemistry and [2]Department of Pharmacology, University of Ottawa, Ottawa, Ontario K1N 6N5, Canada
[3]Department of Botany, Brandon University, Brandon, Manitoba R7A 6A9, Canada
[4]Department of Biology, Queens University, Kingston, Ontario K7L 3N6, Canada

The phototoxic insecticide alpha-terthienyl was ^{14}C labelled for studies of its metabolic and environmental fate. In insects and rats administered the toxin, the label is cleared rapidly by a polysubstrate monooxygenase mediated mechanism. Two metabolites were identified which had no or reduced phototoxicity. Photodegradation of the phototoxin led to the identification of 12 mono and bithiophene carboxylic acid derivatives.

There is a wealth of information of the mode of action of phototoxic materials (1) but relatively little information on the fate of these materials in organisms or the environ ment. In toxicology the risk posed by a substance is the product of exposure and hazard. The exposure can be assessed by the rate of uptake into and clearance from organisms as well as their ability to metabolize the material. The half life in the environment is also important. The hazard can be evaluated by the lethal dose for 50% mortality (LD_{50}) of the substance and its metabolites and degradation products. In the long run information on the mode of metabolism and response of the detoxification system to the toxicant is also important.

Botanical or biopesticides including photoactivated materials are often cited as integrated pest management alternatives for a number of reasons, including their lack of persistence and/or low mammalian toxicity (2), but these properties must be evaluated experimentally before their use can be considered. Information on metabolism is also of interest in basic studies on plant-insect interactions.

The present report reviews recent and/or unpublished research on the fate of the botanical pesticide α-terthienyl (α-T) (1, Figure 1), starting with a review of the larvicidal activity of the parent and related compounds and the synthesis of labelled material used for subsequent toxicokinetic studies and identification of metabolites. New information on the photodegradation products and the mammalian toxicity are also provided.

0097–6156/95/0616–0144$12.00/0

Phototoxic thiophene insecticides.

The botanical pesticide, α-terthienyl, is a potent light-activated insecticide produced as a secondary metabolite by many genera of Asteraceae such as *Echinops, Centauria, Eclipta, Berkleya, Tagetes, Flaveria, Porophylum, Dyssodia* (3). It is highly toxic to a variety mosquito species (Table 1). Efficacy of emulsifiable concentrates and surface spreading formulations in controlling wild populations of *Culex spp.* and *Aedes spp.* has been verified under field conditions in Ontario, Quebec and Manitoba (4-5). Forestry insect pests (6) and agricultural insects are also sensitive to these materials (7). In previous studies (1) the mode of action of the material has been demonstrated to be mainly mediated through type II generation of singlet oxygen rather than superoxide generation. Phototoxicity in target insects results in lipid peroxidation and loss of membrane integrity, for example in the anal gills of mosquito larvae (8). Many derivatives of $\underline{1}$ have been synthesized by the Ottawa (9), Chicago (3) and Pisa (10) groups. A quantitative structure activity model for the phototoxicity of derivatives and analogues of $\underline{1}$ has been developed (11) based on the Hansch model of toxicity as a function of a bilinear or parabolic expression of the partition coefficient. A linear regression model including the bilinear term and a new term describing the rate of singlet oxygen production satisfactorily describes 80% or more of the variation in toxicity of these compounds to mosquito larvae or brine shrimp (11-12). The model also suggests ways that the partition coefficient can be optimized for target efficacy and reduced non-target effects.

Synthesis of ^{14}C labelled α-terthienyl for fate and metabolism studies.

The synthesis of α-terthienyl (α-T) ($\underline{1}$, Figure 1) labelled with ^{14}C starting with ^{14}CH$_3$I was a prerequisite to any studies of the fate of the materials. The synthetic scheme used (13) is shown in Figure 2. The label was introduced into two carbons in the central ring of the molecule to produce alpha-terthienyl with a specific activity of 5.4 mCi/mmole. Considerable effort was expended maximizing the efficiency of each step with "cold" reagents prior to working with the radiolabelled products. Sufficient hot material was obtained to allow us to complete metabolism studies in rats and continue these in insects. An alternate scheme involving the production of a ^{3}H labelled $\underline{1}$ was developed previously (7) and used in some of the studies described here as well.

Toxicokinetics in insects and nontarget invertebrates and mammals.

Once the labelled material was available, studies of the toxicokinetic of $\underline{1}$ in organisms was possible. Uptake of labelled $\underline{1}$ in target mosquito larvae is rapid and reaches a plateau within 24h. The material bioaccumlates strongly in insect tissue presumably because it is highly lipophilic. Hasspieler (8) observed strong fluorescence due to the compound in membranes of larvae, especially midgut and Malpighian tubules. After exposure of mosquito larvae to a single dose of $\underline{1}$, label was rapidly excreted with poly exponential kinetics which are consistent with clearance from several compartments.

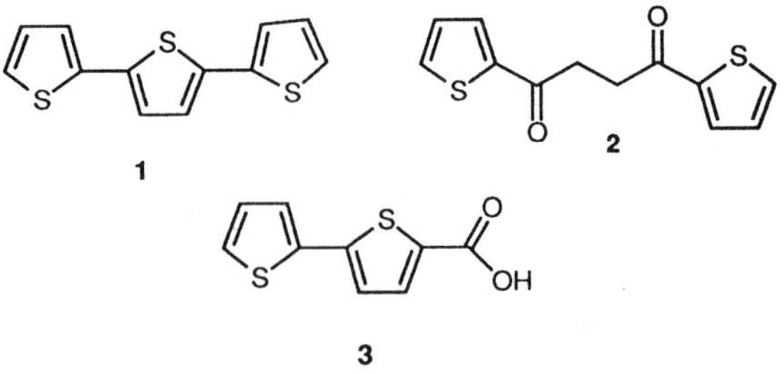

Figure 1. Structures of alpha terthienyl, **1** and two metabolites, **2** and **3** isolated from the urine of rats that were administered **1** orally.

Table 1. Phototoxicity of alpha-terthienyl to mosquito larvae

Mosquito Species	LC_{50} (ppb)	LC_{90} (ppb)
Aedes aegypti	4.22	7.08
Aedes epactius	6.03	18.11
Anopheles stephansi	14.39	28.31
Culex tarsalis (malathion resistant strain)	11.75	15.56
Culex tarsalis (malathion sensitive strain)	15.67	34.28

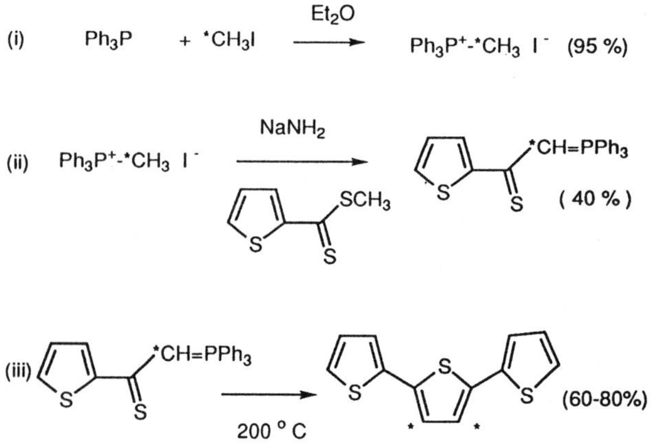

Figure 2. The synthesis of radioactive alpha-terthienyl. The yields of (i), (ii), and (iii) were 95, 40 and 60-80% respectively.

Clearance was even more rapid from non-target *Daphnia magna* with greater than 80% clearance of the label in a few hours (figure 3). The rate constant for clearance from the first compartment, $k_{el} = 1.35$ versus 1.19 for mosquito larvae.

In Lepidoptera, Iyengar et al. (7) found that excretion of a single applied dose was slow in the phototoxin-sensitive insect, *Manduca sexta*, requiring several days. It was much faster in the phototoxin-tolerant *Ostrinia nubilalis* suggesting that clearance rates are significant to toxicity.

As a representative non-target mammal, male Sprague-Dawley rats treated with a sublethal dose of the labeled material (50 mg/kg or 24,000,000 dpm/kg) by gavage (13), showed no significant decreases in growth rate or kidney or liver/ body weight ratios was observed in treated versus control animals. Excretion of the label was maximal at 24 hours in both urine and faeces and most of the material was passed within 2 days of treatment.

Metabolites:

In order to obtain sufficient material for metabolism, studies initially focussed on the rat model. In the faeces, most of the material was unmetabolized and therefore efforts to isolate metabolites focused on the urine. Partition of the combined aqueous urine samples with $CHCl_3$ followed by TLC of the $CHCl_3$ layer indicated that little unmetabolized, unbound α-T was present in the urine, although there was a large amount of label. Hydrolysis of urine metabolites by ß-glucuronidase/arylsulfatase followed by partitioning with EtOAc at pH 9, 7, and 2 yielded 1.51%, 1.83%, and 21.4% of the total radioactivity present in the urine, respectively, but 75.2% of the dpm counts remained in the aqueous layer.

Four radioactive bands (#1-4) accounting for 4, 23, 32, 41 % respectively of the radioactivity of the extract were obtained after thin layer chromatography of the major EtOAc extract (pH 2) on silica gel G developed with $CHCl_3$:MeOH 9:1. These bands were identified respectively as α-terthienyl, two metabolites which were identified as structures **2** and **3** (Figure 1) and a non-migrating material that probably represents conjugated or polymeric material. Metabolites **2** and **3** were analyzed directly by GC-MS, after pretreatment of the acidic **3** with hexamethyldisilazane. The structural assignments made on the basis of the mass spectral data were confirmed by synthesis.

Phototoxicity tests in a simple yeast bioassay indicated that **2** had no phototoxic activity and **3** had reduced phototoxic activity compared to **1**. These results can be explained by the cleavage of the ring structures during metabolism which reduced the conjugation and the light absorbing properties of **2**. The toxicity of the compounds to yeast in the dark was negligible. The toxicity of metabolites in rats is not known, but rats are clearly able to excrete these polar materials readily after oral administration of the parent material even though minimal amounts of the parent are excreted.

Using synthetically produced metabolites, we have now been able to detect the formation of **3** in mosquito larvae, *Aedes atropalpus* and faeces of the lepidopteran, *Manduca sexta* after administration of **1**. This metabolite was identified by HPLC co-

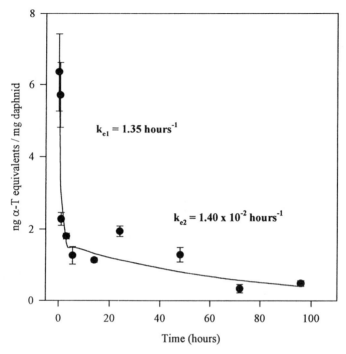

Figure 3. Elimination of ^{3}H-α-T from *Daphnia magna*. Groups of 20 *Daphnia* were exposed for 30 min to 100 ppb/530 dpm/mL, then transferred to dechlorinated water and sampled at various times post-exposure.

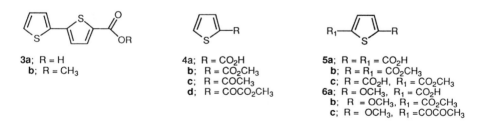

Figure 4. Photolysis products of α-T in methanol upon exposure to solar simulating conditions (900w/m^2).

chromatography of extract and standard while using fluorescence detection of the trace amounts that were isolated from insects.

In mosquito larvae, the involvement of the polysubstrate monooygenases (PSMO's) in metabolism and antioxidant systems as defences against phototoxins was clearly demonstrated by Hasspieler (14). Enhanced toxicity and photooxidative membrane damage, measured by a sensitive fluorescence bioassay was observed when the PSMO inhibitor piperonyl butoxide (PBO) was coadministered with α-T. Reduced damage occurred with PSMO inducers. Recently we have found that the reduction of reduced glutathione levels in insects by synthetic inhibitors or sesquiterpene lactones co-occurring in phototoxic plants, can significantly enhance the photooxidation of target membranes. These experiments on modulation of phototoxic activity suggested improved formulations with PBO that have been tested and shown to be more effective in the field compared to unsynergized material (4-6).

The results suggest that the label passes through the rat and some insect species very rapidly and that the parent material is readily metabolized to several polar metabolites. Those organisms that are unable to clear the metabolites quickly (*M. sexta* and some species of mosquitos) are more sensitive to photosensitization. The trace quantities of $\underline{1}$ in rat urine suggest a putative role for the toxin pump p-glycoprotein which can excrete xenobiotics without metabolism from cells.

The involvement of PSMO's is further indicated by the effect of oral administration of $\underline{1}$ to lepidoptera. Larvae reacted to the presence of the phototoxin by induction of a variety of PSMO or detoxification associated components or activities: cytochrome b_5, NADH cytochrome c reductase, *O*-demethylase, and glutathione *S*-transferase (15).

Photodegradation of $\underline{1}$

In addition to the fate in organisms, an understanding of the environmental fate of phototoxins is required. Despite extensive studies on the singlet oxygen generating capacity of α-T, there appear to be no reports on the environmental fate of α-T, other than our observations of a half-life of approximately 6 hours in sunlight and that a black insoluble material is obtained when α-T suspensions are exposed to sunlight for long periods of time (1). We studied the photolysis of α-T in methanol using a solar simulating xenon lampor sunlight at 900 w/m². In the presence of air, photolysis leads to a large number of identifiable products. The products were first separated into the acidic and neutral components by extraction with 10% Na_2CO_3. Both the neutral and acidic components were analyzed directly by GC-MS, after pretreatment of the acidic components with hexamethyldisilazane. The structural assignments made initially on the basis of the mass spectral data were confirmed by synthesis. The identified products consisted of a number of 2-substituted thiophenes including thiophene 2-carboxylic acid and its methyl ester ($\underline{4}$), thiophene -2,5 dicarboxylic acid, and its mono and dimethyl substituted thiophenes ($\underline{5}$), and a series of 5-methoxy substituted thiophenes, bearing a 2-carboxy, 2-carbomethoxy and 2-COCO$_2$CH$_3$ substituents ($\underline{6}$). Some bis-thiophenecarboxylic acid and its methyl ester was also formed ($\underline{3}$).

The formation of these compounds can be rationalized as stemming from an initial $2+2$ cycloaddition of singlet oxygen on a π-bond of a thiophene ring and subsequent ring opening. The introduction of the methyl substituent to yield 6b and 6c probably results from reaction with the solvent methanol with an α-T derived radical cation. Such intermediates have been shown to be form upon irradiation of α-T by a single electron transfer process (16).

The monothiophene products (**4**, **5** and **6**) were inactive in phototoxic bioassays with yeast or brine shrimp, but the bis-thiophenes **3** retained some of the phototoxic activity of the parent material. Furthermore the photolysis products could conceivably be readily used substrates of microorganisms.

Toxicity to mammals.

Assessment (13) of the toxicity of pure α-T by I.P. administration to Wistar male rats has indicated an accurately determined LD_{50} of 110 mg/kg (fiducial limits 84-153) at 24 h, n = 100. In a trial with Sprague-Dawley rats, the orally administered material was not toxic at the highest dose used (50 mg/kg, n = 16). These results suggest that the pure material is moderately toxic to mammals. However, the ready to use formulation (RTU) with 1g/L of active ingedient was not toxic in acute or subacute trials with Sprague Dawley rats at the maximum oral dose of 5 mL/kg. At high concentrations, 1 also is known to photosensitize skin (1), although it is not mutagenic (13).

Conclusion.

The available research indicates that α-T is a relatively non persistent material, whose photodegradation leads to a variety of relatively simple products with reduced or little photoactivity. Although the material has moderate toxicity to mammals, it is far more active in targeted insects and derivatives with modified hydrophobicity show promise for lower non-target toxicity. This natural source material is rapidly metabolized in and excreted from organisms by PSMO's and photooxidative damage is countered by antioxidant defences.

Acknowledgments. This work was supported by the Natural Science and Engineering Research Council of Canada (Stategic program).

Literature Cited.

1. Arnason, J.T., Philogene, B.J.R. and Towers, G.H.N., In *Herbivores*, Berenbaum M. and Rosenthal, G. Eds; 2nd edition, Academic Press, New York, N.Y. 1991, vol 2, pp 317-342.

2. Isman, M.B. *Pesticide Outlook*, 1994, June, pp. 26-31.

3. Kagan, J., In *Progress in the Chemistry of Organic Natural Products*; Herz, W., Griesbach, H., Kirby G.W. and Tamm, Ch. Eds.; Springer Verlag, New York, N.Y. 1991. Vol. 56 pp. 87-169. 1991

4. Fields, P. G., Arnason, J.T. Philogene, B.J.R., Aucoin, R. Morand, P. and Soucy-Breau, C. *Mem. Ent. Soc. Can.* **1991**, *159*, 29-38.

5. Dodsall, L.M., Galloway, M. and Arnason, J.T. *J. Amer Mosquito Contr. Assoc.* **1992**, *8*, 166-172.

6. Helson B., Kaupp, W., Ceccarelli A. and Arnason, J.T. *Can Ent.* **1995**, submitted.

7. Iyengar, S. Arnason, J.T. Philogene, B.J.R., Morand, P. and Werstiuk, N.H. *Pest. Biochem. Physiol.*, **1990**, *37*, 154-164.

8. Hasspieler, B.M., Arnason, J.T. and Downe, A.E.R. *Pesticide Biochem. Physiol.* **1990**, *38*, 41-47.

10. Rossi, R., Carpita, A. and Lezzi, A. *Tetrahedron*, **1984**, *40* 2773 2790.

11. Marles, R.J., Arnason, J.T., Compadre, R.L., Compadre, C.M., Soucy-Breau, C., Morand, P., Mehta, B., Redmond, R.W. and Scaiano, J.C.. In *Recent Advances in Phytochemistry*, Fischer, N.H., Isman, M.B and Stafford, H.A. Eds., *Modern Phytochemical Methods*, Plenum Press, New York, 1991, Vol 25, pp. 371-395.

12. Marles, R.J., Compadre, R.L., Compadre, C.M., Soucy-Breau, C., Morand, P., Mehta, B., Redmond, R.W., Scaiano, and Arnason, J.T. *Pest. Biochem. and Physiol.* **1991** *41*, 89-100.

13. Marles, R., Durst, T., Kobaisy, M., Abou Zaid, M. Arnason, J.T., Kacew, S., Kanjanapothi, D. and Lozoya, X. *Pharmacology and Toxicology*, **1995,** in press.

14. Hasspieler, B.M., Arnason, J.T. and Downe, A.E.R. *Pest. Biochem. Physiol.* **1991**, *40*, 191-197.

15. Feng, R., Houseman, J.G., Downe, A.E.R. and Arnason, J.T. *J. Chem. Ecol.* **1993**, *19*, 2047-2054.

16. Scaiano, J.C., Evans, C. and Arnason, J.T. *J. Photchem. Photobiol.* **1989**, 3, 411-418.

RECEIVED August 25, 1995

Chapter 13

Porphyric Insecticides

Subcellular Localization of Protoporphyrin IX Accumulation and Its Photodynamic Effects on Mitochondrial Function of the Cabbage Looper (*Trichoplusia Ni.*)

Keywan Lee[1] and Constantin A. Rebeiz

Laboratory of Plant Pigment Biochemistry and Photobiology, 240 A PABL, 1201 West Gregory Avenue, University of Illinois, Urbana, IL 61801–4716

The destructive photodynamic properties of δ -aminolevulinic acid-induced protoporphyrin IX accumulation has been well documented in insect and animal tissues. In this work the responses of isolated organs and tissues of several insect species to Proto accumulation were investigated. In *Trichoplusia ni*, and *Helioverpa* (*Heliothis*) *zea*, significant amounts of Proto accumulated in the midgut, and fat bodies. In *Blattella germanica* (German cockroach), more of the Proto accumulated in the male and female guts than in the abdomen. *Anthonomus grandis* (Cotton boll weevil) abdomens accumulated less Proto than the other three species. Oxygen consumption in *T. ni* midguts enriched in Proto was monitored before and after exposure to 2-hr of illumination. Decrease in O_2 consumption suggested photodynamic damage. To determine the physiological basis of this phenomenon, the subcellular site of protoporphyrin IX accumulation and singlet oxygen generation were investigated. Protoporphyrin IX accumulation occurred mainly in the mitochondria. The accumulated protoporphyrin resulted in the formation of singlet oxygen. ALA + Oph treatment exhibited deleterious effects on mitochondrial activities before illumination, which tended to obscure the possible involvement of singlet oxygen in enzyme Photoinactivation. However, the rate of NADH cytochrome c reductase activity, appeared to decline more rapidly in the light, in protoporphyrin IX-enriched than in control mitochondria. This in turn suggests the possible involvement of singlet oxygen in the enhanced inactivation of mitochondrial cytochrome c reductase activity by light.

[1]Deceased

NOTE: This chapter is part 10 in a series of articles on this topic.

0097–6156/95/0616–0152$12.00/0

Porphyric insecticides are a group of new insecticides with photodynamic properties (1-4). This novel system consists of modulators of the porphyrin-heme biosynthetic pathway, which when used singly or in combination with δ -aminolevulinic acid (ALA), a natural amino acid, induces massive accumulation of protoporphyrin IX (Proto) in treated insects. Protoporphyrin IX, the immediate precursor of heme does

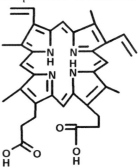

I. δ -Aminolevulinic acid   **II. Protoporphyrin IX**

not accumulate to any large extent in normal tissues. In treated insects, however, massive amounts of Proto accumulate and cause insect death upon exposure to light. In this work the mechanism of photodestruction of ALA + modulator-treated insects is investigated. It is shown that Proto accumulation in treated insects takes place in the mitochondria, where damaging singlet oxygen is generated in the light. Photodamage to the mitochondria appears to result in impaired mitochondrial function.

Materials and Methods

Insects. Adult *Blattella germanica* (German cockroach) were obtained from a colony maintained in a Biotronette® Mark III Environmental Chamber at 28 °C, 50% relative humidity and set for a 8:16 h light:dark photoperiod. The colony was initiated from egg cases purchased form Carolina Biological Supply, and was maintained on a diet of dog food and water. In order to obtain a regular supply of adults, subcolonies were established weekly in separate 42-cm x 27 cm plastic animal cages. Adult *Anthonomus grandis* (cotton boll weevil) were obtained from a colony maintained at the Boll Weevil Research Laboratory, USDA, SEA, Mississippi state University, Mississippi. A single tray of eggs dispersed on boll weevil diet was received weekly and held at room temperature until adult emergence. *Helioverpa (Heliothis) zea* (corn earworm) and *Trichoplusia ni* (cabbage looper) were obtained from a laboratory colony as described in (1).

Chemicals. 2,2'-Dipyridyl (Dpy) and 1,10-phenanthroline (Oph) were purchased from Aldrich (Milwaukee, WI). ALA was purchased from Biosynth Inc. (Skokie, IL). Grace's insect culture medium was purchased from Gibco Laboratories (Grand

III. 2,2'-Dipyridyl

IV. 1,10-Phenanthroline

Island, NY). All other reagents and chemicals, unless otherwise indicated, were purchased from Sigma (St. Louis, MO).

Treatment of Isolated Organs. One week old cockroaches, were starved overnight, and anesthesized with CO_2 prior to dissection. Abdomens of four cockroaches were removed with small surgical scissors, rinsed in cold buffer (0.1 M potassium phosphate, pH 7.0), cut into small pieces, and transferred to a small plastic Petri dish (5 cm in diameter), containing 3 ml of incubation buffer as well as ALA and a modulator at a specific pH. Abdomens of three day old, adult cotton boll weevils were excised and processed in a similar manner. Larvae of *H. zea* and *T. ni* were placed in ice-chilled phosphate buffer pH 7.0, and were dissected under a stereoscopic microscope to remove the tracheal branches and the Malpighian tubules. Midguts were then slit open and cleared of residual food remains, and rinsed in fresh phosphate buffer. Fat bodies were also collected from dissected larvae with a small spatula, washed twice in phosphate buffer and stored in cold fresh buffer. Integuments were removed from the mid-section of the body, cut into small pieces and placed in fresh cold phosphate buffer along with the fat bodies and tracheal branches. Organs and tissues prepared as just described were placed in incubation buffer containing ALA, a modulator or ALA + modulator, and were incubated in darkness for 5-hr. After incubation the organs and/or tissues were homogenized in acetone: 0.1N NH_4OH (9:1 v/v) and after centrifugation the acetone extract was used for tetrapyrrole determination by spectrofluorometry. Concomitant photodynamic damage was assessed by monitoring the decrease in oxygen consumption of the incubated tissue in the light. Oxygen consumption was determined polarographically, using a Clark oxygen electrode.

Subcellular Fractionation of Treated *T. ni*. Forty, 5th-instar larvae of *T. ni.*, were placed on diets containing 4 mM ALA and 3 mM Oph, in eight ounce paper containers with clear plastic lids (Lily-Tulip, Inc., Augusta, GA). The containers were kept in darkness at 28 °C for 17 h. The treated insects were washed with distilled water and homogenized in 50 ml ice-cold medium (0.25 mM sucrose, 10 mM Hepes, 1 mM EGTA, 10 μg/ml PMSF, pH 7.7). The homogenate was filtered through a layer of Miracloth (Calbiochem., La Jolla, CA), and centrifuged at 800g for 10 min in a Beckman JA-20 rotor at 1 °C. The 800g supernatant was decanted and recentrifuged at 10,000g for 10 min at 1 °C. The 10,000g supernatant was saved for further centrifugation. The mitochondrial pellet was washed once by suspension in the same medium, and centrifugation at 10,000g (washed mitochondria). The supernatants were pooled and centrifuged at 10,000*g* for 10 min. The pellet was pooled with the washed mitochondrial preparation. The resulting supernatant was centrifuged at 100,000*g* for 1 h in a Beckman 80 Ti rotor at   1 °C.   This centrifugation separated the post-mitochondrial supernatant into a microsomal pellet and a soluble "cytosol" supernatant. The microsomal pellet was resuspended gently in 0.25 mM sucrose, 10 mM Hepes, 1 mM EGTA, pH 7.7, with a camel hair brush and placed on ice until further use.

Percoll-Purification of Washed Mitochondria. Washed mitochondria (see above) were purified by centrifugation in a Percoll gradient. 35 ml of a 28% Percoll solution [28% (v/v) Percoll:0.25 M sucrose, 10 mM Hepes, 1 mM EDTA, pH 7.7] were placed in two 40 ml centrifuge tubes, and 1.5 ml (about 2-3 mg protein) of washed mitochondria were layered on top of the Percoll medium. Centrifugation at 48,000*g*, and 1°C, for 30 min in a Beckman JA-20 angle rotor resulted in the formation of four bands (Fig. 1). Fractions were collected with a 23 cm Pasteur pipette. All fractions

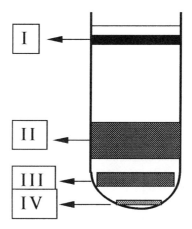

Fig. 1. Subcellular Percoll-Sedimentation Profile of *T. ni.* Fed a Diet Containing 4 mM ALA and 3 mM Oph. II = purified mitochondria.

were mixed with 10 times their volume of washing medium (0.25 M sucrose, 10 mM Hepes, 1 mM EGTA, pH 7.7) and were centrifuged at 39.000*g* for 10 min at 1 °C. The sedimented pellets were resuspended in washing medium and recentrifuged as above. The sedimented fractions were used for Proto determination and marker enzyme assays.

Subcellular Markers. Subcellular marker enzymes were monitored at 25 °C with an Aminco spectrophotometer, Model DW-2, operated in the split beam mode. Activity was linear with respect to time and enzyme concentration. Succinate:cytochrome *c* reductase (SCR) (EC 1.3.99.1), a mitochondrial marker, was monitored in 5 mM potassium phosphate buffet, pH 7.4, 0.05 mM cytochrome *c*, 1 mM NaCN, 10 mM sodium succinate, and 0.2 mM ATP (5). Lactate dehydrogenase (LDH) (EC 1.1.1.27) a cytosol marker, was assayed in 33 mM 2-(N-Morpholino)-ethanesulfonic acid (MES) pH 6.0, 0.125 mM NADH, and 0.33 mM sodium pyruvate (6). NADH cytochrome c reductase, a microsomal marker was assayed in 1 ml reaction cuvette containing 20 mM potassium phosphate, pH 7.4, 0.2 mM NADH, 0.05 mM cytochrome c, and 1 mM NaCN. Five μl of 0.2 mM ethanolic solution of antimycin was added to inhibit mitochondrial activity. Molar extinction coefficients of 6.22 x $10^3.M^{-1}.cm^{-1}$ at 340 nm for NADH and 21.1 x $10^3.M^{-1}.cm^{-1}$ at 550 nm for cytochrome c were used for computation of specific activities.

Detection of Singlet Oxygen Formation in Illuminated Solutions of Proto. The imidazole-RNO (p-nitrosodimethylaniline) method developed by Kralic and El Moshni (7)was used to monitor the formation of singlet oxygen in aqueous solutions of Proto. The singlet-oxygen-mediated oxidation reaction was monitored spectrophotometrically by measuring the increase in absorbance of RNO at 440 nm. The 2 ml reaction mixture consisted of 0.02 or 0.04 mM RNO, 8 mM imidazole, 10 μg Proto, or 100 μl mitochondrial suspension enriched in Proto, and 25 mM potassium phosphate buffer, pH 7.0. The reaction mixture was placed in a test tube (10 mm x 17 cm) and irradiated with 600 W/m^2 of white fluorescent light.

Determination of Mitochondrial Photodynamic Damage. Washed mitochondria prepared from ALA + Oph-treated *T. ni.* were exposed to 900 W/m^2 of white fluorescent light for 15 or 30 min at 25 °C. Then the activities of succinate oxidase, NADH dehydrogenase and NADH-cytochrome c reductase were monitored. Succinate oxidase activity was measured polarographically with a Clark oxygen electrode. The reaction mixture consisted of 2.9 ml 0.25 M sucrose, 0.01 M Hepes, 5 mM Mg Cl₂, 1 mM EGTA, 0.01 M potassium phosphate buffer, pH 7.5, and 0.1 ml of mitochondrial suspension. 100 μl aqueous sodium succinate was added to initiate the reaction. NADH dehydrogenase activity was determined by measuring the decrease in absorption at 600 nm of 2,6-dichlorophenolindophenol (DCIP), in a reaction containing 0.33 mM NADH, 0.1 mM DCIP, 0.5 mM NaCN, and 5-20 μl mitochondrial suspension in 0.1 M potassium phosphate buffer, pH 7.4. NADH-cytochrome c reductase activity was assayed as described above, in the absence of added antimycin.

Tetrapyrrole Determination. Aliquots of various subcellular fractions were mixed with 2 ml of cold acetone:0.1 N NH₄OH (9:1, v/v) and centrifuged at 39,000g for 10 min at 1°C. The acetone extract was delipidated by extraction twice with equal volumes of hexane. The hexane-extracted acetone residue which contained hydrophilic tetrapyrroles such as Proto and Zn-Proto was used for the quantitative spectrofluorometric determination of Proto (8), and detection of generated singlet oxygen. Fluorescence spectra were recorded at room temperature on a fully corrected photon-counting SLM spectrofluorometer, Model 8000C, interfaced with an IBM PS/2 model 60 PC. All spectra were recorded at emission bandwidths of 4 nm.

Protein Determination. Proteins were determined according to Smith et al. (9).

Results and Discussion

To further our understanding of the potential response of insect organs and tissues to porphyric insecticide treatment, in the absence of adsorption, transport, metabolism and compartmentalization, the response of isolated insect organs and tissues to incubation with ALA + Dpy or ALA + Oph was investigated. In these experiments, the following insects were used: adult *B. germanica*, adult *A. grandis*, fifth instar larvae of *H. zea* and fifth instar larvae of *T. ni*.

Table I. Response Of Isolated Organs And Tissues To Incubation With ALA And Dpy Or Oph[a]

			Proto content after 5-h incubation in darkness in the media listed below (nmol per 100 mg protein)					
Exp	*Insect*	*Tissue*	*Buffer*	*ALA*	*Dpy*	*ALA + Dpy*	*Oph*	*ALA + Oph*
A[b]	*T. ni*	Midgut	0.6	0.4	6.1	5.1	8.2	8.8
(n=2-7)			±	±	±	±	±	±
			0.3	0.2	1.8	0.8	0.5	2.7
B1	*T. ni*	Midgut	0.6	0.6	4.4	3.0	8.6	6.5
B2	*T. ni*	Integument	0.2	0.1	-[c]	-	0.8	0.7
B3	*T. ni*	Fat body	0.6	0.6	1.9	5.1	4.7	2.1
C1	*H. zea*	Midgut	1.8	0.7	15.9	20.5	23.2	29.2
C2	*H. zea*	Integument	0.2	0.2	0.6	0.6	0.7	0.8
C3	*H. zea*	Fat body	0.1	0.8	10.3	11.2	8.6	8.3
D1	*B. germanica*	Male gut	3.0	2.0	-	-	4.8	2.3
D2	*B. germanica*	Female gut	2.7	4.8	-	-	6.1	6.9
D3	*B. germanica*	Male abdomen - gut	0.2	0.3	-	-	2.9	4.5
D4	*B. germanica*	Female abdomen - gut	0.4	0.3	-	-	1.5	1.3
E[b]	*A. grandis*	Abdomen + gut	0.5	0.8	1.9	1.7	-	-
(n=2)			±	±	±	±		
			0.03	0.05	0.04	0.04		

[a] Incubation was in 5 mM ALA, 1.5 mM Dpy or in ALA + Dpy or Oph at pH 5.5 for 5-h in darkness.

[b] n, refers to the number of replicates.

[c] Not determined.

The results of isolated organ and tissue investigations for each of the aforementioned four insect species are summarized Table I. In *T. ni*, and *H. zea*, significant Proto accumulation was observed in the midgut, and fat bodies. Proto accumulation occurred when tissues were incubated with Dpy, ALA + Dpy, Oph, and ALA + Oph (Table I). No response to treatment with ALA alone was observed. In cockroaches, more of the Proto appeared to accumulate in the male and female guts than in their abdomen. As in *T. ni* and *H. zea*, the response was elicited by each of the treatments that included Dpy or Oph. Cotton boll weevil abdomens appeared to be less responsive than the abdomens of the other three species.

To determine whether Proto accumulation resulted in photodynamic damage in incubated tissues, *T. ni* midguts were incubated in darkness either in buffer, with ALA, or with Oph + ALA. Oxygen consumption of the tissue was monitored before and after exposure to 2-hr of illumination. It was assumed that decrease in O_2 consumption may indicate photodynamic damage and cell death. It was conjectured that this effect could be direct, *i. e.* rooted in mitochondrial enzyme destruction or indirect, as a result of mitochondrial disfunction. A 30% decrease in O_2 consumption was observed in midguts treated with Oph or with ALA + Oph after 2-hr in the light (data not shown).

Subcellular Distribution of Proto Accumulation. The subcellular site of Proto accumulation was next investigated. Forty, 5th-instar *T. ni* larvae, were placed on diets containing 4 mM ALA and 3 mM Oph, in eight ounce paper containers with

Table II. Subcellular Localization of Proto Accumulation in 5th Instar *T. ni.* larvae Placed On Diets Containing 4 mM ALA and 3 mM Oph for 18 h in Darkness

Subcellular Fraction	Proto (nmol/100 mg protein)	Total Proto (nmol/fraction)	(%)	SCR^a	NCR^b	LDH^c
Homogenate	82	53	100	241	-	249
Mitochondria	270	16	30	921	8.6	28
Microsomes	302	15	28	31	23.7	2
Cytosol	30	12	23	147	-	323

[a] Succinate cytochrome c reductase, a mitochondrial marker (nmol cyt *c* reduced/min/mg protein)
[b] NADH cytochrome c reductase, a microsomal marker (nmol cyt *c* reduced/min/mg protein)
[c] Lactate dehydrogenase, a cytosol marker (µmol NADH oxidized/min/mg protein)

clear plastic lids. Incubation was performed in darkness at 28 °C for 17 h. The treated insects were washed with distilled water and subcellular fractionation was carried out as described in Materials and Methods. Proto accumulation was observed in all subcellular organelles with the highest accumulation in mitochondria and microsomes (Table II). Proto accumulation in the microsomes was not due to mitochondrial contamination as evidenced by SCR activity (Table II). Since the site of Proto accumulation does not necessarily reflect its site of biosynthesis, it is not clear at this stage whether Proto biosynthesis takes place in the mitochondria, microsomes or both organelles. Indeed it has been shown that protoporphyrinogen IX, the immediate precursor of Proto readily translocates from its original site of synthesis to various cellular sites where it is converted to Proto by protoporphyrinogen IX oxidase (10). On the other hand, protoheme which is formed from Proto by ferrous iron insertion into Proto is the prosthetic group of cytochromes which are present in the endoplasmic reticulum as well as in the mitochondria. For example cytochrome P-450, a group of heme protein enzymes involved in detoxification of various xenobiotics, is mainly present in the endoplasmic reticulum.

Protoporphyrin IX Content of Percoll-Purified Mitochondria. In order to ascertain that mitochondria are indeed a major site of Proto accumulation, The Proto content of Percoll-purified mitochondria was evaluated. As shown in Table III, Percoll purified mitochondria (Fraction II, Fig. 1) exhibited the highest Proto content. This, in turn, prompted an investigation of the role that this massive mitochondrial

Table III. Proto Content of Percoll-Purified Mitochondria

Subcellular Fraction	*Proto (nmol/100 mg Protein)*	*SCR*[a]
Washed mitochondria	186	554
Fraction I	67	554
Fraction II	534	991
Fraction III	92	104
Fraction IV	43	23

[a] Succinate cytochrome c reductase, a mitochondrial marker (nmol cyt c reduced/min/mg protein)

Proto accumulation may play in the photochemical death of insects which accumulate Proto. In one experiment attempts were made to determine whether Proto and Proto-enriched mitochondria extracts, were capable of forming singlet oxygen upon light treatment. In the second experiment, the effects of illumination and singlet oxygen generation on Proto-enriched mitochondrial enzymes were investigated.

Singlet Oxygen Generation by Illuminated Proto Solutions and Proto-Enriched Mitochondrial Extracts. Singlet oxygen formation upon illumination of an aqueous solution of Proto was monitored spectrophotometrically by measuring the increase in absorbance of bleached RNO at 440 nm in the presence of imidazole. The reaction was carried out in a total volume of 2 ml at pH 7.0 (see Materials and Methods for more information). To distinguish between the formation of ˙OH radicals and singlet oxygen formation, the reaction was also carried out in the presence of azide and heavy water (D_2O). Azide reduces singlet oxygen lifetime and is therefore an effective quencher of singlet oxygen. On the other hand, D_2O increases singlet oxygen lifetime and the probability of its involvement in oxidative reactions. Thus a reduction of the rate of RNO bleaching in the presence of azide, and an increase in the rate of RNO bleaching in the presence of D_2O are reasonable evidence for the involvement of singlet oxygen in the reaction. As shown in Fig. 2, irradiation of an

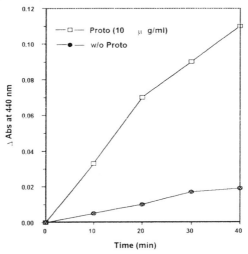

Fig. 2. Time-Dependent Singlet Oxygen Generation in an Aqueous Solution of Proto as Evidenced by RNO Bleaching. ΔAbs, refers to the increase in absorbance at 440 nm due to RNO bleaching.

aqueous solution of Proto with white fluorescent light resulted in a drastic increase in the rate of RNO bleaching as evidenced by the increase in absorbance at 440 nm. The reaction was totally suppressed under nitrogen gas, was reduced in magnitude by addition of azide and increased by substitution of H_2O with D_2O (Table IVA). Altogether these observations indicated that singlet oxygen was being generated in an aqueous solution of Proto upon illumination. This is compatible with the findings of others (11). Essentially the aqueous reaction mixture exhibited the same phenomenology when standard Proto was replaced by 100 μl Proto-enriched mitochondrial suspension (Table IVB). Thus, under illumination Proto-enriched mitochondria appeared to generate singlet oxygen as did an aqueous solution of Proto.

Photoinactivation of Mitochondrial Enzymes in Illuminated Proto-Enriched Mitochondria. Singlet oxygen is known to be toxic to cellular function by inducing lipid peroxidation and enzyme inactivation (12). The detrimental effects of light on oxygen consumption of various insect tissues that accumulate Proto, and the generation of singlet oxygen in Proto-enriched mitochondria, suggested that one of the toxic effects of Proto accumulation in insects may be mediated via inactivation of mitochondrial function. To test this hypothesis, the activity of several mitochondrial enzymes was evaluated in ALA and Oph-treated and untreated insects after

Table IV. Effect of Various Additives on the Generation of Singlet Oxygen in an Aqueous Solution of Proto[a] and in an Aqueous Reaction Mixture Containing Proto-enriched Mitochondria

Experiment	Treatment	ΔAbsorbance at 440 nm during 20 min illumination (% of control)
A[a]	Control	0.037 (100)
A	- Imidazole	0.011 (30)
A	+ Azide (10 mM)	0.023 (62)
A	+ N_2 gas	0 (0)
A	70 % D_2O	0.067 (181)
B[b]	Control	0.015 (100)
B	- Imidazole	0.0075 (50)
B	+ Azide (10 mM)	0.002 (13)
B	+ N_2 gas	0 (0)
B	70 % D_2O	0.067 (127)

[a]The reaction mixture contained 0.4 mM RNO, 8 mM imidazole, and 10 μg Proto.
[b]The reaction mixture contained 0.2 mM RNO, 8 mM imidazole, and 100 μl of the mitochondrial suspension which contained 120 pmol of Proto.

illumination. ALA + Oph treated insects accumulated massive amounts of Proto, in various subcellular compartments including the mitochondria as described above. After mitochondrial isolation, the washed mitochondria were irradiated with 600 W/m^2 of white fluorescent light, for various periods and the activity of several mitochondrial enzymes were determined.

As depicted in Fig. 3, the activities of succinooxidase, NADH dehydrogenase and NADH cyt c reductase, decreased in a time-dependent fashion in control and treated insects. The decrease in enzyme activities in untreated insects probably results from photosensitization of endogenous mitochondrial photosensitizers (13). It was also observed that ALA + Oph treatment, lowered enzyme activities by 30 to 50 % before the onset of illumination. In other words ALA + Oph treatment exhibited deleterious effects on mitochondrial activities even before illumination. This may be caused by

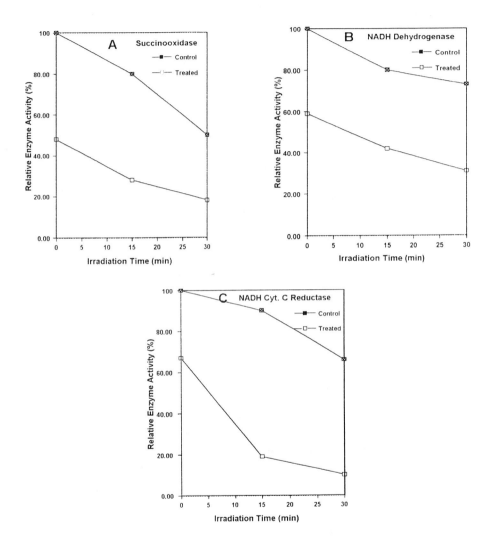

Fig. 3.   Photoinactivation of Three Mitochondrial Enzymes.   Washed mitochondria were prepared from untreated (control) insects and insects treated with 4 mM ALA and 3 mM Oph as described in Material and Methods. The washed mitochondria were irradiated with 600 W/m^2 of white fluorescent light for the times indicated on the graphs before assaying for enzyme activity.

the premature release of O_2^- and ˙OH radicals from the active site of damaged cytochrome c oxidase (14), and tended to obscure the possible involvement of singlet oxygen in enzyme Photoinactivation. However, the rate of NADH cyt c reductase activity, declined more drastically in mitochondria prepared from ALA + Oph treated than in controls. This in turn suggests the possible involvement of singlet oxygen in the enhanced inactivation of cytochrome c reductase activity by light.

Abbreviations. ALA: δ-aminolevulinic acid; Dpy: 2,2'-dipyridyl; Oph: 1,10-phenanthroline; Proto: protoporphyrin IX; Succinooxidase =. SCR, succinate:cytochrome c oxidase; NCR, NADH cytochrome c oxidase, LDF, lactic dehydrogenase; RNO, p-nitrosodimethylaniline.

Acknowledgments. Supported by funds from the Illinois Agricultural Experimental Station and by the John P. Trebellas Photobiotechnology Research Endowment to C.A.R.

Literature Cited

1. Rebeiz, C. A., Juvik, J. A., and Rebeiz, C. C. *Pesticide Biochem. Physiol.* **1988,** 30, 11-27.
2. Rebeiz, C. A., *J. Photochem. Photobiol. B: Biol.* **1993,** 18, 97-99.
3. Gut, L. J., Juvik, J. A., and Rebeiz, C. A. *Amer. Chem. Soc. Symp. Ser.* **1994,** 559, 206-232
4. Rebeiz, C. A., Gut, L. J., Lee, K., Juvik, J. A., Rebeiz, C. C. and Bouton, C. E. *Crit. Rev. Plant Sci.* **1995,** In Press.
5. Bergman A., Gardestrom P., and Ericson I. *Plant Physiol.* **1980,** 66:442-445.
6. Davies, D. D., and Davies, S. *Biochem. J.* **1972,** 129:831-839.
7. Kralic, I., and El Mohsni, S. *Photochem. Photobiol.* **1978,** 28:577-581.
8. Rebeiz, C. A., Mattheis, J. R., Smith, B. B., Rebeiz, C. and Dayton, D. F. *Arch. Biochem. Biophys.* **1975,** 171:549-567.
9. Smith P. K., Krohn, R. I., Hermanson, G. I., Mallia, A. K., Gartner, F. H., Provenzano E. K., Fujimoto E. K., Goeke, N. M., Olson B. J., and Klenk, D. C. *Anal. Biochem.* **1985,** 150:76-85.
10. Lee H. J., Duke, M. V. and Duke S. O. *Plant Physiol.* **1993,** 102:881-889.
11. Hopf, F. R., and Whitten, D. G. In *The porphyrins - Volume 2,* D. Dolphin, Ed.; Academic Press, New York, **1978,** pp. 161-195.
12. Spikes, J. D., and Straight, R. C. *Amer. Chem. Soc. Symp. Ser.* **1987,** 339:98-108.
13. Jung . J. and Kim, Y. J. *Photochem. Photobiol.* **1990,** 52:1011-1015.
14. Rebeiz, C. A., Juvik, J. A., Rebeiz, C. C., C. E. Bouton, and Gut, L. J. *Pest. Biochem. Physiol.* **1990,** 36:201-207.

RECEIVED August 24, 1995

Chapter 14

Thiarubrines: Novel Dithiacyclohexadiene Polyyne Photosensitizers from Higher Plants

Shona M. Ellis[1], Felipe Balza[1], Peter Constabel[1], J. B. Hudson[2], and G. H. Neil Towers[1]

[1]Department of Botany and [2]Department of Pathology, University of British Columbia, Vancouver, British Columbia V6T 1Z4, Canada

The thiarubrines are highly toxic sulfur heterocycles occurring predominantly in resin canals in the cortex and periderm of roots in certain species of Asteraceae. They are readily converted to thiophenes on exposure to light. In addition to light-mediated antiviral, antibacterial, nematocidal, and insecticidal activities they possess activity against fungi which does not require light. They are particularly effective against the human pathogenic yeast, Candida albicans.

Introduction to the Thiarubrines

Polyynes are widely distributed in the Asteraceae. These acetylenic compounds are also commonly found in the Campanulaceae, Apiaceae, Araliaceae as well as in the Basidiomycetes (1). A number of sulfur derivatives including dithiacyclohexadiene polyynes and thiophenes occur exclusively in the Asteraceae primarily in the Heliantheae. The dithiacyclohexadiene polyynes, which we have called thiarubrines (Figure 1), are found in the roots and, in some species e.g. Ambrosia chamissonis, in stems and leaves as well. Chaenactis douglasii and Eriophyllum lanatum produce the positional isomers thiarubrine A, 3-(1-propynyl)-6-(5-hexen-3-yn-1-ynyl)-1,2-dithiacyclohexa-3,5-diene and thiarubrine B, 3-(pent-3-yn-1-ynyl)-6-(3-buten-1-ynyl)-1,2-dithiacyclohexa-3,5-diene (5).

A

B

C

D

E

F

G

H

I

J

K

L

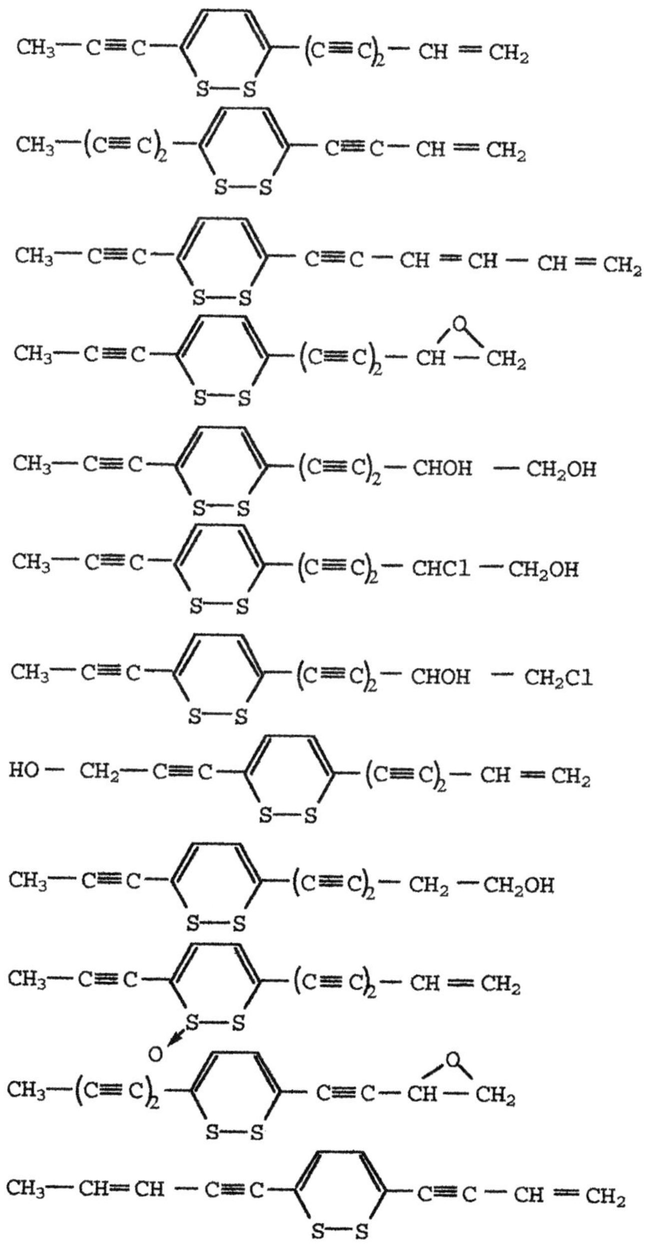

Figure 1. The Chemical Structures of the Thiarubrines

Rudbeckia hirta produces geometric isomers of thiarubrine C *(22)*. Table I summarizes the distribution of the thiarubrines throughout the Asteraceae.

Ambrosia chamissonis produces the widest array of thiarubrines of any plant analyzed to date *(11, 23)*. Thiarubrines A and B, are usually the major compounds. Other thiarubrines, based on the thiarubrine A skeleton, have been identified. Thiarubrine D, 3-(1-propynyl)-6-(5,6-epoxyhex-3-yn-1-ynyl)-1,2-dithiacyclohexa-3,5-diene, and thiarubrine E, 3-(1-propynyl)-6-(5,6-dihydroxyhex-3-yn-1-ynyl)-1,2-dithiacyclohexa-3,5-diene, a vicinal diol, are unique to this group and in relatively high quantity accompanied by their corresponding thiophenes. Compounds present in lower concentrations include two chlorohydrins, thiarubrines F and G (3-(1-propynyl)-6-(6-chloro-5-hydroxyhex-3-yn-1-ynyl)-1,2-dithiacyclohexa-3,5-diene), a primary alcohol, thiarubrine H, (3-(hydroxyprop-1-ynyl)-6-(5-hexen-3-yn-1-ynyl)-1,2-dithiacylohex-3,5-diene), and another primary alcohol, thiarubrine I, (3-(1-propynyl)-6-(6-hydroxyhex-3-yn-1-ynyl)-1,2-dithiacyclohexadiene) *(13)* and their corresponding thiophenes. Thiosulphinate J, (3-(1-propynyl)-6-(5-hexen-3-yn-1-ynyl)-cyclohexa-3,5-diene-1,2-thiosulphinate occurs without a corresponding thiophene *(12)*.

The thiarubrines are readily converted into thiophenes via ring contraction and sulphur extrusion on irradiation (Figure 2). Thiarubrine A yields the thiophene (2-(1-propynyl)-5-(5-hexen-3-yn-1-ynyl)-thiophene) and B produces the thiophene (2-(pent-3-yn-1ynyl)-5-(3-buten-1-ynyl)-thiophene). The thiarubrine solutions becomes yellow upon irradiation because of the presence of elemental sulfur. The thiophenes are colorless. The electronic state of the sulfur has not been established.

In Bohlmann's comprehensive book *(1)* on plant acetylenes UV-visible spectra are presented for a large number of polyynes and much can be gleaned about chromophores of these compounds from these data.

Biosynthesis

There is little known about the biosynthesis of these 13 carbon sulfur compounds although it is clear that they originate from polyynes which, in turn, are known to be derived from fatty acids. Neither the mechanism of triple bond formation nor the biochemical source of the sulfur atoms is known. Tracer studies by Bohlmann *et al.* *(1)* indicate that the terminal carbon of a fatty acid is usually lost in the generation of the odd carbon polyynes. This was corroborated by Gomez-Barrios *et al.* *(24)* using ^{13}C-acetate in tracer studies.   Even though the thiophenes can be produced by irradiation of the corresponding thiarubrine, there appears to be a separate biosynthetic pathway for these two classes of sulfur heterocycles as demonstrated in *Chaenactis douglasii* using ^{35}S incorporation *(25)*. Bohlmann *(1)* proposed that the thiarubrines exist in equilibrium with their thioketone isomers which could explain the red coloration.   This explanation is not widely accepted *(5)*.

The thiarubrines are obviously biosynthetically related and thiarubrine A is probably the precursor of the hydroxylated and chlorinated   thiarubrines.

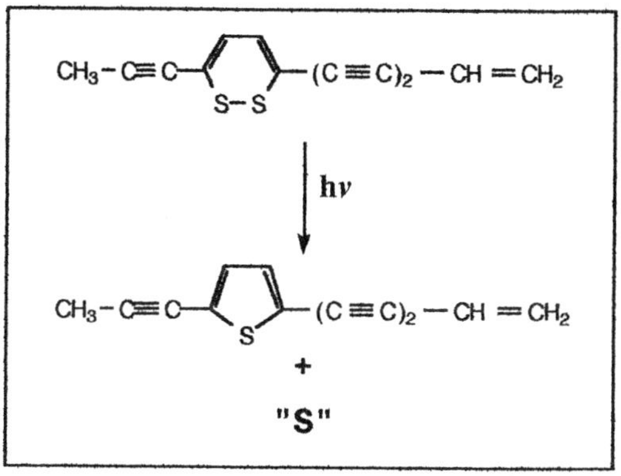

Figure 2.   Conversion of Thiarubrine   A   to   Thiophene   A   after Irradiation   with   UV-A   Light.

Isolation,   Purification   and   Identification

The   light   sensitive   nature   of   the   thiarubrines   require   that extractions   and   other   chemical   manipulations   be   performed   in subdued light *(11,12,13)*.   The plant samples are   frozen,   freeze dried   and   then   ground   in   a   mortar   or   a   Waring   blender   and extracted with a solvent.   The solvent of choice for the non-polar thiarubrines   is   petroleum   ether.   When   extracting   a   mixture   of non-polar   and   polar   thiarubrines   methanol:acetonitrile   (7:3)   is used.   HPLC   analyses   using   a   preparative   MCH-10   (C_{18},   10mm

Table 1.   Distribution of Thiarubrines in the Asteraceae
           -based on the taxonomy of Bremer (2).

Species	Thiarubrine	Reference

Tribe   Eupatorieae

Subtribe   Mikaniinae

Mikania officinalis	B	(1)

Tribe   Helenieae

Subtribe   Baeriinae

Eriophyllum caespitosum	B	(3)
E. lanatum	A, B	(4,5)
E. staechadifolium	A	(4)
Lasthenia chrysostoma	L	(6)
L. coronaria	L	(6)

Subtribe   Chaenactidinae

Chaenactis douglasii	A, B	(5)
C. glabriuscula	B	(1)
Palafoxia hookeriana	A	(1)
P. texana	B	(1)
Picradeniopsis woodhousei	L	(7)
Schkuhria abrotanoides	A	(1)
S. advena	A	(3)
S. multiflora	A	(8)
S. pinnata	A	(3,9)
S. senecioides	A	(3)

Tribe   Heliantheae

Subtribe   Ambrosiinae

Ambrosia artemisifolia	A	(1)
A. chamissonis	A, B, D, E, F, G, H, I, J	(5,10,11, 12,13)

Table 1. *Continued*

Species	Thiarubrine	Reference
A. confertiflora	A	(Isabel Lopez per. com.)
A. cumanensis	A	(10)
A. eliator	A	(3)
A. psilostachya	A	(Isabel Lopez per. com.)
A. trifida	A. K	(3,14)
A. trifoliata	A	(3)
Iva xanthifolia	A	(3)

Subtribe   Ecliptinae

Aspilia mossambicensis	A	(15)
(now placed with the *Wedelia* group)		
Oyedeae boliviana	B	(16)
Verbesina alata	B	(1)
V. boliviana	B	(17)
V. cinerea	B	(17)
V. latisquamata	A	(18)
V. occidentalis	A	(19)
Zexmenia hispida	B	(20)

Subtribe   Melampodiinae

Melampodium divaricatum	C (E)	(3, 21)
M. longifolium	C (E)	(3)

Subtribe   Milleriinae

Milleria quinquefolia	B	(1)

Subtribe   Rudbeckiinae

Rudbeckia. bicolor	C (E)	(3)
R. hirta	C	(3)
R. newmannii	C	(1)
R. speciosa	C (E)	(3)
R. sullivantii	C	(1)

particles) column, run isocratically with 28:72 or 50:50 acetonitrile:water is particularly useful. A photodiode array detector allows for distinctions between thiarubrines and thiophenes as well as other polyynes. The UV spectra of thiarubrines have diagnostic features and are particularly useful. They are not easy to tell apart however. The thiophenes on the other hand, although similar, have distinctive UV-Vis spectra. A useful method is to re-run samples, after exposure to UV radiation, when all thiarubrines have been converted to thiophenes.

Column chromatography, using dichloromethane as the initial solvent with the gradual addition of acetone, may be carried out to isolate larger quantities. In some cases chromatography on a second column is necessary to eliminate "oil-like" materials which accompany the thiarubrines extracted from roots.

Thin layer chromatography may be used to monitor fractions. Here again caution is necessary to exclude light. Thiophenes may be successfully isolated using this technique as well.

Final purificiation of thiarubrines and thiophenes is best achieved by HPLC with an MCH-10 column (8mm x 300mm).

The usual spectroscopic techniques are employed for the identification of the compounds: Ultraviolet-Visible spectrometry, Gas Chromatography-Mass Spectrometry of hydroxylated compounds, and solid probe Mass Spectrometry (MS). Balza et al. (11,12) presented proton Nuclear Magnetic Resonance (NMR) data in the structural identification of the compounds and also assigned ^{13}C NMR values (12).

Quantification of the thiarubrines can be carried out by means of UV-Vis spectrometry. The molar absorptivity (ε) is the same for thiarubrines A and B (ε=3000 at 490 nm, ε=10,300 at 340 nm). These values are assumed to be the same for the other thiarubrines.

Bioactivity Against Organisms

Thiarubrines exhibit toxicity to a number of organisms including mammals (13, 26 - 32). We have found that they induce pulmonary oedema in mice (unpublished) and may therefore be of limited value as therapeutic agents unless applied topically.

Insecticidal and Antifeedant Activities. There has been relatively little investigation of the activity of the thiarubrines on insects. We have found that, commercial dogfood preparations, when laced with thiarubrine A, in very low concentrations (10μg/g of dogfood) are a potent feeding deterrent to the cockroach *Blatta orientalis* (Towers and Abramowski unpublished). The thiophene α-terthienyl was adminstered orally to tobacco cutworms *(Manduca sexta) (26)*. Food laced with 50 μ g α-terthienyl per gram of food, was eaten by the insects which did not develop to adulthood.when irradiated with UV-A. Tissue damage resulted from topical application and UV-A treatment. These pronounced activities suggest that, when they occur, these compounds are important features of the natural defenses of plants.

Nematocidal Activity. The soil nematode *Coenorhabditis elegans*, was killed in the dark when exposed to 5 ppm and killed in the light (30 minute irradiation) when exposed to 0.03 ppm thiarubrine A *(27)*. It was once proposed that wild chimpanzees in Tanzania use the leaves of *Aspilia mossambicensis* as a nematocidal agent, but, although the roots of this species contain small amounts of thiarubrines, leaves have been found, from only one sample, to contain these compounds *(15)*. Page *et al. (28)* and Huffman *et al.* (unpublished), using sensitive HPLC techniques with freshly collected leaves from plants eaten by chimpanzees, found neither thiarubrines nor thiophenes in any of the samples.

Antifungal Activity. The preliminary antifungal and antibacterial test of choice is the disk-diffusion method *(13,27)*. Paper disks, impregnated with test compound are placed on a thin lawn of an organism or its spores. After 24 hours (sometimes longer in the case of fungi) growth is evaluated. A zone of clearance (area of no growth) around the disk indicates either death or fungistatic activity. Activity towards fungi does not require light. Fungi tested include the yeast *Candida albicans*, the dermatophytic fungi *Trichophyton mentagrophytes* and *Microsporum gypseum.* as well as *Fusarium trincticum* and *Verticillium lateritium*, plant pathogenic imperfect fungi. In all cases the thiarubrines exhibited pronounced activity at 0.1 mg or less per disk. *T. mentagrophytes* was susceptible at 0.01 mg/disk. Disk-diffusion tests on *Candida albicans* revealed the thiarubrines to be powerful fungicides at concentrations of 0.01 mg/disk in all thiarubrines tested except for thiarubrines A and B which were active at higher concentrations. The clearing zones were found to be larger when these two compounds were exposed to light. The converse was true for the other thiarubrines which exhibited smaller clearing zones upon irradiation. This implies that the activity of the thiarubrine is not based on its conversion to the thiophene. This class of phytochemicals has thus a very complex mode of action depending on whether light is involved in the activity analysis. Broth dilution tests with *Aspergillus fumigatus* and *Candida albicans* indicated that *A. fumigatus* was susceptible to the thiarubrines at 125-250 ng/ml. The MIC of thiarubrines A and B with *C. albicans* was 500 ng/ml whereas the MICs of the other thiarubrines was much lower (7.8-31.3 ng/ml) except for thiarubrine D whose MIC was 62.5-250 ng/ml. Fungizone, a commercial fungicide, in our hands gave an MIC of 160 ng/ml.

A very important factor to consider when testing chemicals against microorganisms is that the activity is sometimes dependent on the growth medium *(13,27)*.

Antibacterial Activity. In all of tests with bacteria it was found that the thiarubrines A to I are required in relatively high concentrations for toxicity eg. 1 μg or more per disk *(13,27,29)*. Ultraviolet light is essential for antibacterial activity. The gram-negative *Pseudomonas flavescens* and *P. aeruginosa* are resistant to the thiarubrines at this concentration. The gram-positive *Bacillus subtilis* and *Staphlococcus aureus,* on the other hand are susceptible. The polar thiarubrines are active at lower concentrations than the more non polar thiarubrines A and B. *Escherichia coli,* once believed to be non-pathogenic to humans, has been recently implicated in many serious infections. This species was found to be resistant to all thiarubrines in dark except for thiarubrine D. In light thiarubrines A and B exhibited good activity *(13)*.

Thiophenes also exhibit light-dependent antibacterial activity, but at higher concentrations than the thiarubrines *(28)*.

Antiviral Activity. Antiviral activity was light-dependent with cases of thiarubrines A and D in micromolar concentrations but not as good as that of the thiophenes *(30,31)*. Although they absorb at 480-490 nm their activity was slight at this wavelength. UV-A (360 nm) was required for optimum activity. The thiarubrines were found to be most active against membrane-bound viruses indicating that the target is probably some component of the membrane and that either singlet oxygen generation or a radical mechanism is involved. The viruses were unable to replicate in the presence of these compounds. Their activity was found to be not as efficient as α-Terthienyl. Anti-HIV activity required light *(32)*.

Cytotoxicity Tests. When evaluating chemicals as potential antibiotics it is important to evaluate their toxicity toward mammalian cells. We have tested thiarubrines against mouse mastocytoma cells and found that the more polar thiarubrines are effective at lower concentrations than the nonpolar ones (thiarubrine A). Thiarubrines F, H, and I show the best potential as antifungal compounds. They are toxic to the fungi at much lower concentrations than they are toxic to mammalian cells *(13)*.

The Mode of Action of the Thiarubrines.

Thiarubrines exhibit some of their activities e.g. towards fungi, without a light requirement. The corresponding thiophenes, however, are inactive against fungi unless they are irradiated during the experiment. This suggests that thiarubrines *per se* are antifungal and that the thiophenes, produced on irradiation are active agents in light (Figure 3). Much has been written about the light-mediated activities of the thiophenes belonging to the polyyne series of phytochemicals and this will not be discussed here.

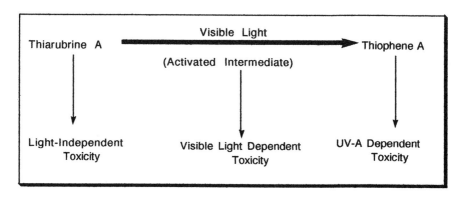

Figure 3   Proposed Biological Activity of Thiarubrine A
            (adapted from ref. *29*).

The different thiarubrines exhibit a range of toxicities. The stability of the compounds varies, for example, the diol is much more stable in aqueous conditions and storage over time than is thiarubrine A *(33)*. Differences in pH may also affect biological activities. When comparing the light-independent activity of thiarubrine A with that of thiarubrine C *(29)* it was found that they were essentially the same. The difference between the two molecules is that a triple bond in the former is replaced by a double bond in the latter which suggests that the biological activity may reside in the dithiane ring.

Block *et al. (34)* have successfully synthesized thiarubrine B which we have tested and found to be as effective as the natural compound.

Tissue   Culture

Tissue culture has become a very important technique in many areas of biology and biotechnology.   In times of diminishing forests and vegetation in general, it would seem to be a potentially viable alternative source of useful chemicals.   This application has not met with the success one would have expected.   There are a number of problems with this technology: low   yields   of   desired   compounds,   problems   with   the establishment of continuous cultures, many compounds are not excreted into the medium.   The production of thiarubrines   in tissue culture has been studied extensively in our laboratory. Cosio *et al. (35)* examined tumourous cultures of *Chaenactis douglasii* generated with *Agrobacterium tumefaciens* and found thiarubrines were being produced.   Anatomical sections however clearly showed tissue differentiation in calli with the appearance of thiarubrines in canals in peripheral tissue. Thiarubrines and thiophenes are not produced in suspension cultures or even normal (non-transgenic) callus cultures of *Chaenactis douglasii (36)* and *Ambrosia chamissonis (13)*. Normal and transgenic root cultures, on the other hand, produce the thiarubrines and thiophenes characteristically found in the roots of the respective plants. If transgenic callus or root cultures are exposed to light during growth and development bleaching occurs and the production of thiarubrines is shut down. The red color of these tissues reappear but the new pigments are anthocyanins and not thiarubrines.

Hairy roots are readily generated with *Agrobacterium rhizogenes.* Like *Agrobacterium tumefaciens* this bacterium is attracted to the wound site of a plant, becomes attached and transfers a piece of DNA into the plant cell where it becomes incorporated into the nuclear genome of the plant. Unlike *A. tumefaciens, A. rhizogenes* causes a proliferation of roots from the infected cell.   These roots grow very rapidly and do not require growth regulators in the medium.   An excellent paper by Zambryski *et al.* *(37)* describes the process and the differences between the two bacterial species.

Hairy root cultures of *Chaenactis douglasii (36)* and *Ambrosia chamissonis (13)* have been established.   In the case of *Chaenactis* the thiarubrine content of the hairy root cultures was found to be higher than in normal roots cultures.   This was not the case for *Ambrosia chamissonis* in which both normal and transgenic cultures produced approximately 1 mg/ g dry weight of thiarubrine A   which is less than that produced by soil grown roots.

Manipulations of culture conditions may also enhance the production of important chemicals, through precursor feeding as well as through the stimulation of production with   elicitors. Since many secondary compounds play a defensive role in plant tissues, exposure to a pathogen may bring about an increase in the production of these compounds.   When root cultures of *Ambrosia chamissonis* are exposed to cell wall preparations of the fungal pathogen, *Phytophthora megasperma* f. sp. *glycine*, instead of an elevation of the overall thiarubrine content, the concentration of thiarubrine E was enhanced markedly *(13)*. In antimicrobial tests thiarubrine E is more potent than thiarubrine A.   In other words there was increased synthesis of thiarubrine E at the expense of the less active thiarubrine A.   This type of chemical response does not appear to have been described in the voluminous literature on elicitation.

Localization of Thiarubrines in Tissues

Thiarubrines are located in canals which run with the longitudinal axis of roots and, where they occur, parallel to the long axis of stems and leaves. They are easily seen in entire small roots and in hand sections of species which produce them. The canals are in the same strategic locations previously noted for similar canals in other species of the Asteraceae. It has been suggested that polyynes are sequestered in these canals.   In *Ambrosia chamissonis* roots larger canals also occur in the bark. In younger non-woody roots the canals run longitudinally in the outer cortex and in the immature periderm.   In cultured primary roots thiarubrines are localized in canals between what appears to be a double endodermis.   Interestingly when root cultures of *Ambrosia* are exposed to light the cells of the endodermis accumulate anthocyanins *(13)*.   This suggests a need for a protective mechanism against the effects of the light.

Species in the Asteraceae, such as *Chaenactis douglasii* *(35)* and *Verbisina* spp. do not contain thiarubrines or thiophenes in their above ground parts *(17)*. Roots are generally the major sources of these chemicals and the analysis of the other parts of the plants have often been ignored. The instability of these compounds to light may have prejudiced researchers against looking for them in leaves or stems. In *Ambrosia chamissonis* young leaves and stems are visibly streaked with thiarubrines. Upon sectioning leaves and stems it is apparent that the thiarubrines are in canals closely associated with the vascular bundles. Canals in the leaf are found in close proximity to tracheary elements. The localization of the canals in woody stems is not discernible using the hand sectioning methods employed.

The overall thiarubrine content of the above ground portions in *Ambrosia* is less than that of the roots. The relative concentrations of the thiophenes is of course much higher. This is possibly the result of photoconversion. In the case of thiarubrine B this does not appear to be the case. Thiarubrine B is found in much higher concentrations in woody stems of *Ambrosia* than would be predicted for normal degradation when compared with the other thiophenes *(13)*. This may be the result of differential stability of this compound, but more likely it is due to the elevated synthesis of thiophene B independent of thiarubrine B. High levels of polyynes, e.g. phenylheptatryine, appear to exist in a stable condition in leaves of tropical species e.g. *Bidens*, in full sunlight.

Ambrosia chamissonis has in the past been segregated into many species based on leaf morphology. Recently these taxa have been placed into one heteromorphic species *(38)*. Upon examination of the chemistry of a number of collections it was apparent that there were not only differences in the chemical profile of thiarubrines between different morphological types, but also between samples collected at different sites. Whether these represent responses to environmental conditions remains to be seen.

Concluding Remarks

As we have indicated the thiarubrines exhibit strong toxicities and there are both light-dependent and light-independent activities. The light-independent activity is the subject of patents for the treatment of candidiasis in humans *(33)*.

Acknowledgements:

We thank the Natural Sciences and Engineering Research Council of Canada for support of this work.

Literature Cited

1. Bohlmann, F.; Burkhardt, T.; and Zdero, C. *Naturally Occurring Acetylenes.* Academic Press, New York, 1973
2. Bremer, K *Asteraceae Cladistics and Classification*, Timber Press, Portland, Oregon; 1994
3. Bohlmann, F.; Kleine, K. M. *Chem. Ber.* 1 9 6 5, *9 8*, 3081-3086.
4. Bohlmann, F.; Zdero, C.; Jakupovic, J.; Robinson, H.; King, R. M. *Phytochemistry* 1 9 8 1, *2 0*, 2239-2244.
5. Norton. R. A.; Finlayson, A. J.; Towers, G. H. N. *Phytochemistry* 1 9 8 5, *2 4*, 356-357.
6. Bohlmann, F.; Zdero, C. *Phytochemistry* 1 9 7 8, *1 7*, 2032-2033.
7. Bohlmann, F.; Zdero, C.; Grenz, M. *Phytochemistry* 1 9 7 6, *1 5*, 1309-1310.
8. Bohlmann, F.; Jakupovic, J.; Robinson, H.; King, R. M. *Phytochemistry* 1 9 8 0,19, 881-884.
9. Bohlmann, F.; Zdero, C. *Phytochemistry* 1 9 7 7, *1 6*, 780-781.
10. Bohlmann, F.; Zdero, C.; Lonitz, M. *Phytochemistry* 1 9 7 7, *1 6*, 575-577.
11. Balza, F.; Lopez, I.; Rodriguez, E.; Towers, G. H. N. *Phytochemistry* 1 9 8 9 , *2 8*, 3523-3524.
12. Balza, F.; Towers, G. H. N. *Phytochemistry* 1 9 9 0, *2 9*, 2901-2904.
13. Ellis, S. M. *MSc. Thesis*, Univeristy of British Columbia, 1993.
14. Lu, T.; Parodi, F. J.; Vargas, D.; Quijano, L.; Mertooetomo, E. R.; Hjortso, M. A.; Fischer, N. H. *Phytochemistry* 1 9 9 3, *3 3*, 113-116.
15. Rodriguez, E. R., Aregullin, M., Uehara, S., Nishida, T., Wrangham, R., Abramowsji, Z., Finlayson, A., and Towers, G. H. N. *Experimentia* 1 9 8 5 , *4 1*, 419-420.
16. Bohlmann, F.; Zdero, C. *Phytochemistry* 1 9 7 9, *1 8*, 492-493.
17. Bohlmann, F., Grenz, M., Gupta, R. K., Dhar, A. K., Ahmed, M., King, R. M., and Robinson, H. *Phytochemistry* 1 9 8 0, *1 9*, 2391-2397.
18. Bohlmann, F.; Lonitz, M. *Chem. Ber.* 1 9 7 8, *1 1 1*, 254-263.
19. Bohlmann, F.; Lonitz, M. *Phytochemistry*, 1 9 7 8, *1 7*, 453-455.

20. Bohlmann, F.; Lonitz, M. *Chem., Ber.* **1 9 7 8**, *1 1 1*, 843-852.
21. Bohlmann, F.; Le Van, N. *Phytochemistry*, **1 9 7 7**, *1 6*, 1765-1768.
22. Constabel, C. P.; Balza, F.; Towers, G. H. N. *Phytochemistry* **1 9 8 8**, *2 7*, 3533-3535.
23. Balza, F., and Towers, G. H. N. In *Methods in Plant Biochemistry-Alkaloids and Sulphur Compounds*, Editor, P. M. Dey, J. B. Harbourne; Academic Press: Toronto, Can., 1993, Vol. 8; pp 551-572.
24. Gomez-Barrios, M. L.; Parodi, F. J.; Vargas, D.; Quijano, L.; Hjortso, M. A.; Flores, H. E.; Fischer, N. H. *Phytochemistry* **1 9 9 2**, 31, 2703-2707.
25. Constabel, C. P.; Towers, G. H. N. *Phytochemistry* **1 9 8 9**, *2 8*, 93-95.
26. Downum, K. R.; Rosenthal, G. A.; Towers, G. H. N. *Pesticied Biochemistry and Physiology* **1 9 8 8**, *2 2*, 104-109.
27. Towers, G. H. N.; Abramowski, Z.; Finlayson, A. J.; Zucconi, A. *Planta Medica* **1 9 8 5**, *5 1*, 225-229.
28. Page, J. E.; Balza, F.; Nishida, T.; Towers, G. H. N. *Phytochemistry* **1 9 9 2**, *3 1*, 3437-3439.
29. Constabel, C. P.; Towers, G. H. N. *Planta Medica* **1 9 8 9**, *5 5*, 35-37.
30. Hudson, J. B.; Graham, E. A.; Fong, R., Finlayson, A. J.; Towers, G. H. N. *Planta Medica* **1 9 8 5**, *1*, 51-54.
31. Hudson, J.B. *Antiviral Research* **1 9 8 9**, *1 2*, 55-74.
32. Hudson, J. B.; Balza, F.; Harris, L.; Towers, G. H. N. *PPhotochem. and Photobiol.* **1 9 9 3**, *5 7*, 675-680.
33. Towers, G. H. N.; Bruening, R. C.; Balza, F.; Abramowski, Z. A.; Lopez-Bazzochi, I. *Thiarubrine antifungal agents.* United States Patent No. 5,202,348, 1993.
34. Block, E.; Guo, C.; Thiruvazhi, M.; Toscano, P. J. *J. of American Chemical Soc.* **1 9 9 4**, *1 1 6*, 9403-9404.
35. Cosio, E. G.; Norton, R. A.; Towers, E.; Finlayson, A. J.; Rodriguez, E.; Towers, G. H. N. *J. Plant Physiol.* **1 9 8 6**, *1 2 4*, 155-164.
36. Constabel, C. P.; Towers, G. H. N. *J. of Plant Physiology* **1 9 8 9**, *1 3 3*, 67-72.
37. Zambryski, P.; Tempe, J.; Schell, J. *Cell* **1 9 8 9**, *5 6*, 193-201.
38. Payne, W. W. *J. of the Arnold Aboretum* **1 9 6 4**, *XLV*, 401-438.

RECEIVED August 24, 1995

Chapter 15

Alternative Mechanisms of Psoralen Phototoxicity

Karen A. Marley[1], Richard A. Larson[1], and Richard Davenport[2]

[1]Institute for Environmental Studies and [2]School of Life Sciences, University of Illinois, Urbana, IL 61801

Traditional studies of the mechanism of phototoxicity of psoralens (furocoumarins) have emphasized their photoaddition reactions with DNA. Photooxidation products of furocumarins, although they have been observed in some studies, have been given little attention. By the use of improved extraction and separation techniques, we recently identified a group of salicylaldehyde derivatives from psoralen photooxidation *in vitro*. The occurrence of these phenolic aldehydes could possibly account for some of the observed photobiological effects now attributed to the parent compounds. Psoralen and 8-methoxypsoralen are quite susceptible to photolysis in polar solvents such as water, giving many cleavage products including aldehydes, carboxylic acids, and hydrogen peroxide. Studies with photolyzed psoralens and isolated photoproducts such as aromatic aldehydes reveal them to be toxic, even in the absence of light, to *Xenopus* embryos.

Naturally occurring substances having a variety of structures have been found to display insecticidal effects in sunlight. Many other phototoxic effects have also been attributed to these compounds. In particular, furocoumarins (also known as psoralens) induce blistering, "accelerated" sunburn and hyperpigmentation when humans or animals come into contact with furocoumarin-containing plants (*1*). Structure-activity studies have established that psoralen, the parent compound, is the most phototoxic furocoumarin and that ultraviolet light in the region of 320-360 nm is the most effective region of irradiation (*2*).

Psoralen

The concept of phototoxicity implies some biochemical target within the cell (or at the cell surface) that interacts in some fashion with the excited state of the toxic agent, leading to a deleterious change. However, in no case of phototoxicity

Table I. Products identified from the aqueous photolysis of psoralen

Compound	Structure	Status	Identification Method
5-formyl-6-hydroxy-benzofuran (FHBF)		New product (*15*)	from SPE; HPLC-PDAD, gc-ms, IR, NMR
6-formyl-7-hydroxy-coumarin (FHC)		Previously identified (*20*)	from SPE; HPLC-PDAD, gc-ms
furocoumaric acid (FCA)		Previously identified (*33*)	from polar residue (gc-ms); methylated deriv by gc-ms
2,5-dihydroxy-1,4-benzenedicarbox-aldehyde		New photoproduct; Acid previously identified as an alkaline oxidation product (*21*)	from SPE (gc-ms); also found corresponding acid as methylated deriv by gc-ms
2,4-dihydroxy-benzaldehyde		New product	From polar residue (gc-ms); also found corresponding acid as methylated deriv by gc-ms
salicylaldehyde		New product	From polar residue (gc-ms); also found corresponding acid as methylated deriv by gc-ms

Source: Reprinted with permission from The Spectrum (in press). Copyright 1995 Center for Photochemical Sciences, Bowling Green State University, Bowling Green, Ohio.

has one particular target or group of targets been definitively identified. Experiments *in vivo* with furocoumarins, for example, have demonstrated that they react with DNA, proteins, and lipids to a significant degree (*3,4*).

Various mechanisms have been proposed for the action of furocoumarin phototoxicity. In summary, these are: 1) intercalation of the furocoumarin molecule into the DNA helix and photoiniated [2+2] cyclobutane adduct formation leading to crosslinked DNA; 2) [2+2] cycloadducts with lipids; 3) formation of reactive oxygen, either $^1O_2$ and/or $HO_2 \cdot$; 4) formation of reactive intermediates or "transients", such as free radicals, peroxides, epoxides or dioxetanes. (For review articles, see 4-6). The majority of *in vitro* experiments conducted by photobiologists have highlighted psoralen reactions with DNA. However, DNA adducts are usually formed only in a dry or solid state with a quantum yield of < 1% (*7*). Quantum yields of the reactive oxygen species are also quite low (*8*).

Although the importance of light induction has been well-established, the subsequent reactivity of furocoumarins without light has also been noted. Several studies have demonstrated that pre-illuminated furocoumarins (sometimes called "photooxided psoralens") can hemolyze red blood cells (*9*) and oxidize lipids (*10*). Other laboratory investigations of the inflammatory response initiated by psoralens note that these effects are not immediate -- usually there is a delay (or induction period) of 6-12 hours with maximum effect noted at 24-48 hours. In contrast, rose bengal, a well-known $^1O_2$ producer, produces erythemal effects within three hours (*11*). Additional notes in various experimental protocols also indicate that enhanced toxicity can be observed if the reaction mixtures are held for a short period of time following illumination (e.g., see *12*). Among the structural classes of furocoumarin products that may be formed, synthetic furocoumarin-dioxetanes can form adducts with DNA (*13*) and synthetic furocoumarin-peroxides can cleave DNA (*14*). However, there is no good evidence for these exact structural types in mixtures of photooxidized psoralens.

Recently we demonstrated that 5-formyl-6-hydroxybenzofuran (FHBF) (Table I) is a major early photodecomposition product of psoralen in aqueous solution (*15*). This aldehyde is structurally related to phototoxic keto benzofuran derivatives that occur in the plant genus *Encelia* (*16*). Preliminary testing of this compound suggests that it also may have phototoxic properties. In the aqueous photolysis of psoralen, additional aldehydes are formed, including short-chain aldehydes, such as acetaldehyde, as well as phenolic aldehydes; minor products include the corresponding acid derivatives; H_2O_2 is also formed. None of the previous papers on furocoumarin photooxidation has identified either FHBF or these aldehydes (and acids) as reaction products.

Results and Discussion

Psoralens are somewhat hydrophobic compounds and might be expected to partition into lipid-rich regions of the cell. However, much metabolic activity (such as transport through membranes and vitamin C-vitamin E interactions (*17*)) occurs at lipid-water interfaces, and aqueous processes can be expected to play important roles. Most solution photochemistry by previous investigators on furocoumarins has been conducted with 8-methoxypsoralen (8-MOP), at millimolar concentrations in methylene chloride, of which it was noted that photolysis was slow and "intractable" (*18-19*). However, in our laboratory, using simulated sunlight, the half-life of a 20 μM air-saturated aqueous solution of 8-MOP was found to be approximately 6 hours; for 20 μM psoralen the half-life was 20 minutes. Because the product mixtures were less complex and the yields much higher with psoralen than with 8-MOP, most of the following work describes psoralen photoproducts, although some of the 8-MOP analogues have also been detected.

Reverse phase HPLC analysis of the reaction mixtures immediately following illumination showed that one major comparatively non-polar product (eluting after psoralen) was formed. However this product was not present if the analysis was delayed following irradiation (e.g., if analyzed after an overnight delay) or during various isolation procedures. Finally, a combination of solid phase extraction and semi-prep HPLC purification allowed the identification of 5-formyl-6-hydroxybenzofuran (FHBF), a lactone ring-opened product (*15*). In addition, a relatively minor polar product was identified as 6-formyl-7-hydroxycoumarin (FHC) as previously described (*20*).

Further investigation of the polar residue showed that some phenolic aldehydes and the corresponding acids, all of which are structurally related to FHBF or FHC, were also formed (Table I). Of these compounds, 2,5-dihydroxy-1,4-benzenedicarboxylic acid has been identified as a product of alkaline permanganate-induced oxidative degradation of psoralen (*21*). None of the remaining compounds have been described as a photolytic or oxidative product of psoralen. Short-chain aldehydes were also detected in the reaction mixtures; they are most likely the result of ring-cleavage reactions.

During the photolysis of psoralen, the formation of H_2O_2 was also observed. Small amounts of H_2O_2 were produced in aqueous solutions; however, when better H-donor solvents were used such as ethanol or isopropanol, the yields of H_2O_2 increased by factors of 10 and 100, respectively. Previous work has shown that these solvents participate in H_2O_2 production by a mechanism of hydrogen atom abstraction by the excited state of a carbonyl group [e.g., for the phototoxic aldehyde, citral (*22*)]. In fact, high yields of H_2O_2 continued long after psoralen was photodegraded (Figure 1) thus indicating that aldehyde products could also be precursors of H_2O_2. Preliminary work on the model compounds salicylaldehyde and isophthalaldehyde also confirm high yields of H_2O_2 in H-donor solvents (manuscript in prepration). Independently, a group of *para* phenolic aldehydes, such as vanillin, have also been shown to yield H_2O_2 under aqueous photolysis conditions (Anastasio, personal communication).

The participation of carbonyl compounds in furocoumarin phototoxicity may occur by at least three pathways:

1) reaction of the carbonyl group with electron-rich cellular constituents, such as sulfhydryl and amino groups of proteins and amino groups of guanine, through Schiff base formation:

$$R\text{-CHO} + H_2N\text{-R'} \rightarrow RCH=NR'$$

protein crosslinking or DNA adduct formation may result.

2) reactions requiring the presence of H_2O_2, as in the addition of the HOOH to a carbonyl group to form a hydroxyhydroperoxide (*23*):

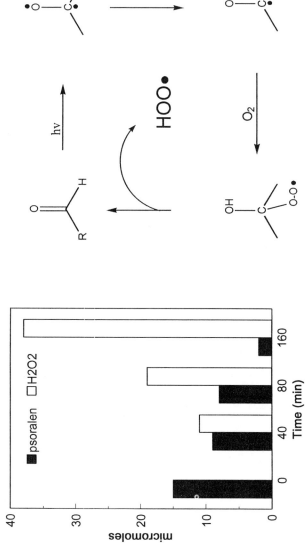

Figure 1. Loss of psoralen and formation of hydrogen peroxide in a H-donating solvent (ethanol). (Reproduced with permission from The Spectrum (in press). Copyright 1995 Center for Photochemical Sciences, Bowling Green State University, Bowling Green, Ohio.)

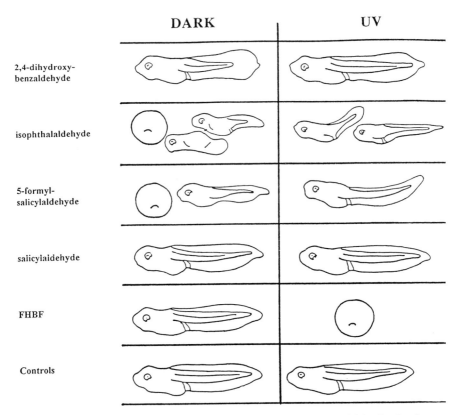

Figure 2. Xenopus laevis blastulae exposed to phenolic aldehydes in the presence and absence of UV.

Reactions with this type of compound could lead to protein oxidation, lipid oxidation or DNA strand breaks.

3) continued photoreactions, such as an H atom abstraction pathway where the H donor is a lipid, amino acid or DNA, leading to lipid oxidation, protein degradation or DNA strand breakage. Oxetane formation via the Paterno-Büchi reaction may occur when the carbonyl compound is photolyzed in the presences of alkenes, dienes or alkynes. Cycloadditon of the alkene to the 1,4-diradical of the carbonyl gives a 4-membered ring with an oxygen heterocycle:

Lipid or DNA adducts might result from reactions of this type.

Biological Effects of Phenolic Aldehydes. Embryos of *Xenopus laevis* (African clawed frog) at early blastula stage were used to screen for teratogenicity and other developmental effects of phenolic aldehydes (Figure 2). 2,4-Dihydroxybenzaldehyde resulted in the development of abnormal tadpoles; abnormalities were evident in the tail region, affecting only cells involved in the spinal cord and tail elongation. This effect occurred both in the light and in the dark. Both 5-formylsalicyaldehyde and isophthalaldehyde suppressed normal development in the absence of light; severely abnormal larvae developed under UVA illumination. FHBF allowed production of apparently normal larvae in the dark, but arrested development at the blastula stage after exposure to light. These results demonstrate that, while all of these compounds are teratogens, their dependency upon UV varied. 2,4-Dihydroxybenzaldehyde showed no appreciable degree of dependency, while FHBF appeared to be completely light dependent. Both isophthalaldehyde and 5-formylsalicylaldehyde were inhibited to a marked degree in the light, possibly due to photolysis of the compounds during the relatively long exposure to UV. Embryos were also exposed to salicylaldehyde; no abnormalities were observed.

Although low molecular weight aldehydes are well established as acutely toxic compounds, phenolic aldehydes have not been as widely studied, although various biological effects have been described, such as anti-mutagenicity (*24*) and cell growth inhibition (*25*). Salicylaldehyde has some demonstrated toxicity but is used extensively in the perfume and cosmetic industry and is generally regarded as safe (*26*). Recent studies, though, indicate that salicylaldehyde is antifungal against several *Fusarium* species responsible for dry rot in potatoes (*27*); and that willow leaf beetles use salicylaldehyde as a defensive secretion against generalist insect predators (*28,29*).

Conclusions

Even though psoralen phototoxicity has been extensively linked to cycloadduct formation with DNA, little attention has been given to its photooxidation or photolysis products. Carbonyl compounds were formed by photolysis of psoralen in aqueous solution that have either some known toxicity or the potential for toxic action. It would be too simple to say that these phenolic aldehydes would have the same mode of action or reactivity as that of known toxic aldehydes such as crotonaldehyde or acrolein, since those aliphatic compounds probably react by pathways that are not available to, or less favorable in, aromatic derivatives. In addition, the reactivity of any aldehyde can be altered by the presence of other potential reactants such as H_2O_2 or metals. Nevertheless, our preliminary observations of potentially significant biological effects, as well as scattered literature data on the toxicity of these substances, suggests that investigation of their biological properties should be undertaken. While many conjectures have been made as to some "key event" in the phototoxicity of psoralens, there now exists the possibility for additional pathways.

Methods and Materials

Reagents, sample preparation, illumination procedures, solid phase extraction (SPE) methods and chromatographic analyses have been previously described (15). H_2O_2 was determined by a peroxidase-based method (30).

The polar residue (that portion of aqueous material remaining after SPE) was reduced to dryness *in vacuo*, redissolved in methanol, derivatized with diazomethane (31) reevaporated using a gentle stream of argon and finally dissolved in 0.1 mL of of methylene chloride and analyzed by GC-MS.

Xenopus laevis embryos at early blastula stage (32) were placed in dilute aqueous solutions (10 μg/mL) of the compounds tested. One batch was irradiated with a UV-A "black-light" (with maximal output near 354 nm) at 1000 μW/cm^2 for 12 hours. A second batch was allowed to develop in the dark. At the end of 12 hours embryos were removed, placed in fresh culture medium, and allowed to continue development at 23^0 C. Development was terminated at 96 hr at which time embryos were fixed in borax-formalin. Compounds tested were 2,4-dihydroxybenzaldehyde, 5-formylsalicylaldehyde, isophthalaldehyde and salicylaldehyde (all of which were obtained from Sigma Chemical Co., St. Louis MO) , and the psoralen photolysis product 5-formyl-6-hydroxybenzofuran (FHBF) (isolated by solid phase extraction) (15).

Literature Cited

1. Murray, R. D.; Mendez, J.; Brown, S. A. *The Natural Coumarins: Occurrence, Chemistry and Biochemistry*; Wiley (Interscience), New York, 1982; pp. 291-292.
2. Kuske, H. *Dermatologica* **1940**, *82*, 273-338.
3. Beijersbergen van Henegouwen, G. M. J.; Wijn, E. T.; Schoonderwoerd, S. A.; Dall'Acqua, F. *J. Photochem. Photobiol. B: Biol.* **1989**, *3*, 631-635.
4. Midden, W. R. In *Psoralen DNA Photobiology*; Gasparro, F. P., Ed.; CRC Press, Boca Raton, FL, 1988, Vol. 2; pp. 1-49.

5. Potapenko, A. Ya. *J. Photochem. Photobiol. B:Biol.* **1991**, *9*, 1-34.

6. Larson, R. A.; Marley, K. A. In *Environmental Oxidants*; Nriagu, J. O.; Simmons, M. S., Eds.; Wiley Series in Advances in Environmental Science and Technology; John Wiley and Sons, New York, NY, 1994, Vol. 28; pp. 269-317.

7. Cadet, J.; Vigny, P.; Midden, W. R. *J. Photochem. Photobiol. B: Biol.* **1990**, *6*, 197-206.

8. Joshi, P. C., Pathak, M. A. *Biochem. Biophys. Res. Commun.* **1983**, *112*, 638-646.

9. Lysenko, E. P.; Federov, P. I.; Remisov, A. N.; Potapenko, A. Ya.; Wunderlich, S.; Pliquett, F. *Stud. Biophys.* **1988**, *124*, 225-234.

10. Potapenko, A. Ya.; Moshinin, M. V.; Krasnovsky, A. A., Jr.; Sukhorukov, V. L. *Z. Naturforsch.* **1982**, *37c*, 70-74.

11. Kumar, J. R.; Ranadive, N. S.; Menon, I. A.; Haberman, H. F. *J. Photochem. Photobiol. B: Biol.* **1992**, *14*, 125-137.

12. Beijersbergen van Henegouwen, G. M. J.; Wijn, E. T.; Schoonderwoerd, S. A.; Dall'Acqua, F. *Z. Naturforsch.* **1989**, *44c*, 819-823.

13. Adam, W.; Mosandl, T.; Dall'Acqua, F.; Vedaldi, D. *J. Photochem. Photobiol. B: Biol.* **1991**, *8*, 431-437.

14. Adam, W.; Cadet, J.; Dall'Acqua, F.; Epe, B.; Ramaiah, D.; Sahamoller, C. R. *Angew. Chem. Int. Ed. Engl.* **1995**, *34*, 107-110.

15. Marley, K. A.; Larson, R. A. *Photochem. Photobiol.* **1994**, *59*, 503-505.

16. Proksch, P.; Rodriguez, E. *Phytochemistry* **1983**, *22*, 2335-2348.

17. Burton, G. W.; Foster, D. O.; Perley, B.; Salter, T. F.; Smith, I. C. P.; Ingold, K. U. *Phil. Trans. Royal Soc. Lond. B* **1985**, *31*, 565-578.

18. Wasserman, H. H.; Berdahl, D. R. *Isr. J. Chem.* **1982**, *23*, 409-414.

19. Logani, M. K.; Austin, W. A.; Shah, B.; Davies, R. A. *Photochem. Photobiol.* **1982**, *35*, 569-573.

20. Rodighiero, G.; Musajo, L. Dall'Acqua, F.; Caprale, G. *Gazz. Chim. Ital.* **1964**, *94* 1073-1083.

21. *Polycyclic Compounds Containing Two Hetero Atoms in Different Rings. Five- and Six-Membered Heterocycles Containing Three Hetero Atoms and Their Benzo Derivatives;* Elderfield, R. C. Ed.; Heterocyclic Compounds; John Wiley and Sons, Inc., New York NY, 1961; Vol. 7, pp. 11-29.

22. Asthana, A.; Larson, R. A.; Marley, K. A.; Tuveson, R. W. *Photochem. Photobiol.* **1992**, *56*, 211-222.

23. Peyton, G. R.; Gee, C. S.; Smith, M. A.; Bandy, J.; Maloney, S. W. In *Biohazards of Drinking Water Treatment*, Larson, R. A., Ed. Lewis Publishers, Chelsea, MI, 1988, pp. 185-200.

24. Watanbe, K.; Ohta, T.; Shirasu, Y. *Agric. Biol. Chem.* **1988**, *52*, 1041-1045.

25. Dedonder, A.; van Sumere, C. F. *Z. Pflanzenphysiol.* **1971**, *65*, 70-80.

26. Opdyke, D. L. J. *Food Cosmet. Toxicol.* **1979**, *17*(Suppl). 903-905.

27. Vaughn, S. F.; Spencer, G. F. *J. Agric. Food Chem.* **1994**, *42*, 200-203.

28. Kearsley, M. J. C.; Whitham, T. G. *Oecologia* **1992**, *92*, 556-562.

29. Rank, N. E. *Oecologia* **1994**, *97*, 342-353.

30. Bader, H.; Sturzennegger, V.; Hoigné, J. *Water Res.* **1988**, *22*, 1109-1115.

31. Ngan, F. and Toofan, M. *J. Chromatogr. Sci.* **1991**, *29*, 8-10.

32. Dawson, D. A.; Wilke, T. S.; *Environ. Toxicol. Chem.* **1991**, *10*, 941-948; Stover, E., L.; Rayburn, J. R.; Hull, M. Bantle, J. A.; *Teratogen. Carcinogen. Mutagen.* **1993**, *13*, 35-45.
33. Décout, J. and Lhomme, J.. *Photobiochem. Photobiophys.* **1985**, *10*, 113-120.

RECEIVED September 12, 1995

Chapter 16

Plant Photosensitizers Active Against Viral Pests

J. B. Hudson[1], R. J. Marles[2], J. T. Arnason[3], and G. H. Neil Towers[4]

[1]Department of Pathology and Laboratory Medicine, University of British Columbia, Vancouver, British Columbia V6T 1Z4, Canada
[2]Department of Botany, Brandon University, Brandon, Manitoba R7A 6A9, Canada
[3]Department of Biology, University of Ottawa, Ottawa, Ontario K1N 6N5, Canada
[4]Department of Botany, University of British Columbia, Vancouver, British Columbia V6T 1Z4, Canada

Many bioactive photosensitizers have been isolated from plants, and some of these possess antiviral activities which depend on UVA or visible light. Among the more potent examples are α-terthienyl and other terthiophenes, which are active against membrane-containing viruses, probably via singlet oxygen mediated mechanisms. The antiviral activities of terthiophenes are, however, profoundly influenced by the nature of the side chains, as revealed by studies on numerous natural and synthetic derivatives. Many compex quinonoids have been isolated from plants, and recently some of them, such as hypericin and hypocrellin, have been shown to possess light-dependent antiviral activities. But these activities are also strongly influenced by chemical structure. The fact that these antiviral compounds often possess other antimicrobial activities as well makes them attractive candidates for therapeutic applications to virus-induced diseases, which are often compounded by secondary microbial infections.

Viruses are ubiquitous. They have probably evolved in parallel with animal, plant and microbial hosts for millions of years, and are likely to presist as threats to the health of their hosts for at least the foreseeable future. Prevention of viral disease by vaccination is practical for only a few cases, while therapeutic approaches to date have led to few successes. One human virus, smallpox, has been eradicated, and there are prospects for the eradication of a few more viruses; but in general viruses have resisted our attempts to control or manage them. Furthermore, when "novel" viruses emerge, we are often helpless, as shown by our experience with the AIDS pandemic.

For various reasons millions of people worldwide prefer to treat diseases, including many that are known to be caused by viruses, by means of so-called "medicinal" plants and animals. The concept of plants as a source of antiviral compounds is not new, although it has been, and still is, often derided by some scientists who believe that only 'man made' materials can benefit mankind. In fact, there are innumerable precedents for the assertion that potentially useful compounds can be found in the plant kingdom; and by concentrating on the traditional "medicinal plants" we can largely eliminate potentially toxic materials. Moreover, plants contain a wealth of chemicals that challenge the imagination of the average chemist, including some compounds that have defied laboratory synthesis.

Plant Antivirals

The subject of plant-derived antiviral compounds has been reviewed in recent years (1-5), and the list of potentially useful materials continues to grow year by year as more investigators try their luck with plants from different parts of the world.

The question then arises: what are we actually looking for in these plants? Are we looking for compounds that inactivate one or more viruses, or that interfere with the replication of one or more viruses; or are we searching for multifunctional compounds that have a variety of bioactivities - e.g., immune stimulation, tissue repair, etc. It is conceivable that the benefits attributed to plant extracts may be due to several bioactive compounds acting in synergy.

Another factor to take into account is the relative importance of photosensitizers. Many phytochemicals with documented bioactivities are photosensitizers and, accordingly, their bioactivities either depend upon or are enhanced by light of appropriate wavelengths (3). It is probably significant that many traditional treatments were supplemented by sunlight. Photosensitizers have the potential for controllable applications in viral diseases, by analogy with photodynamic therapy in cancer.

Evaluation of Antivirals

How do we carry out an analysis of plant materials for antiviral photosensitizers? In theory, an antiviral compound could work:
i)   by protecting cells from subsequent virus infection, i.e. by an interferon-like mechanism;
ii)   by a direct virucidal effect;
iii)   by interfering with the virus replication cycle, i.e. at some stage between virus adsorption to the cell and the emergence of progeny viruses from the infected cell.
These effects could be relatively specific for one or a few virus targets, or more general. Many such activities have been described (4-5).

In addition an antiviral compound could work by enhancing one or more components of the immune system, e.g., via an effect on cytokines; but this effect may only be manifest *in vivo*.

Many different viruses have been used as models to evaluate antivirals. We have used the examples depicted in Table I because they collectively represent all the likely targets for antivirals and also represent important models for human infections.

Table I. Virus Models

Virus	Genome	Membrane	Assays
Sindbis	RNA	+	plaque assays, cpe end-point
Polio type 1	RNA	-	plaque assays, cpe end-point
HSV-1, Mouse CMV	DNA	+	plaque assays, cpe end-point
HIV-1	RNA	+	cpe end-point, p24 antigen
some bacteriophages	DNA/RNA	-	plaque assays

The three types of mechanism mentioned above can all be accommodated in fairly simple cell culture systems, and, for photosensitizers, the appropriate type of light (usually UVA or broad spectrum visible light) can be administered when required.

It is premature to attempt a systematic classification of plant-derived antivirals, for too few plant families have been investigated adequately, and, likewise, too few chemical types. Some families are clearly rich in certain compounds that possess impressive antiviral activities, e.g. the Asteraceae, which contain abundant thiophenes and related compounds. But there are likely to be many other useful materials that have not yet been evaluated. Table II summarizes the types of plant photosensitizers with known antiviral activities.

Two groups of compounds will be dealt with in more detail: thiophenes and quinonoids.

Thiophenes

The thiophenes are widely distributed among species of the Asteraceae, where they occur as mono-, bi-, or terthiophenes, with a variety of side chains. Some thiophene-containing plants, such as *Tagetes* species, have been used in traditional medicine in many countries for controlling skin and eye infections and intestinal parasites. Their chemistry and biological activities have been extensively reviewed (6). Many of their bioactivities, including antiviral activities, require long wavelength UV (UVA).

Several years ago it was shown that α-terthienyl (αT) had extremely potent phototoxic effects on animal viruses with membranes (7). Thus as little as 10^{-10}

Table II. Summary of Antiviral Plant Photosensitizers

Class of compound	Example	Origin	Spectrum of activity	Light requirement
1. Quinonoid compounds	Hypericin	*Hypericum* spp.	Viruses with membranes	Visible light
	Hypocrellins	*Hypocrella bambuase*	Viruses with membranes	Visible light
2. Thiophenes	α-Terthienyl	*Tagetes* spp.	Viruses with membranes	UVA
3. Dithiins	Thiarubrine A	*Ambrosia* spp.	Viruses with membranes	UVA
4. Polyynes	Phenylheptatriyne	*Bidens* spp.	Viruses with membranes	UVA
5. Furyl compounds	Furanocoumarins (psoralens)	*Psoralea* spp.	Broad—especially DNA viruses	UVA
	Furanochromones (visnagin)	*Ammi visnaga*	Broad—especially DNA viruses	UVA
	Furoisocoumarin (coriandrin)	*Coriandrum sativum*	Viruses with membranes	UVA
6. Alkaloids	Harmine	*Peganum harmala*	Broad	UVA

M of αT had significant antiviral activity, but only in the presence of UVA. Much greater concentrations were needed to demonstrate cytotoxic activity, or activity against viruses without membranes (*7*). However αT + UVA-treated virus (murine CMV) remained intact and could penetrate cultured cells normally, and the viral genome entered the nucleus, although viral genes were not expressed in the cells; consequently the virus could not replicate. The key structural requirements for good antiviral activity (against membrane-containing viruses) are the possession of two or three thiophene rings in a linear configuration, with retention of conjugation, and one side chain consisting of an acetylenic group or certain other electron-withdrawing groups. Thus a number of functional groups such as alcohols, carboxylic-acids, cyano, methyl and silyl groups, are compatible with efficient activity (Table III). In contrast the addition of phenyl groups or halides, or the introduction of angular configurations between thiophene rings, are all inhibitory (*8*).

Some of the inactive compounds absorb very poorly in the UVA region and therefore would not be excited by the UVA lamps. However, no correlation was found between quantitative UVA absorbance and antiviral potency for the active compounds. Analyses of singlet oxygen quantum yields were made for many of these compounds; but in some cases there was a lack of correlation with antiviral potency. It may be that efficient activity requires the appropriate chemical configuration for proper insertion into the membranes of the virus (with some degree of hydrophobicity), together with sufficient UVA absorbance to generate a high quantum yield of singlet oxygen.

In general, even when the same host cells were used, Sindbis virus (SINV) was significantly more sensitive than mouse cytomegalovirus (MCMV) was to a given compound. This may reflect differences between the viral membrane compositions, which could be anticipated since SINV derives its membrane lipids from host cell cytoplasmic membranes and MCMV derives them from the host nuclear membrane. Furthermore the membrane associated viral proteins are quite different. However the more active compounds were active against both viruses; there were very few qualitative discrepancies. The most active compounds, including αT itself, were also found to be very potent against HIV-1 (Table III), in the presence of UVA, as determined by the reduction in the ability of this virus to produce characteristic cpe or to synthesize p24 protein (*9-10*).

When tested against a sensitive cell line that is commonly used for cytotoxicity tests, most of the thiophenes, and especially αT, showed phototoxicity. There was some correlation between cytotoxicity and antiviral activity, as might be expected in view of the similarity of their membranes; but there were exceptions. Thus some antiviral thiophenes possessed markedly less photocytotoxicity than αT (*8*). This is encouraging from the therapeutic aspect, especially since αT itself is not particularly toxic to animals kept in normal room light.

Most of the thiophenes showed little or no cytotoxicity in the absence of UVA, at least in the micromolar concentration range. α-Terthienyl did not cause

Table III. Effect of Serum on Anti-HIV Activities

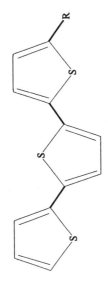

Compound #	Side-Chain (R)	Rate*	CLOGP†	Antiviral activity‡ +serum	-serum
1 (α-terthienyl, αT)	H	11.9	5.7	3	3
2	COOH	11.6	5.49	0	3
3	CH₂COOH	14.51	4.97	0	4
4	CH₂CH₂COOH	11.74	5.27	0	4
5	CH₂CH₂COOOCH₃	6.67	5.64	0	3
6	CO-(C₆H₄)-o-COOH	6.24	5.97	0	3
7	CH=CHCOOH	11.27	5.70	2	3
8	CH-CHCOOCH₃	2.71	6.02	0	3
9	CH₂CH(NH₂)COOCH₂CH₃	9.57	5.04	2	3
10	CHO	7.79	5.15	2	3
11	I	4.05	6.87	2	3
12	CN	10.18	5.24	3	4
13	NO₂	0	5.54	1	1

* Rate = rate of singlet oxygen production in $\mu mol/s/m^2$.
† CLOGP = calculated log octanol/water partition coefficient. These calculations were made as described in ref. 8.
‡ Relative antiviral activity: 0 = none; 1 to 4 = increasing activity (potency). Activities determined with and without serum (10, 23)

chromosome aberrations in cultured cells, and was not mutagenic in E. coli (*6*). Fish seemed to be able to tolerate larvicidal doses of αT, and mice and rats survived substantial doses, depending on the route of inoculation. However, αT produced photodermatitis in animal skin as well as allergic contact dermatitis (*3*). Some of the less cytotoxic terthiophenes may be safer to use *in vivo* than αT; thus more animal tests need to be done.

The mechanism of action of αT is thought to be a relatively simple type II photosensitization mechanism which requires singlet oxygen exclusively, as indicated by various experiments involving the use of inhibitors and aerobic-anaerobic atmospheres. In liposome models, it appears that αT inserts into membrane regions rich in unsaturated fatty acids and in the presence of UVA generates $^1O_2$, and hence membrane damage. This may also explain phototoxicity in insects and vertebrates when vital tissues that have accumulated αT are exposed to UVA (*11*). Alternative targets such as proteins or nucleic acids have also been suggested (*2, 6*).

In regard to viruses, the selective effect of the thiophenes against membrane-containing viruses implies a lipid target. On the other hand the fact that viruses without membranes may be susceptible at high thiophene concentrations suggests that proteins, or nucleic acids, may serve as targets whenever they are accessible.

We have recently observed that the antiviral activities of terthiophenes were substantially decreased in the presence of serum components, which are sometimes added to virus stocks to conserve viability. It appears that constituents of serum can bind the compound and block its antiviral activity (Table III) (*9, 12*); consequently optimal antiviral activity *in vitro* requires simple buffered salt solutions. This might be seen initially as a disadvantage in connection with therapeutic applications in the presence of tissue fluids; but on the other hand it has been pointed out that serum and other fluids often contain singlet oxygen quenchers, which could conceivably decrease unwanted side effects *in vivo*. Furthermore, photodynamic dyes used in cancer chemotherapy, such as porphyrins, require specific lipoprotein components of plasma to facilitate their penetration into target cells (*13*). Consequently we believe that α-terthienyl and other potent terthiophenes should be evaluated in model animal virus infections.

Quinonoids

Complex quinones are widely distributed among plant families, and many of them are likely to be photosensitizers, although until recently there had been few attempts to evaluate them as antivirals.

That situation changed dramatically when Meruelo and associates reported in 1988/89 that hypericin (Figure 1) and, to a lesser extent, pseudohypericin, were capable of inhibiting the growth of several murine retroviruses, both in cell culture and in mice (*14-15*). Hypericin could act therapeutically in infected mice and could also prevent subsequent viral disease. A comprehensive evaluation of

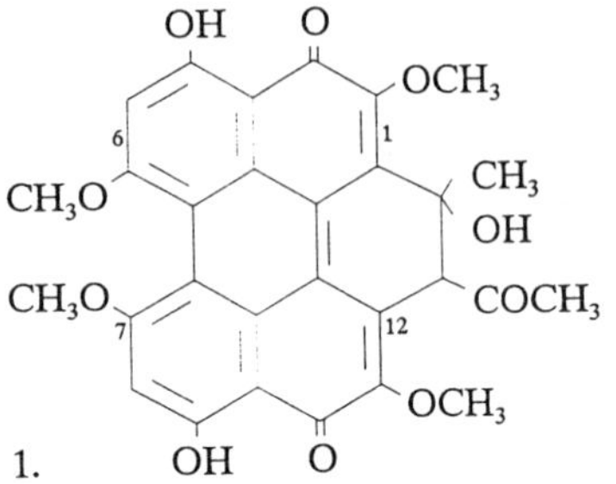

1.

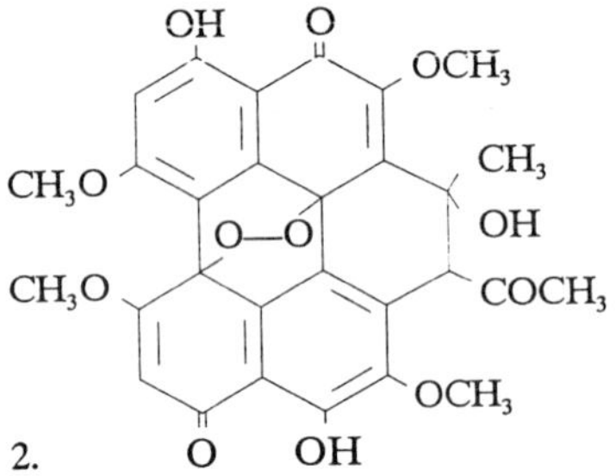

2.

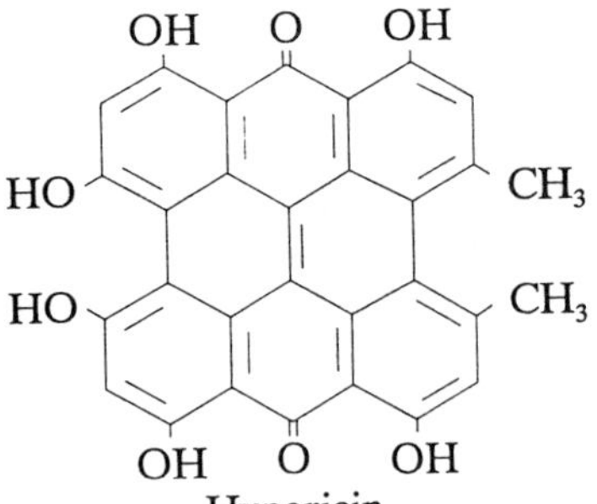

Hypericin

Figure 1. Chemical Structures of 1. Hypocrellin A and 2. Peroxyhypocrellin, in comparison with Hypericin.

physiological and biochemical parameters in the treated mice revealed no side-effects. It appeared that in cell culture the compound interfered in the process of development or maturation of the virions, although it was not clear if this mechanism could explain the antiviral effect *in vivo*. The replication of HIV in cell culture was also inhibited by hypericin.

Additional studies with other membrane-containing viruses confirmed the antiviral effect of hypericin in cell cultures and in animals (*16*). However, optimal antiviral activity *in vitro* required visible light (*16-20*). This specific requirement for visible light could not be replaced by an equivalent dose of UVA, even though hypericin does have an absorption peak in this region. Most of the other studies referred to ignored the contribution of light, and this may have been responsible for occasional discrepancies between different reports.

Nevertheless hypericin also possesses a "dark" antiviral effect which is revealed at much higher concentrations (Figure 2). Another variable in the hypericin-virus reactions is the presence of enhancing or inhibitory molecules. Anderson et al (*21*) pointed out the importance of adequate "solubilization" of the compound, which they achieved by adding a small amount of neutral detergent to the reaction mixtures. A low concentration of serum also ensured optimal antiviral activity, possibly by "presenting" the compound in the appropriate manner (*12*). In contrast higher concentrations of serum were progressively inhibitory. Thus the influence of serum on antiviral activity is different from the effect on terthiophenes (see above, Table IV), and this illustrates the attention that must be given to reaction parameters when evaluating phytochemicals.

Table IV. Effect of Serum on Antiviral Effect of Hypericin
(0.1μg/ml)

% Serum Added	% pfu Remaining
10	53
2	10.7
0.5	0.42
0.1	0043
0	16

Since hypericin-containing extracts have been administered in the past to humans, for unrelated reasons, it seems reasonable to be optimistic for the prospects of antiviral therapy by hypericin. Nevertheless, the importance of light must not be ignored, for severe phototoxicity has often been described in livestock that consumed *Hypericum* plants in sunny environments (*22*).

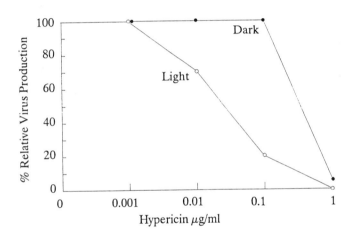

Figure 2. Anti-HIV effect at different hypericin concentrations, ±light. Aliquots of HIV-1 (10^3TCD$_{50}$) were incubated with various concentrations of hypericin in light (open circles) or dark (closed circles), under standard conditions, and the mixtures were inoculated into CEM cell cultures. At 6 days p.i. supernatants were removed for p24 assays. 10μg/ml was cytotoxic in light and dark.

Various structural analogues of hypericin and other anthraquinones have been examined for antiviral activities; but to date none have been as effective as hypericin itself (*16*), even when tested under optimal conditions of light exposure and serum concentration (Hudson et al, unpublished data).

A structurally similar perylene, hypocrellin A (Figure 1), isolated from the Chinese medicinal fungus, *Hypocrella bambuase*, was found to be almost as potent an antiviral as hypericin, but its activity also required light (*23*), which was not surprising because photodynamic properties have been documented for several hypocrellins (*24*). However, hypocrellin A has considerably less "dark" antiviral activity than hypericin. In contrast the peroxy derivative of hypocrellin A, generated by visible light irradiation of the hypocrellin, had relatively little activity. Hypocrellins have been used in Chinese traditional medicine for many years, especially for skin afflictions.

A summary of antiviral plant photosensitizers is given in Table II.

Plant Extracts

In total, several thousand plant extracts have been investigated for antiviral activity, with surprisingly little overlap in species between the studies (*2*). Most of the investigations were performed by means of only one or two test protocols for the purpose of looking for a particular type of activity. Nevertheless the successes were sufficient to encourage other workers in the future to use all the test modes instead of just one.

Aqueous and organic extractions have in general proved equally fruitful; thus it is not feasible at present to conclude that any one method of extraction is preferable.

In view of the significant proportion of plant extracts that have yielded positive results, and the variety in test protocols, test viruses, and plant extraction methods, it seems reasonable to conclude tentatively that there are probably numerous kinds of antivirals in these materials; hence further characterization of the active ingredients of some of these plants should reveal more useful compounds, including photosensitizers. Many of these may turn out to be identical to or structurally related to the antivirals just described; but there may also be novel phytochemicals. Therefore the search should continue.

Literature Cited

1. Vanden Berghe, D. A., Vlietinck, A. J. and Van Hoof, L. *Bull Inst. Pasteur* **1986,** *84*, 101-147.
2. Hudson, J. B. *Antiviral compounds from plants*. CRC Press, Boca Raton, FL, 1990; 200 pp.
3. Hudson, J. B. and Towers, G. H. N. *Pharmacol. Ther.* **1991,** *49*, 181-222.

4.   Vlietinck, A. J. and Vanden Berghe, D. A. *J. Ethnopharm.* **1991,** *31,* 141-153.
5.   Hudson, J. B. In *Antiviral Proteins in Higher Plants*; Chessin, M., Ed. CRC Press, Boca Raton, FL, 1995, 161-174.
6.   Kagan, J. *Prog. Chem. Organ. Natural Products,* **1991,** *56,* 87-169.
7.   Hudson, J. B. and Towers, G. H. N. *Photochem. Photobiol.,* **1988,** *48,* 289-296.
8.   Marles, R. J., Hudson, J. B., Graham, E. A., Soucy-Breau, C., Morand, P., Compadre, R. L., Compadre, C. M., Towers, G. H. N. and Arnason, J. T. *Photochem. Photobiol.,* **1991,** *56,* 479-487.
9.   Hudson, J. B., Harris, L., Teeple, A. and Towers, G. H. N. *Antiviral Res.,* **1993,** *20,* 33-43.
10.   Hudson, J. B., Harris, L., Marles, R. J. and Arnason, J. T. *Photochem Photobiol,* 58, 246-250, 1993.
11.   Arnason, J. T, Philogene BJR, and Towers, G. H. N. In *Herbivores*; 2nd ed. Rosenthal G, Berenbaum M. (eds) *Academic Press,* New York, 1990.
12.   Hudson, J. B., Graham, E. A., Towers, G. H. N. *Planta Medica,* 60:329-332;1994.
13.   Allison, B. A., Pritchard, P.H, Richter, A. M. and Levy, J. G. *Photochem. Photobiol.,* **1990,** *52,* 501-506.
14.   Meruelo, D., Lavie, G. and Lavie, D. *Proc. Natl. Acad. Sci.,* **1988,** *85,* 5230-5234.
15.   Lavie, G., Valentine, F., Levin, B., Mazur, Y., Gallo, G., Lavie, D., Weiner, D. and Meruelo, D. *Proc. Natl. Acad. Sci.,* **1989,** *86,* 5963-5967.
16.   Schinazi, R. F., Mead, J. R. and Feorino, P. M. *AIDS Res. and Hum. Retroviruses,* **1992,** *8,* 963-990.
17.   Carpenter, S. and Kraus, G. A. *Photochem. Photobiol.,* **1991,** *53,* 169-174.
18.   Hudson, J. B., Lopez-Bazzochi, I. and Towers, G. H. N. *Antiviral Res.,* **1991,** *15,* 101-112.
19.   Hudson, J. B., Harris, L. and Towers, G. H. N. *Antiviral Res.,* **1993,** *20,* 173-178.
20.   Stevenson, N. R. and Lenard, J. *Antiviral Res.,* **1993,** *21,* 119-127.
21.   Anderson, D. O., Weber, N. D., Wood, S. G., Hughes, B. G., Murray, B. K. and North, J. A. *Antiviral Res.,* **1991,** *16,* 185-196.
22.   Ivie, G. W. *J. Natl. Cancer Inst.,* **1982,** *69,* 259-266.
23.   Hudson, J. B., Harris, L., Zhou, J., Chen, J., Yip, L., Towers, G. H. N. *Photochem. Photobiol.,* **1994,** *60,* 253-255.
24.   Zhenjun, D., Lown, J. W. *Photochem. Photobiol.,* **1990,** *52,* 609-616.

RECEIVED August 24, 1995

Chapter 17

Fungal Resistance to Photosensitizers That Generate Singlet Oxygen

Margaret E. Daub, Anne E. Jenns, and Marilyn Ehrenshaft

Department of Plant Pathology, North Carolina State University, Raleigh, NC 27695-7616

Fungi in the genus *Cercospora* synthesize a photoactivated perylenequinone toxin, cercosporin, which plays an important role in the ability of these organisms to parasitize plants. In the presence of light, cercosporin generates singlet oxygen, and is toxic to plants, mice, bacteria, and many fungi. *Cercospora* species, however, are resistant to high concentrations of cercosporin. Targeted gene disruption of the *Cercospora* phytoene dehydrogenase gene revealed no role for carotenoids in cercosporin resistance. By contrast, resistance is highly correlated with the ability of these fungi to transiently reduce and detoxify cercosporin. Mutants which are deficient in resistance to cercosporin have been isolated. These mutants are also sensitive to five other singlet oxygen generating photosensitizers. The mutants are being used to identify genes for photosensitizer resistance using mutant complementation. Once identified, these gene(s) can be used to determine whether resistance is targeted against the photosensitizer molecule or singlet oxygen itself.

Cercosporin (**I**) is a photosensitizing perylenequinone phytotoxin produced by members of the genus *Cercospora* (*1-3*). *Cercospora* species are a highly successful group of fungal pathogens that cause disease on a diversity of host plants including corn, sugar beet, tobacco, coffee, soybean, and banana, as well as many ornamental and weed species. *Cercospora* species cause major economic problems due not only to their world-wide distribution and wide host range, but also because naturally-occurring resistance to the disease has not been identified in many host species. One of the reasons for the success of this group of pathogens appears to be their production of cercosporin. Cercosporin is produced by many members of the genus and has near universal toxicity to plants. Also, production of cercosporin appears critical for

0097–6156/95/0616–0201$12.00/0

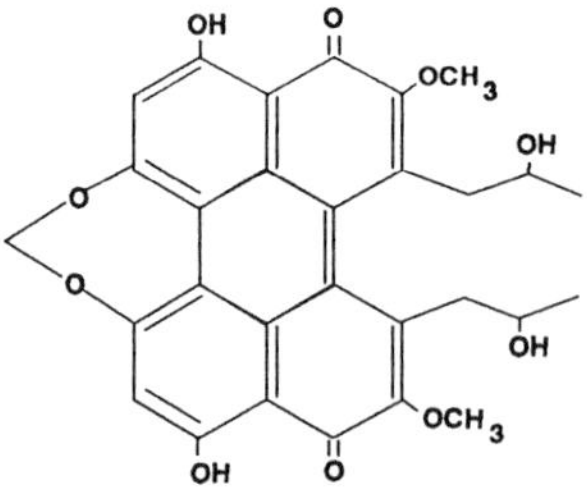

(I) Cercosporin

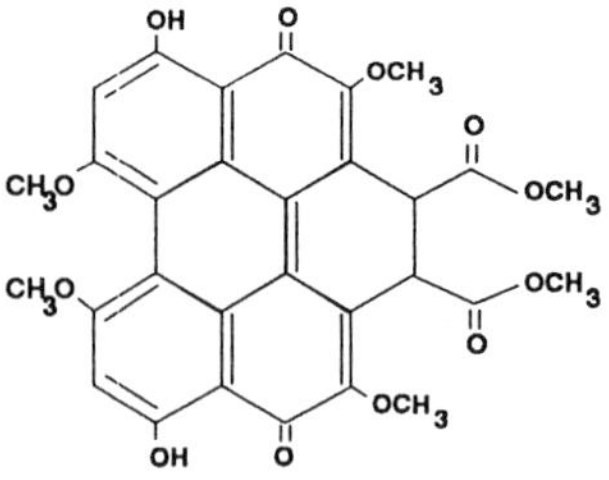

(II) Elsinochrome A

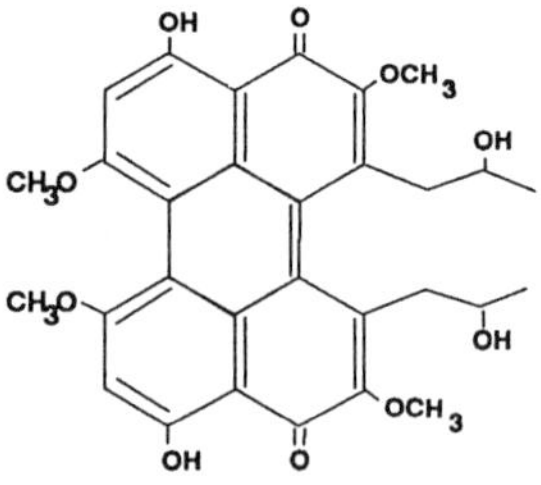

(III) Phleichrome

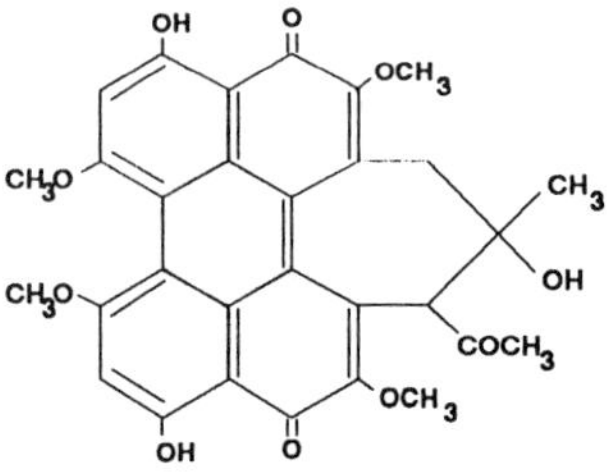

(IV) Hypocrellin A

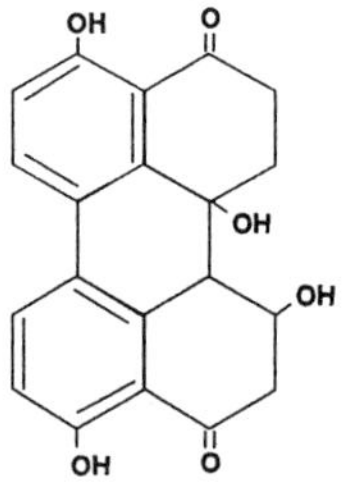

(V) Altertoxin I

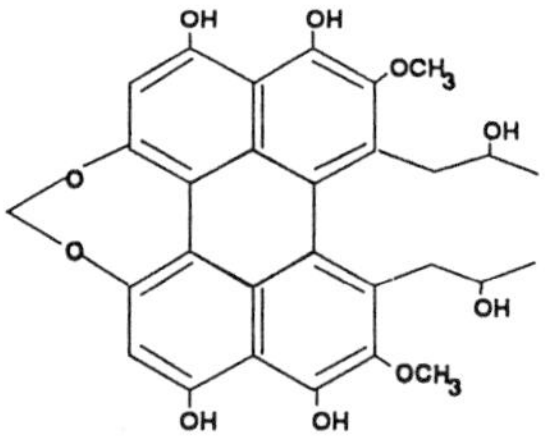

(VI) Reduced Cercosporin

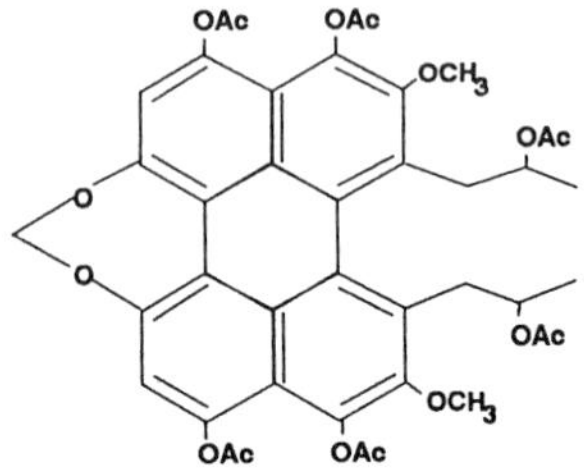

(VII) Hexaacetyl-dihydrocercosporin

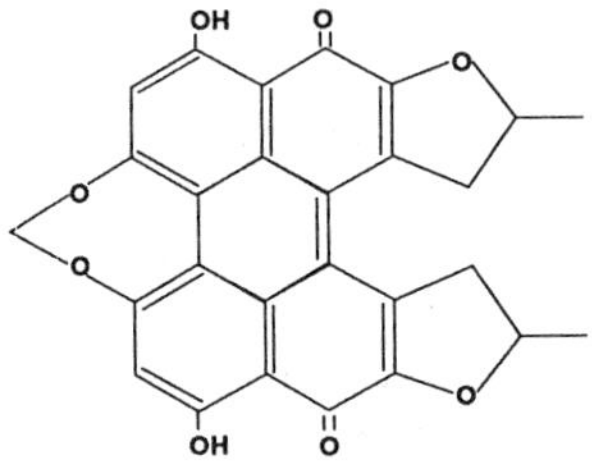

(VIII) Noranhydro-cercosporin

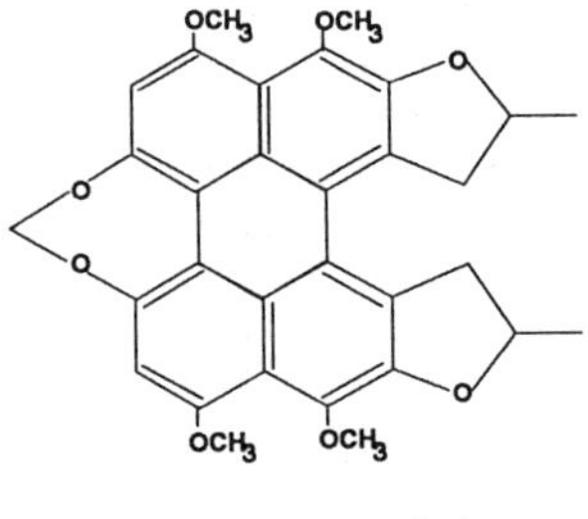

(IX) Tetramethyl-noranhydrodihydrocercosporin

Structures I-IX

successful pathogenesis as fungal mutants deficient in cercosporin synthesis are unable to parasitize their host plants (*4*), and light has been shown to be critically important in the development of disease symptoms on susceptible hosts (*2,3*).

Cercosporin was the first toxin synthesized by plant pathogens to be recognized as a photosensitizer, but numerous other plant pathogenic fungi also produce perylenequinone toxins. Elsinochromes such as elsinochrome A (**II**) are produced by species of *Elsinoe* and *Sphaceloma* (*5*). Perylenequinones such as phleichrome (**III**) have been isolated from cucumbers infected with *Cladosporium cucumerinum* and from *Cl. phlei* and *Cl. herbarum* (*6-8*). Shiraiachromes have been isolated from the bamboo pathogen *Shiraia bambusicola* (*9*). Hypocrellins, isolated from another bamboo pathogen, *Hypocrella bambusae*, have been used as medicinal agents in China for many years, and are currently being investigated for use in photodynamic tumor therapy (*10*). Numerous perylenequinones have been isolated from *Alternaria* species including alteichin, altertoxins I (**V**), II, and III, alterlosin I and II, and stemphyltoxin III (also isolated from *Stemphylium botryosum*) (*11-15*). Altertoxins have been shown to produce superoxide (O_2^-) when exposed to light (*12*), and altertoxins and stemphyltoxin have been reported to be mutagenic (*11, 14*). Although only cercosporin has been studied in any detail for its involvement in plant disease, the production of photoactivated perylenequinones by such a diversity of plant pathogens suggests that photosensitization may be a more common plant pathogenesis factor than has been previously recognized.

Cercosporin is toxic to mice, bacteria, many fungi, and, with one exception, all plants that have been tested (*16-20*). It has also been shown to be cytotoxic to human tumor cells and to inactivate protein kinase C (*21*). Attempts to obtain resistant plants and fungi through mutagenesis and selection of cells in culture have not been successful (*2*, Daub, unpublished results). Virtually the only organisms which show resistance to cercosporin are the *Cercospora* fungi themselves and some related fungi that produce similar toxins. These observations suggest that cercosporin has almost universal toxicity to cells, and that resistance is due to active defense mechanisms present in the few resistant organisms. Here we describe our work on identifying the mechanisms used by *Cercospora* fungi to resist cercosporin, and our efforts to isolate the gene(s) which encode cercosporin resistance. Our ultimate goal is to understand the mechanism by which these genes function to provide resistance to cercosporin and to singlet oxygen ($^1O_2$), and to investigate the possible utility of these genes in the engineering of plants for resistance to diseases caused by *Cercospora* species and other perylenequinone-producing fungi.

Cercosporin

Isolation and Characterization of Cercosporin. Cercosporin was first isolated in 1953 by Deutschmann (*22*) from mycelium of *Cercospora kikuchii*, a soybean pathogen. The compound was independently isolated in 1957 by Kuyama and Tamura from the same species, and named cercosporin (*23-24*). Its structure was determined independently by Lousberg et al. (*25*) and Yamazaki and Ogawa (*26*). Its stereochemistry and absorption and fluorescence properties have also been studied (*26-28*). Cercosporin is a derivative of 4,9-dihydroxyperylene-3,10-quinone. It is red in color and turns green when dissolved in dilute alkali. It is soluble in dilute alkali and in many organic solvents, but is only sparingly soluble in water. When secreted by the fungus during growth in culture, it is readily visible due to the accumulation of cercosporin crystals in the culture medium. A possible pathway for biosynthesis in the fungus has also been proposed (29).

Cercosporin Phototoxicity. Cercosporin's photosensitizing activity was first reported by Yamazaki et al. (20). They demonstrated that cercosporin toxicity to mice and bacteria was light dependent, and that oxygen was required for the cercosporin-induced photooxygenation of dimethylfuran. Macri and Vianello (30) and Daub (1, 31) demonstrated cercosporin's photodynamic activity on plant tissues. Cercosporin causes rapid leakage of ions from plant tissues and bursting of plant protoplasts, indicative of membrane damage. Antioxidants and $^1O_2$ quenchers inhibit this membrane breakdown. Membrane damage was shown to be due to peroxidation of the membrane lipids. Cells damaged by cercosporin show an accumulation of lipid peroxidation products (31, 32), and a marked increase in the ratio of saturated to unsaturated fatty acids (33). Cercosporin treatment also causes marked decreases in the fluidity of plant protoplast membranes and an apparent increase in the membrane phase transformation temperature, as measured by electron spin resonance (ESR) spectroscopy with fatty acid spin labels (33). All of these changes are known to occur in membranes damaged by lipid peroxidation (34, 35). These results support ultrastructural studies which demonstrated membrane damage in both cercosporin-treated and *Cercospora*-infected sugar beet leaf tissue (36, 37).

Active Oxygen Production by Cercosporin. The generation of $^1O_2$ and O_2^- by cercosporin has been studied by several laboratories. Singlet oxygen production by cercosporin has been measured directly (38, 39), as well as through analysis of cholesterol oxidation products (40) and by measurement of BHMF oxidation (41). Production of O_2^- by cercosporin has also been demonstrated when cercosporin is incubated in the presence of reducing agents such as methionine, ergothioneine, and urate (40, 41).

Although both $^1O_2$ and O_2^- can be produced by cercosporin *in vitro*, evidence suggests that $^1O_2$ formation is the major mechanism by which cercosporin exerts its toxicity. Cercosporin is a potent producer of $^1O_2$ (quantum yield of $^1O_2$ formation = 0.81 ± 0.07) (38), but only a weak producer of O_2^- (40). Cercosporin killing of suspension-cultured plant cells is inhibited in the presence of two different $^1O_2$ quenchers (Dabco and the carotenoid bixin), demonstrating the importance of $^1O_2$ production *in vivo* (1). Evidence for a role of O_2^- is less certain. Paraquat-resistant cell culture mutants having elevated levels of catalase and superoxide dismutase (42, 43) show no or only low levels of increased resistance to cercosporin (43, 44), suggesting that O_2^- may not be important in cercosporin toxicity *in vivo*.

Resistance Mechanisms

Photosensitizer Resistance. Photosensitizers are commonly-occurring compounds. They are natural products of plants, microorganisms, and protozoans, are used as dyes and reagents in laboratories and increasingly as agricultural pesticides and as pharmaceuticals (*45-49*). In spite of their common occurrence, surprisingly little is known about how organisms defend themselves against them. Plants are major producers of photosensitizers in nature (*50-51*), yet mechanisms of protection against autotoxicity have not been described. Among protozoans and insects, behavioral, light-avoidance responses are a common method of protection (*52-53*). Some insects contain pigments that block transmission of activating wavelengths, or are able to metabolize different photosensitizers (*53*). The protozoan *Blepharisma*, which contains the endogenous photosensitizing pigment blepharismin, normally protects itself by photophobic responses, but can also irreversibly oxidize blepharismin to an inactive form when exposed to high light intensities (*52*).

Molecules that quench or block the formation of active oxygen species by photosensitizers have also been identified. Many compounds are known which quench $^1O_2$ (*54*), but only a few of these are found in biological systems, and even fewer have been shown to play a role in cell defense. The best characterized mechanism of protection against photosensitizers is the production of carotenoids by the target organism (*55, 56*). Carotenoids quench both $^1O_2$ and the triplet state of photosensitizers, and are the most efficient quenchers identified which exist in biological systems (*54, 57, 58*). Other $^1O_2$ quenchers found in biological systems include thiol compounds, the amino acids histidine, methionine, and tryptophan, and some peptides, amines, and phenols (*54, 59-61*). The actual role of these molecules in the resistance of living organisms to photosensitizers, however, has not been well documented. Enzymes such as superoxide dismutase and peroxidases have been shown to be important in defense against elevated oxygen levels and agents such as ozone and paraquat (*62-64*), but there is little evidence for a role for these enzymes in defense against most photosensitizers.

Fungal Resistance to Cercosporin. In spite of cercosporin's almost universal toxicity, *Cercospora* fungi are resistant to its effects. Cercosporin can accumulate to concentrations as high as 1 mM in light-grown cultures with no measurable decrease in fungal growth (*65*). *Cercospora* species are also resistant to a number of other $^1O_2$-generating photosensitizers, including hematoporphyrin, methylene blue, and eosin Y (*18, 66, 67*). Thus *Cercospora* fungi obviously have evolved mechanisms that allow them to tolerate high concentrations of $^1O_2$-generating photosensitizers without a major expenditure of energy.

The goal in this laboratory over the last several years has been to define the resistance mechanisms used by *Cercospora* species to protect themselves against cercosporin. Several factors were found to be unimportant in resistance. Membrane fatty acids are the target of cercosporin, however cultures of *Cercospora* species contain high concentrations of linoleic acid (18:2) which is susceptible to peroxidation, and the fatty acid composition of cultures grown under cercosporin-producing and non-producing conditions does not change (*33*). *Cercospora* species do not differ from cercosporin-sensitive fungi in superoxide dismutase or catalase activity or general antioxidant activity (*18*). Also, although reducing agents and thiols such as ascorbate, cysteine, and reduced glutathione protect sensitive fungi against cercosporin, *Cercospora* species do not differ from sensitive fungi in endogenous levels of these agents or in levels of total soluble or protein thiols (*68, 69*) By contrast, two mechanisms were identified which appeared to be involved in cercosporin resistance. These are: 1) the presence of carotenoids, and 2) the ability of the fungus to chemically reduce and detoxify the cercosporin molecule.

Carotenoids. As stated previously, carotenoids are highly effective quenchers of $^1O_2$ and of triplet states of photosensitizers and are the most commonly identified means of defense of many organisms against photosensitizers (*55-58*). Preliminary data suggested that carotenoids may play a role in cercosporin resistance. Killing of plant cells by cercosporin can be inhibited by the carotenoid carboxylic acid, bixin (*1*). A recent study of resistance of rice to *Cercospora oryzae* infection correlated carotenoid content with the resistance shown by a wild rice plant to cercosporin and *Cercospora* infection (*16*). *Cercospora* species produce high levels of β-carotene (*70*). Also, carotenoid-deficient mutants of *Neurospora crassa* and *Phycomyces blakesleeanus* were shown to be significantly more sensitive to cercosporin than are carotenoid-producing isolates of the same species (*70*). Direct proof of the role of carotenoids in cercosporin resistance of *Cercospora* species, however, was lacking.

To test the role of carotenoids in *Cercospora* resistance to cercosporin, we used targeted gene disruption to create mutants deficient in the enzyme phytoene dehydrogenase (pdh), which catalyzes the dehydration reactions leading from the colorless carotenoid precursor phytoene to the colored carotenoid pigments. Probes to the *N. crassa* phytoene dehydrogenase (*al-1*) gene (*71, 72*) were used to isolate the corresponding pdh gene from *Cercospora nicotianae* (*73*). A gene conferring hygromycin resistance was cloned into the middle of the *C. nicotianae* pdh gene, creating a disrupted version which was then transformed into the wild type fungus (*74*). Mycelial fungi transform by integration of DNA into the chromosomes, usually by homologous recombination. Integration via a single cross-over event yields transformants with both a wild type copy and a disrupted copy of the gene, but a double cross-over event allows for replacement of the endogenous gene with the disrupted copy. Transformants were selected for resistance to hygromycin and then screened for production of β-carotene, the sole carotenoid produced by this fungus (*70*). Approximately one out of ten transformants were deficient for β-carotene production (Figure 1). Southern analysis indicated that the β-carotene-producing transformants contained both a wild type and a disrupted version of the gene, whereas in the β-carotene-deficient mutants, the disruption construct had replaced the endogenous copy of the gene (*74*).

β-carotene-deficient mutants were then screened for resistance to cercosporin, both by inducing them to synthesize cercosporin and by growing them on medium amended with cercosporin (*74*). Surprisingly, the β-carotene-deficient mutants were no more sensitive to cercosporin than were β-carotene-producing isolates (Table 1). The disruption mutants were also no more sensitive to five other $^1O_2$-generating photosensitizers (rose bengal, hematoporphyrin, methylene blue, toluidine blue, eosin yellow) (Table 1). Carotenoid disruption mutants were also created in mutant strains of *C. nicotianae* sensitive to cercosporin (strains CS8 and CS10, see below). These disruptants were also no more sensitive to cercosporin than the isolates from which they were derived. From these data we concluded that carotenoids are not involved in the resistance of *Cercospora* species to cercosporin or to other photosensitizers. These results were surprising, given the extensive role carotenoids play in photosensitizer resistance.

Table 1. Photosensitizer sensitivity of carotenoid-producing and carotenoid-deficient strains of *Cercospora nicotianae*.

Strain[a]	β-carotene production	Photosensitizer					
		Cercosporin 10μM	Rose bengal 10μM	Eosin Y 10μM	Hematoporphyrin 50μM	Methylene blue 100μM	Toluidine blue 100μM
WT	+	105[b]	53	100	71	108	108
VH1	+	100	59	100	76	94	108
DH1	-	100	67	107	73	121	119
DH9	-	108	63	95	62	105	121
DH25	-	98	59	94	68	113	100
DH37	-	105	59	106	78	108	113
DH48	-	113	47	95	78	100	95
CS10	+	66	57	107	66	100	100
CS10.15	-	70	57	100	55	100	93
CS10.20	-	58	64	100	63	96	133

[a]WT: wild type. VH1: Vector-transformed control. DH1, DH9, DH25, DH37, DH48: carotenoid-minus disruption mutants of the wild type strain. CS10: cercosporin-sensitive mutant. CS10.15 and CS10.20: carotenoid-disruption mutants of CS10.

[b]Growth on photosensitizer-containing medium as a percent of growth on non-amended medium.

SOURCE: Adapted from ref. 74.

Reductive Detoxification. Recent studies suggest that the major mechanism of resistance of *Cercospora* species to cercosporin is due to the ability of the fungus to transiently detoxify cercosporin by reducing it (*39, 66, 68*). This hypothesis was initially investigated due to two observations. First, fluorescence microscopy indicated that hyphae of cercosporin-producing *Cercospora* cultures emitted a green fluorescence rather than the red fluorescence typical of cercosporin, suggesting that the cercosporin molecule may be modified when in contact with hyphae. Second, it was known that some photosensitizers, when reduced, are converted to colorless "leuco" forms which do not absorb light and thus are not photoactive. Photosensitizer reduction, therefore, could be a mechanism of biological detoxification.

Initial studies to test this hypothesis focused on estimating the cell surface reducing ability of cercosporin-resistant and sensitive fungi (*68*). Reducing power was assayed by measuring the ability of various fungi to reduce tetrazolium dyes spanning a wide range of redox potentials. Species resistant to cercosporin (*Cercospora* species and *Alternaria alternata*) were capable of reducing a significantly greater number of dyes than were sensitive species (*N. crassa, Aspergillus flavus*, and *Penicillium* species).

Next, the properties of reduced cercosporin were investigated (*39*). Cercosporin can be reduced by the addition of a strong reducing agent such as dithionite or zinc dust, however reduced cercosporin (**VI**) is highly unstable, and instantly reoxidizes upon aeration or extraction away from the reducing agent. In order to test the toxicity of reduced cercosporin, two stably-reduced derivatives, a reduced-acetylated derivative of cercosporin (hexaacetyl-dihydrocercosporin, **VII**), and a reduced-methylated derivative of noranhydrocercosporin (**VIII**) (tetramethyl-noranhydrodihydrocercosporin, **IX**), were synthesized. Reduced cercosporin and the reduced methylated and acetylated derivatives are green in color (as compared to the red color of cercosporin), are highly fluorescent, and absorb approximately half the amount of light as cercosporin and noranhydrocercosporin. Dissipation of absorbed light energy by fluorescence is in direct competition with transfer of energy to oxygen to produce $^1O_2$, suggesting that in addition to absorbing less light, the derivatives would be less efficient in transferring the energy from the light they do absorb to oxygen. This hypothesis was confirmed by direct measurement of $^1O_2$ generation from the reduced derivatives. Per quantum of light absorbed, the reduced methylated and acetylated derivatives, respectively, produced only 26 and 16% of the $^1O_2$ produced by oxidized forms. The reduced derivatives were also shown to be less toxic as assayed both by inhibition of growth of sensitive fungi and by a lipid peroxidation assay.

In order to demonstrate that *Cercospora* species do in fact reduce cercosporin, we took advantage of differences in the fluorescence properties of cercosporin and reduced cercosporin (*66*). Using bandpass filters which allowed differentiation between fluorescence emission from the two compounds by fluorescence microscopy, we determined that cercosporin in contact with fungal hyphae was in a reduced form. Cercosporin in contact with hyphae emitted green fluorescence, detectable with a 515-545 nm bandpass filter, characteristic of reduced cercosporin. Crystals of active (non-reduced) cercosporin were clustered near the hyphal strands, indicating that diffusion away from the hyphae allowed for spontaneous reoxidation. Hyphae killed by heat, UV light, or exposure to chloroform vapor were unable to maintain cercosporin in a reduced form. Red fluorescence detectable with a 575-635 nm bandpass filter was present in this case. Thus maintenance of reduced cercosporin is an active function. Further, the cercosporin-sensitive fungi *N. crassa* and *A. flavus* were unable to reduce cercosporin. Additional support for the reduction hypothesis came from studies with the $^1O_2$-generating photosensitizer rose bengal (*66*). Rose bengal is the only photosensitizer identified to date that shows significant toxicity to *Cercospora* species. $^1O_2$ quantum yields for cercosporin and rose bengal are similar (*38, 75*), thus differential toxicity is not due to differences in $^1O_2$ production, a conclusion supported by the observation that rose bengal and cercosporin are equally toxic to sensitive fungi. One major difference between these two compounds, however, is their ease of reduction. Rose bengal is significantly harder to reduce than is cercosporin. Measured redox potentials are -0.72 V and -0.16 V vs. NHE for rose bengal and cercosporin, respectively (*66, 76*). Further, microscopic observation similar to that conducted for cercosporin indicated that *Cercospora* species were not able to reduce rose bengal.

Based on these results, we constructed a model which proposes that *Cercospora* species protect themselves against cercosporin by reducing and detoxifying molecules in contact with hyphae (Figure 2). Cercosporin molecules which diffuse away from the hyphae are able to spontaneously re-oxidize, and revert to the photoactive form involved in infection of host plants. This mode of resistance would be highly effective for an organism which requires production of an active photosensitizer for survival in nature.

Identification of Cercosporin-Resistance Genes

Isolation of Cercosporin-Sensitive Mutants. In order to isolate genes involved in cercosporin reduction and resistance, mutants of *C. nicotianae* were isolated which are sensitive to cercosporin (*67*). The mutants were isolated from UV-mutagenized mycelial protoplasts, and screened for cercosporin sensitivity by replica-plating colonies on cercosporin-containing medium. Six cercosporin-sensitive (CS) mutants were isolated, and these fell into two phenotypic classes. Five of the mutants (CS2, CS6, CS7, CS8, and CS9, designated class 1) are totally inhibited when grown on cercosporin-containing medium at concentrations as low as 1 μM. The sixth mutant (CS10, designated class 2), is partially inhibited by 10 μM cercosporin, but not at lower concentrations. When assayed by fluorescence microscopy, the class 1 mutants were found to be incapable of reducing cercosporin (Figure 3), supporting our hypothesis that cercosporin reduction is correlated with resistance. The partially-sensitive CS10, however, was normal in cercosporin-reducing ability (Figure 3).

The mutants were also tested for resistance to five other $^1O_2$-generating photosensitizers, methylene blue, toluidine blue, rose bengal, eosin Y, and hematoporphyrin. Wild type *C. nicotianae* is highly resistant to all these photosensitizers, with the exception of rose bengal which shows significant toxicity to this fungus. Surprisingly, the class 1 sensitive mutants (CS2, CS6, CS7, CS8, and CS9) were completely inhibited by all of the photosensitizers (Table 2). This level of sensitivity was unexpected, as even the most cercosporin-sensitive naturally-occurring fungal species show only 50-80% growth inhibition by eosin Y and no inhibition at all by methylene blue at the concentrations used (*18, 66*). Thus the class 1 mutants appear to be mutant in a gene which mediates resistance to a range of $^1O_2$-generating photosensitizers, and such a mutation results in levels of photosensitizer sensitivity not occurring in fungi in nature. By contrast, the

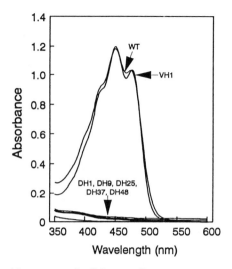

Figure 1. Carotenoid content of wild type *Cercospora nicotianae* and phytoene dehydrogenase disruption mutants. WT, wild type; VH1, wild type transformed with the vector plasmid alone; DH1, DH9, DH25, DH37, and DH48, carotenoid disruption mutants. Reproduced with permission from ref. 74. Copyright 1995 American Phytopathological Society.

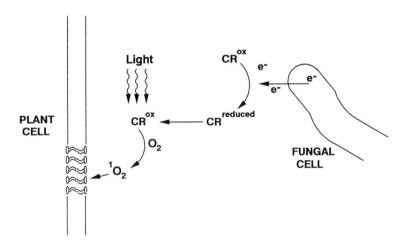

Figure 2. Model for resistance of fungal hyphae to cercosporin by transient reductive detoxification. Fungal hyphae produce reducing power (e⁻) which results in the reduction and detoxification of the cercosporin molecule (cerc^{red}) in contact with fungal hyphae. Cercosporin which diffuses away from the fungal cell spontaneously reoxidizes to the photoactive form (cerc^{ox}) responsible for $^1O_2$ formation and toxicity to plant cells. Reproduced with permission from ref. 68. Copyright 1992 American Society for Microbiology.

Table 2. Sensitivity of wild type *Cercospora nicotianae* and cercosporin-sensitive (CS) mutants to photosensitizers.

	Cercosporin		Methylene Blue	Toluidine Blue	Eosin Y		Rose Bengal		Hemato-porphyrin	
Conc. (µM)	10	50	100	100	10	50	10	50	10	50
Isolate										
wild type	100[a]	77*[b]	106	100	96	86	63*	50*	98	74*
CS2	5*	0*	0*	0*	0*	0*	0*	0*	0*	0*
CS6	0*	0*	0*	0*	0*	0*	0*	0*	0*	0*
CS7	0*	0*	0*	0*	0*	0*	0*	0*	0*	0*
CS8	0*	0*	0*	0*	0*	0*	0*	0*	0*	0*
CS9	0*	0*	0*	0*	0*	0*	0*	0*	0*	0*
CS10	56*	38*	102	114	90	76	62*	54*	98	71*

[a] Growth on photosensitizer-containing medium as a percent of growth on non-amended medium.

[b] *Indicates growth on photosensitizer significantly different from growth on non-amended medium (P≤0.05).

SOURCE: Adapted from ref. 67.

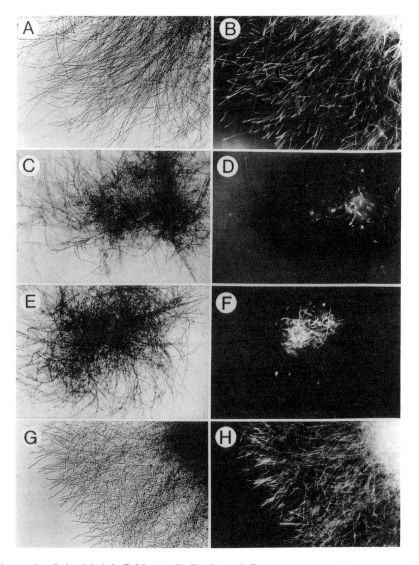

Figure 3. Paired brightfield (A, C, E, G) and fluorescence (B, D, F, H) light micrographs of *Cercospora nicotianae* wild type and cercosporin-sensitive mutant cultures.   Cultures were grown under conditions that suppress endogenous cercosporin synthesis, and 10 μM cercosporin was added to cultures 2-4 hr. prior to assay. Fluorescence pictures were taken with a 515-545 nm bandpass barrier filter, which allows visualization of fluorescence emitted by reduced cercosporin. (A,B) *C. nicotianae* wild type; (C,D) mutant CS7; (E,F) mutant CS8; (G,H) mutant CS10.   Wild type and mutant CS10 are able to reduce cercosporin. Mutants CS7 and CS8 show limited cercosporin reduction in the center of the colony, but actively growing young hyphae are not able to reduce cercosporin. Reproduced with permission from ref. 67. Copyright 1995 American Society for Photobiology.

sensitivity of mutant CS10 was specific to cercosporin (Table 2). Response of CS10 to the other photosensitizers was identical to that of wild type.

The characterization of the class 1 set of mutants raises a number of questions concerning the hypothesis of resistance resulting from transient reduction and detoxification of cercosporin. The observation that these mutants are unable to reduce cercosporin lends supports to the hypothesis. However, the complete sensitivity of the mutants to the five other photosensitizers would seem to argue against the presence of a specific cercosporin reductase and suggest that we have mutated a gene that confers resistance to $^1O_2$. It is possible that cercosporin reduction is an effect of resistance gene expression rather than the mechanism. Isolation and characterization of genes complementing the class 1 mutation will help to answer this question.

Isolation of Genes Encoding Resistance to Cercosporin and Other Photosensitizers. Current efforts are directed at isolating genes which mediate cercosporin resistance by mutant complementation. A genomic library of the *C. nicotianae* wild type strain was constructed in a cosmid vector containing a gene which confers resistance to bialaphos. Clones are being transformed into mutants CS8 and CS10. Transformants are being selected for resistance to bialaphos and then screened for resistance to cercosporin. Clones which complement the CS8 and CS10 mutations will be characterized, and the resistance-conferring genes will be sequenced and identified. The isolated genes and gene products will be useful for a number of studies. Our goal is to investigate the mechanism by which the gene products function to provide resistance, by determining whether they act on $^1O_2$ or on photosensitizer molecules. We will also work to express these genes in sensitive organisms to determine if they have utility in the genetic engineering of plants and other organisms for $^1O_2$ or photosensitizer resistance.

Summary and Conclusions

Plant pathogenic fungi which produce photoactivated perylenequinone toxins have evolved a potent mechanism for parasitism of host plants. The same wavelengths of light required by plants for photosynthesis also activate the perylenequinones to produce $^1O_2$, leading to destruction of plant cell membranes and leakage of nutrients required by the fungi for growth within the host. Fungi which produce these toxins also have evolved a unique and potent resistance mechanism that allows them to tolerate high concentrations of $^1O_2$-generating photosensitizers. Whether this resistance acts by altering the photosensitizer molecules themselves or is directed against $^1O_2$ has yet to be determined. Isolation of the genes encoding this resistance will help answer this question. Resistance genes may also have utility in the genetic engineering of organisms for photosensitizer and $^1O_2$ resistance.

Acknowledgments

This work was supported by grant MCB 9205578 from the National Science Foundation.

Literature Cited

1. Daub, M. E. *Phytopathology* **1982**, *72*, 370-74.
2. Daub, M. E. In: *Light-Activated Pesticides*; Heitz, J. R.; Downum, K. R., Ed.; ACS Symposium Series No. 339; American Chemical Society: Washington, D. C., **1987**, pp. 271-80.
3. Daub, M. E.; Ehrenshaft, M. *Physiol. Plant* **1993**, *89*, 227-36.
4. Upchurch, R. G.; Walker, D. C.; Rollins, J. A., Ehrenshaft, M.; Daub, M. E. *Appl. Environ. Microbiol.* **1991**, *57*, 2940-45.
5. Weiss, U.; Merlini, L.; Nasini, G. In: *Progress in the Chemistry of Organic Natural Products*; Herz, W.; Grisebach, H.; Kirby, G. W.; Tamm, C. H., Ed.; Springer-Verlag: Vienna, **1987**, *Vol. 52*; pp. 1-71.
6. Overseem, J. C.; Sijpesteijn, A. K. *Phytochemistry* **1967**, *6*, 99-105.
7. Robeson, D. J.; Jalal, M. A. F. *Biosci. Biotech. Biochem.* **1992**, *56*, 949-52.
8. Yoshihara, T.; Shimanuki, T.; Araki, T.; Sakamura, S. *Agric. Biol. Chem.* **1975**, *39*, 1683-84.
9. Wu, H.; Lao, X. F.; Wang, Q. W.; Lu, R. R. *J. Nat. Products* **1989**, *5*, 948-51.
10. Zhenjun, D.; Lown, J. W. *Photochem. Photobiol.* **1990**, *52*, 609-16.
11. Davis, V. M.; Stack, M. E. *Appl. Environ. Microbiol.* **1991**, *57*, 180-82.
12. Hartman, P. E.; Suzuki, C. K.; Stack, M. E. *Appl. Environ. Microbiol.* **1989**, *55*, 7-14.
13. Robeson, D.; Strobel, G.; Matusumoto, G. K.; Fisher, E. L.; Chen, M. H.; Clardy, J. *Experientia* **1984**, *40*, 1248-50.
14. Stack, M. E.; Mazzola, E. P.; Page, S. W.; Pohland, A. E.; Highet, R. S.; Tempesta, M. S.; Corely, D. G. *J. Nat. Products* **1986**, *49*, 866-71.
15. Stierle, A. C.; Cardellina II, J. H. *J. Nat. Products* **1989**, *52*, 42-47.
16. Batchvarova, R. B.; Reddy, V. S.; Bennett, J. *Phytopathology* **1992**, *82*, 642-46.
17. Balis, C.; Payne, M. G. *Phytopathology* **1971**, *61*, 1477-84.
18. Daub, M. E. *Phytopathology* **1987**, *77*, 1515-20.
19. Fajola, A. O. *Physiol. Plant Pathol.* **1978**, *13*, 157-164.
20. Yamazaki, S.; Okube, A.; Akiyama, Y.; Fuwa, K. *Agric. Biol. Chem.* **1975**, *39*, 287-88.
21. Tamaoki, T.; Nakano, H. *Bio/Technology*, **1990**, *8*, 732-35.
22. Deutschmann, F. *Phytopathol. Z.* **1953**, *20*, 297-310.
23. Kuyama, S.; Tamura, T. *J. Am. Chem. Soc.* **1957**, *79*, 5725-26.
24. Kuyama, S.; Tamura, T. *J. Am. Chem. Soc.* **1957**, *79*, 5726-29.
25. Lousberg, R. J. J. Ch.; Weiss, U.; Salmink, C. A., Arnone, A.; Merlini, L.; Nasini, G. *Chem. Commun.* **1971**, *71*, 1463-64.
26. Yamazaki, S.; Ogawa, T. *Agric. Biol. Chem.* **1972**, *36*, 1707-18.
27. Nasini, G. L.; Merlini, L.; Andretti, G. D.; Bocelli, G.; Sgarabotto, P. Tetrahedron **1982**, *38,* 2787-96.
28. Wolfbeis, O. S.; Fürlinger, *Mikrochim. Acta* **1983**, *3*, 385-98.
29. Okubo, A.; Yamazaki, S.; Fuwa, K. *Agric. Biol. Chem.* **1975**, *39*, 1173-75.
30. Macri, F.; Vianello, A. *Plant Cell Environ.* **1979**, *2*, 267-271.
31. Daub, M. E. *Plant Physiol.* **1982**, *69*, 1361-64.

32. Cavallini, A.; Bindoli, A.; Macri, F.; Vianello, A. *Chem. Biol. Interact.* **1979**, *28*, 139-46.
33. Daub, M. E.; Briggs, S. P.   *Plant Physiol.* **1983**, *71*, 763-66.
34. Fukuzawa, K.; Chida, H.; Tokumura, A.; Tsukatani, H. *Arch. Biochem. Biophys.* **1981**, *206*, 173-80.
35. Pauls, K. P.; Thompson, J. E. *Physiol. Plant* **1981**, *53*, 255-62.
36. Steinkamp, M. P.; Martin, S. S.; Hoefert, L. L.; Ruppel, F. G. *Phytopathology* **1981**, *71*, 1272-81.
37. Steinkamp, M. P.; Martin, S. S.; Hoefert, L. L.; Ruppel, E. G. *Physiol. Plant Pathol.* **1979**, *15*, 13-16.
38. Dobrowolski, D. C.; Foote, C. S. *Angewante Chemic.* **1983**, *95*, 729-30.
39. Leisman, G. B.; Daub, M. E. *Photochem. Photobiol.* **1992**, *55*, 373-79.
40. Daub, M. E.; Hangarter, R. P. *Plant Physiol.* **1983**, *73*, 855-57.
41. Hartman, P. E., Dixon, W. J.; Dahl, T. A.; Daub, M. E. *Photochem. Photobiol.* **1988**, *47*, 699-703.
42. Hughes, K. W.; Holton, R. W. In: *Vitro* **1981**, *17*, 211 (Abstr).
43. Furusawa, I.; Tanaka, K.; Thanutong, P.; Mizuguchi, A.; Yazaki, M.; Asada, K. *Plant Cell Physiol.* **1984**, *25*, 1247-54.
44. Hughes, K.; Negrotto, D.; Daub, M.; Meeusen, R.   *Environ. Exp. Bot.* **1984**, *24*, 151-157.
45. Ben-Hur, E. *Photomedicine*; CRC Press, Inc: Boca Raton, FL., 1987.
46. Downum, K. R. *New Phytol.* **1992**, *122*, 401-20.
47. Heitz, J. R.; Downum, K. R. *Light-Activated Pesticides*; ACS Symposium Series No. 339; American Chemical Society: Washington, DC, **1987**.
48. Hudson, J. B.; Towers, G. H. N. *Pharmac. Ther.* **1991**, *49*, 181-222.
49. Song, P. S.; Poff, K. L. In: *The Science of Photobiology*; Smith, K. C., Ed; Plenum Press: New York, **1989**, pp. 305-46.
50. Knox, J. P.; Dodge, A. D. *Phytochemistry* **1985**, *24*, 889-896.
51. Towers, G. H. N. *Can. J. Bot.* **1984**, *62*, 2900-11.
52. Giese, A. C. *Photochem. Photobiol. Rev.* **1981**, *6*, 139-80.
53. Berenbaum, M. R. In: *Light-Activated Pesticides*; Heitz, J. R.; Downum, K. R., Ed.; ACS Symposium Series No. 339; American Chemical Society: Washington, DC, **1987**, pp. 206-16.
54. Bellus, D. *Adv. Photochem.* **1979**, *11*, 105-205.
55. Foote, C. S. In: *Free Radicals in Biology;* Pryor, W. A., Ed.; Academic Press: New York, **1976**, *Vol. II*; pp. 85-133.
56. Krinsky, N. I. *Pure Appl. Chem.* **1979**, *51*, 649-60.
57. Foote, C. S.; Denny, R. W.; Weaver, L.; Chang, Y.; Peters, J. *Ann. N.Y. Acad. Sci.* **1970**, *171*, 139-48.
58. Truscott, T. G. *J. Photochem. Photobiol. B.* **1990**, *6*, 359-71.
59. Hartman, P. E. *Meth. Enzymol.* **1990**, *186*, 310-18.
60. Lindig, B. A.; Rogers, M. A. J. *Photochem. Photobiol.* **1981**, *33*, 627-34.
61. Rougee, M.; Bensasson, R. V.; Land, E. J.; Pariente, R. *Photochem. Photobiol.* **1988**, *47*, 485-89.
62. Bors, W.; Saran, M.; Tait, D. *Oxygen Radicals in Chemistry and Biology*; Walter de Gruyter: Berlin, **1984**.

63. Harper, D. B; Harvey, B. M. R. *Plant Cell Environ.* **1978**, *1*, 211-15.
64. Lee, E. H.; Bennett, J. W.; *Plant Physiol.* **1982**, *69*, 1444-49.
65. Rollins, J. A.; Ehrenshaft, M.; Upchurch, R. G. *Can. J. Microbiol.* **1993**, *39*, 118-24.
66. Daub, M. E.; Leisman, G. B.; Clark, R. A; Bowden, E. F. *Proc. Natl. Acad. Sci. USA* **1992**, *89*, 9588-92.
67. Jenns, A. E.; Daub, M. E. *Photochem. Photobiol.* **1995**, *61*, 488-93.
68. Sollod, C. C.; Jenns, A. E.; Daub, M. E. *Appl. Environ. Microbiol.* **1992**, *58*, 444-49.
69. Jenns, A. E.; Daub, M. E. *Phytopathology* **1995**, *85*, In Press.
70. Daub, M. E.; Payne, G. A. *Phytopathology* **1989**, *79*, 180-85.
71. Schmidhauser, T. J.; Lauter, F. R.; Russo, V. E. A.; Yanofsky, C. *Mol. Cell Biol.* **1990**, *10*, 5064-70.
72. Schmidhauser, T. J.; Yanofsky, C. *Fung. Genet. Newsl.* **1989**, *36*, 26.
73. Ehrenshaft, M.; Daub, M. E. *Appl. Environ. Microbiol.* **1994**, 2766-71.
74. Ehrenshaft, M.; Jenns, A. E.; Daub, M. E. *Molec. Plant Microbe Interact.* **1995**, *8*, 569-75.
75. Redmond, R. W. *Photochem. Photobiol.* **1991**, *54*, 547-56.
76. Clark, R. A.; Stephens, T. R.; Bowden, E. F. *J. Electroanal. Chem.*, **1995**, In Press.

RECEIVED August 24, 1995

Chapter 18

Photodynamic Effects of Several Metabolic Tetrapyrroles on Isolated Chloroplasts

Simin Amindari, W. E. Splittstoesser, and Constantin A. Rebeiz[1]

Laboratory of Plant Pigment Biochemistry and Photobiology, 240 A PABL, 1201 West Gregory Avenue, University of Illinois, Urbana, IL 61801-4716

While δ-aminolevulinic acid (ALA)-dependent photodynamic destruction of insect and animal tissues is mainly photosensitized by protoporphyrin IX (Proto), additional Mg-containing tetrapyrroles are involved in the photodynamic destruction of plant tissues. To gain better understanding of the destructive photodynamic effects of these plant tetrapyrroles, the effects of divinyl (DV) Proto, DV Mg-Proto and its monomethyl ester and DV and monovinyl (MV) protochlorophyllides (Pchlides) on isolated chloroplasts was compared. Incubation of isolated cucumber chloroplasts with the tetrapyrroles, in the light, exhibited different effects on the pigments and pigment-protein complexes of the plastids. Only one of the five exogenous tetrapyrroles failed to trigger chloroplast destruction in the light, namely divinyl DV Mg-Proto. Esterification of DV Mg-Proto to yield DV Mg-Proto monomethyl ester (Mpe) rendered this tetrapyrrole extremely destructive. While overall destructive effects were manifested by Chl *a* and *b* disappearance and the appearance of Chl degradation products, such as chlorophyllide *a*, and *b* and pheophytin and pheophorbide *a*, more specific effects on the pigment-protein complexes became evident from *in organello* 77 °K fluorescence spectroscopy. DV Proto, an early intermediate in Chl *a* biosynthesis, affected the photosystem (PS) II antenna Chl *a* pigment-protein complexes, but had no effect on the PS I antenna complex and the Chl *a/b* light harvesting antenna complex (LHCII). On the other hand DV Mpe and DV Pchlide *a*, destroyed completely all the thylakoid pigment-protein complexes. As for DV-Pchlide *a*, it exhibited its strongest effect on the disorganization of the PS I antenna LHCI-730 complex. Altogether these results indicate that individual tetrapyrroles have distinct and different disruptive effects on the structure of

[1]Corresponding author

0097-6156/95/0616-0217$14.50/0

thylakoid membranes in the light. Specific effects appear to be related to the position of particular tetrapyrrole in the Chl a biosynthetic chain and its electrostatic properties.

Tetrapyrrole-dependent photodynamic herbicides (TDPH) are compounds which force green plants to accumulate undesirable amounts of metabolic intermediates of the chlorophyll (Chl) and heme metabolic pathways, namely tetrapyrroles (1-7). In the light the accumulated tetrapyrroles photosensitize the formation of singlet oxygen which kills treated plants by oxidation of their cellular constituents. Tetrapyrrole-dependent photodynamic herbicides usually consist of a 5-carbon amino acid, ALA, the precursor of all tetrapyrroles in plant and animal cells, and one of several chemicals referred to as modulators. δ-Aminolevulinic acid and the modulators act in concert. The amino acid serves as a building block of tetrapyrrole accumulation, while the modulator alters quantitatively and qualitatively the pattern of tetrapyrrole accumulation. The tetrapyrrole-dependent connotation is meant to differentiate between this class of photodynamic herbicides and other light activated herbicides such as paraquat which are not dependent on tetrapyrrole metabolism for herbicidal activity. During the past 12 years, the scope of TDPH research has expanded considerably, as some established herbicides and a plethora of new compounds which act via the TDPH phenomenon have been discovered (8).

Originally photodynamic herbicides were assumed to be non selective in their mode of action. Further experimentation under controlled laboratory and field conditions indicated that various ALA and modulator combinations exhibited a significant degree of photodynamic herbicidal selectivity. This selectivity appeared to be rooted (a) in the different tetrapyrrole accumulating capabilities of various plant tissues, (b) in the differential susceptibility of various greening groups of plants to the accumulation of various tetrapyrroles, and (c) in the differential response of various greening groups to photodynamic herbicide modulators (3,7), as green plants have been classified into three different greening groups (9,10), depending upon which Chl a biosynthetic route (11) predominates during the dark and light phases of the photoperiod.

The dependence of TDPH susceptibility upon the greening group affiliation of treated plants as well as upon the nature of accumulated tetrapyrroles suggested that it may be possible to chemically modulate the activity of TDPH. It was conjectured that this may be achieved with the use of chemicals that may modulate the Chl biosynthetic pathway by forcing ALA-treated plants belonging to certain greening groups to accumulate the" wrong" type of MV or DV tetrapyrrole, while inducing other plant species belonging to other greening groups to accumulate the "right" type of MV or DV tetrapyrrole. A systematic search led to the identification of a large number of chemicals which acted in concert with ALA and which exhibited a definite modulating propensity toward the Chl a biosynthetic pathway (3,4). These chemicals were therefore designated as TDPH modulators. They were classified into four groups depending on their effects on the Chl a biosynthetic pathway (3,4).

During the past 12 years our Laboratory has investigated the effects of various photodynamic herbicide modulators on different greening groups of plants *in vivo.* In most cases one modulator caused the accumulation of several different tetrapyrroles. This in turn made it difficult to determine which tetrapyrrole was responsible for photodynamic damage. In this work, this issue is addressed by performing investigations with isolated chloroplasts which are incubated with various exogenous metabolic tetrapyrroles. Photodynamic damage was evaluated by determining changes in the Chl, chlorophyllide (Chlide) pheophytin and pheophorbide content of treated chloroplasts, as well as assessing alterations to the Chl-protein complexes of thylakoid membranes. An effort is also made to relate specific photodynamic effects to the electrostatic properties of various tetrapyrroles.

Material And Methods

Plant Material. Cucumber (*Cucumis sativus*, var. Beit alpha) was purchased from L. L. Olds Seed Co. Madison, Wisconsin. Corn (*Zea mays*) seeds were purchased from FS Growmark Inc. Bloomington, IL.
Chemicals. ALA was purchased from Biosynth. Skokie, IL. 2,2-dipyridyl (Dpy) was purchased from Aldrich, Milwaukee, WI. DV Proto and DV Mg-Proto were

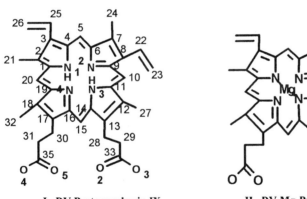

I. DV Protoporphyrin IX   **II. DV Mg-Protoporphyrin IX**

III. DV Mg-Proto Monomethyl Ester

IV. DV Protochlorophyllide *a*   **V. MV Protochlorophyllide** *a*

purchased from Porphyrin Products, Logan UT. DV Mpe, DV-Pchlide *a*, and MV-Pchlide *a* were prepared as described below..

Preparation of MV-Pchlide *a*. MV Pchlide *a* was prepared from etiolated corn leaves incubated with ALA and Dpy in darkness. Corn, a dark monovinyl/light divinyl plant species, accumulate massive amounts of MV Pchlide *a* in darkness.

Corn seeds, were planted in moist vermiculite in plastic trays (37x30x8 cm). Each tray was covered with another tray, wrapped in aluminum foil and kept in a dark growth chamber at 28 °C. Five-day old etiolated corn leaves were harvested under a safe green light. The leaves were incubated at 28 °C for 20 hr in darkness, in deep petri dishes (9x9 cm) each containing 5 g of tissue and 10 ml of an aqueous solution composed of 4.5 mM ALA, 3.7 mM Dpy and 228 mM methanol. At the end of incubation, 2.5 g batches of tissue were transferred to a 50 ml centrifuge tube containing 17 ml acetone:0.1 N NH_4OH (9:1 v/v). The tissue was homogenized at 1 °C for 2 min, with a Brinckman polytrone homogenizer. The homogenate was centrifuged at 39,000g for 10 min at 1 °C. The supernatant was collected and extracted with an equal volume of hexane and again with a one-third volume of hexane. Fully esterified tetrapyrroles such as Pchlide *a* ester and Mg-Proto-diester were transferred to hexane, whereas dicarboxylic tetrapyrroles such as Proto and Mg-Proto, and monocarboxylic tetrapyrroles such as Pchlide *a* and Mg-Proto monomethyl ester (Mpe) remained in the hexane-extracted acetone residue. To the hexane extracted acetone residue fraction was added 1/70 volume of 0.37 M KH_2PO_4 buffer (pH 7.0) and 1/17 volume of saturated NaCl, followed by 1/5 volume of peroxide-free diethyl ether. After mixing, the ether layer was collected with a Pasteur pipette. The hexane extracted acetone residue was re-extracted 7 times with ether. In this manner the Pchlide *a*, Mg-Proto and Mpe pools were transferred to ether. The ether extracts were combined and concentrated under N_2 gas until the solution became cloudy and the pigments began to precipitate. This solution was mixed with one to two volumes of peroxide-free ether and the emulsion was centrifuged briefly to separate the ether phase from the aqueous phase. The ether epiphase was collected and concentrated under N$_2$ gas.   The concentrated ether extract from 5 g of tissue, was chromatographed on 10 thin layer plates of Silica Gel H (5x20 cm), developed in toluene:ethyl acetate:ethanol (8:2:2 v/v/v) at 4 °C.   In this solvent MV Pchlide *a* migrates with an Rf of about 0.21 and moves below Mpe.   After development and while the plates were still wet, the MV Pchlide *a* band was scraped into a beaker containing methanol:acetone (4:1 v/v).   The slurry was transferred to a conical centrifuge tube and centrifuged briefly to separate the silica from MV Pchlide *a*. The methanol:acetone extract containing MV Pchlide *a* was mixed with an equal volume of ether and washed in 0.37 M phosphate buffer, pH 7.0.   The washed ether extract was transferred to a small flat-bottom plastic vial, dried under N$_2$ gas and stored at -80 °C until use.   Thirty grams of tissue yielded about 100 nmoles of MV Pchlide *a*.

Preparation of DV-Pchlide *a*.   DV Pchlide *a* was prepared form etiolated cucumber cotyledons which were poised in the divinyl Pchlide *a* biosynthetic mode by light/dark pre-treatments (12).   Cucumber is a dark divinyl/light divinyl plant species, and after light/dark pre-treatments, accumulate only DV Pchlide *a*.

Eight ml of distilled water were added to a large petri dish (15x2 cm).   Five grams of 4-day-old etiolated cucumber cotyledons were excised without hypocotyl hooks, under a safe green light and were spread as a monolayer in the petri dish. The cotyledons were then subjected to a 2.5 ms flash of actinic white light followed by 45 min of darkness.   This process was repeated one more time, after which a third 2.5 ms flash was given.   After the third flash the water in the petri dish was replaced with 8 ml of an aqueous solution containing 3.7 mM 2,2'-dipyridyl, 4.5 mM ALA and 228 mM methanol.   The petri dish was wrapped in aluminum foil and incubated in darkness at room teMperature for 2 h.   After incubation, the tissue was used for the extraction of DV-Pchlide *a* exactly as described for the preparation of MV Pchlide *a*.   Twenty five grams of tissue yielded about 100 nmoles of DV-Pchlide *a*.

Preparation of DV-Mpe.   Divinyl Mpe was prepared from 4-day old etiolated cucumber cotyledons, incubated with ALA and Dpy in darkness.   After such a treatment, cucumber cotyledons accumulate massive amounts of DV Mpe.

Cucumber seeds were planted in moist vermiculite in glass trays (50x30x5 cm), covered with other glass trays, wrapped in aluminum foil, and kept in darkness overnight, at 28 °C.   Etiolated cotyledons were excised without hypocotyl hooks and were incubated at 28 °C for 20 hr in darkness, in deep petri dishes (9x9 cm) each containing 5 g of tissue and 10 ml of an aqueous solution containing 4.5 mM ALA, 3.7 mM Dpy and 228 mM methanol.   DV-Mpe was extracted exactly as described for MV-Pchlide *a*.   The concentrated ether extract from 5 g of tissue, was chromatographed on 10 thin layer plates of Silica Gel H (5x20 cm), developed in toluene:ethyl acetate:ethanol (8:2:2 v/v/v) at 4 °C.   In this solvent Mpe migrated with an Rf of 0.31, ahead of Pchlide *a*.   Mpe was eluted exactly as described for MV Pchlide *a*.   Twenty g of tissue yielded about 100 nmoles of DV-Mpe.

Isolation of Chloroplasts. Cucumber seeds were planted in moist vermiculite held in glass trays (50x30x5 cm). Germination was carried out in a growth chamber illuminated by six 1000-W metal halide lamps (21.1 mW cm^{-2}) under a 14-h dark / 10-h light photoperiod. The teMperature ranged from 27 °C in the light to 21 °C in darkness. The seedlings were watered with Hoagland's nutrient solution. Forty grams of green, 3-day old cotyledons were hand-ground in two pre-chilled ceramic mortar each containing 75 ml of cold homogenization medium which consisted of 500 mM sucrose, 15 mM Hepes, 30 mM Tes, 1 mM MgCl$_2$, 1 mM EDTA, 0.2% BSA, and 5 mM L-cysteine at a room temperature pH of 7.7 (13). The homogenate was filtered through two layers of Miracloth (Calbiochem) and was centrifuged at 200g for 5 min in a Beckman JA-20 angle rotor at 1 °C. The supernatant was decanted and centrifuged at 1500g for 20 min at 1 °C. The pelleted chloroplasts were gently resuspended in 5 ml of homogenization medium using a small paintbrush. The resuspended chloroplasts were further purified by layering over 35 ml of homogenization medium containing 35% Percoll, in two 50 ml centrifuge tube and centrifuged at 6000g for 5 min in a Beckman JS-13 swinging bucket rotor at 1°C. Intact chloroplasts were recovered as a pellet, whereas broken chloroplasts and other cell organelles formed a band at the top of the tubes. The chloroplast pellet from each tube was resuspended in 1ml incubation medium (see below).

Chloroplast Incubation with ALA. To flat bottom glass tubes (2.5x10 cm) was added 0.9 ml incubation buffer containing 0.33 ml chloroplast suspension, 500 mM sucrose, 200 mM Tris, 20 mM MgCl$_2$, 2.5 mM EDTA, 8 mM methionine, 40 mM NAD, 1.25 mM methanol, 20 mM ATP and 0.1% (w/v) BSA, at a room teMperature pH of 7.7. 0.1 ml of 60 mM ALA was added to some tubes while 0.1 ml of H$_2$O was added to controls. Incubation was carried out for 2 h in a shaking water bath (51 oscillation/min) at 28 °C in darkness.

Incubation of Isolated Chloroplasts With Exogenous Tetrapyrroles. To each of two 15 ml conical centrifuge tubes was added a total volume of 2 ml incubation buffer containing 0.66 ml chloroplast suspension, 1.33 ml incubation medium, at a room temperature pH of 7.7 medium (see above). To one of the tubes which was used as control, 10 µl of 80 % acetone was added. To the second tube was added 100 nmoles of exogenous tetrapyrroles in 10 µl of 80 % acetone. Immediately after mixing, an aliquot of 0.8 ml was transferred to a 50 ml centrifuge tube containing 4 ml acetone:0.1 N NH$_4$OH (9:1 v/v) for pigment extraction and determination (0 hr control). Another aliquot of 0.2 ml of the chloroplast incubation mixture, was mixed with 0.4 ml of glycerol for monitoring the state of the chloroplast by 77 °K spectrofluorometry, before the onset of incubation. Each remaining 1 ml of chloroplast incubation mixture was transferred to a separate flat bottomed 2.5x10 cm tube. The tubes were covered with Saran Wrap to prevent evaporation, and were incubated under 14 mW cm^{-2} of metal halide light for 2h. at 28 °C in a water bath operated at 51 oscillations per min. After incubation, 0.8 ml from each tube, were extracted with 4 ml acetone:0.1 N NH$_4$OH (9:1 v/v). The acetone extract was collected after centrifugation at 39,000*g* for 10 min at 1 °C and processed as described below for pigment analysis. The remaining 0.2 ml of each incubation were mixed with 0.4 ml of glycerol for monitoring the state of the chloroplast membranes by 77 °K spectrofluorometry.

Determination of Tetrapyrrole Content. The acetone extracts were delipidated by extraction with one volume of hexane, then with 1/3 volume of hexane. The combined hexane extracts containing Chl and other apolar pigments were used for the determination of Chl *a* and *b*, and pheophytin *a*, and *b* (Chls without Mg) as described below. The remaining hexane-extracted acetone fraction containing dicarboxylic and monocarboxylic tetrapyrroles was used for the determination of Mpe, Pchlide *a*, and chlorophyllide (Chlide) *a*, (Chl without phytol), as well as pheophorbide *a* (Chl without phytol and without Mg). Pigments were determined quantitatively by spectrofluorometry at room temperature. To determine the amount of Chls and pheophytins, 0.1 ml of the hexane fraction was dried under N_2 gas and the residue was re-dissolved in 80% acetone prior to quantitative spectrofluorometric determinations (14). The amounts of Proto, Mpe, Pchlide *a*, Chlide *a* and *b* and pheophorbide *a* was determined by room teMperature spectrofluorometry on an aliquot of the hexane-extracted acetone residue (14,15). The rest of the hexane-extracted acetone residue fraction was extracted with ether as described for MV Pchlide *a*, and the ether extract washed with 0.37 M phosphate buffer (pH 7.0) prior to the determination of the proportion of MV and DV tetrapyrroles by spectrofluorometry at 77 °K (16).

Spectroscopy. Emission and excitation fluorescence spectra were monitored on a fully corrected photon-counting SLM spectrofluorometer model 8000C, interfaced with an IBM model XT microcomputer. Room temperature determinations were performed in cylindrical microcells 3 mm in diameter. All room temperature spectra were recorded at emission and excitation bandwidths of 4 nm. 77 °K fluorescence spectra were recorded in cylindrical glass tubes of 3 mm outside diameter (17) at emission and excitation bandwidths that varied from 0.5 to 4 nm depending on signal intensity. The photon count was integrated for 0.5 sec at each 1 nm increment.

Protein Determination. The acetone-insoluble protein pellet that was left after acetone precipitation and centrifugation was used for protein analysis. The protein pellets were suspended in distilled water, and total protein was determined by BCA on an aliquot of the suspension after delipidation (18).

Statistical Analysis. Differences between paired treatments were analyzed via a paired-comparison T test using SAS software, version 6.10).

Calculation of Electrostatic Exclusive Volumes. Exclusive electrostatic volumes were determined on a Digital Equipment Co. workstation Model 3520, operating on a VMS platform. Chem-X software, (Chemical Design Limited,

Oxford, England, October 1994 version) was used to build 3-dimensional chemical structures of the various tetrapyrroles. Optimization of three-dimensional chemical structures, was performed with Mopac (QCPE, version 6.0) using the PM3 hamiltonian.

Calculation of electrostatic potential energy levels at various sites of a given molecule was performed using Chem-X software. Chem-X calculates electrostatic potential energy levels by treating the charge on each atom in a molecule as a point charge positioned at the center of the atom. Subsequently, a positive unit charge equivalent to that of a proton is placed at each grid point and the electrostatic interaction between groups of atoms and the unit charge is calculated. This is followed by drawing isopotential contour lines. In our calculations, the number of grid points per molecule was set at a density of one point per angstrom (*i. e.* at a resolution of 1 Å). Values of 10 Kcal/mole for positive potential energy levels and -10 Kcal/mole for negative potential energy levels were selected. Since the interaction of a positive charge with a positive region of the molecule generates positive energy levels, these energy levels were interpreted as repelling. Likewise, since the interaction of a positive charge with a negative region of the molecule generates negative energy levels, these energy levels were considered as binding or attracting energy levels. In this manner, the attraction or repulsion at various loci of a particular molecule toward a positive or negative charge was well defined by the negative and positive potential energy contour lines respectively, which in fact delineate positive and negative charge binding or repelling electrostatic volumes surrounding various sections of a molecule.

Quantitative and positional differences among the electrostatic fields of various tetrapyrroles were calculated by determining the exclusive negative charge binding and repelling volumes (i. e. non-overlapping volumes between two molecules) for selected molecule pairs. The calculation of total electrostatic exclusive volumes was achieved via a Chem-X module that calculates total exclusive volumes for pairs of molecules from the electrostatic volumes of each individual molecule. The distribution of negative charge binding and negative charge repelling electrostatic volumes around a particular molecule was evaluated from the total exclusive volume and from the relative distribution of various negative charge binding and negative charge repelling exclusive volumes around the molecule. The exclusive volumes were best visualized by rotating the electrostatic exclusive volume map along the X and Y axes.

Results

Experimental Strategy. The experimental strategy consisted of incubating Percoll-purified chloroplasts with various tetrapyrrole intermediates of the Chl biosynthetic pathway in the light. This was followed by monitoring the effects of the treatment (a) on the Chl *a* and *b* pools (b) on the generation of Chl degradation products, and (c) on the state of the chloroplast membranes. An effort was also made to relate the observed effects to the electrostatic properties of the used tetrapyrroles.

Determination of the Tetrapyrrole Biosynthetic Activity of Isolated Chloroplasts. We first determined whether the chloroplast isolation procedure resulted in plastids which were active in tetrapyrrole metabolism. This was achieved by evaluating the conversion of added ALA to various tetrapyrroles by the isolated chloroplasts in darkness. As shown in Table I, after 2h of dark incubation , ALA was converted to Proto, Mpe and Pchlide *a*. This in turn indicated that the isolated chloroplasts were metabolically active.

Table I.   Tetrapyrrole Biosynthetic Activity of Isolated Chloroplasts. Incubated with 6 mM ALA in Darkness for 2 hr

Incubation period (hr))	*Proto*	*Mpe*	*Pchlide a*
		nmoles/100 mg protein/2hr[a]	
0	0.00	1.95	64.62
2	182.07	100.40	73.3

[a] Mean of two experiments.

Fluorescence Properties of Chl and Freshly Isolated Chloroplast at 77°K. Fluorescence spectroscopy at 77 °K was used extensively to probe the effects of various metabolic tetrapyrroles on the state of organization of the chloroplast membranes. It was deemed important therefore, to discuss the 77 °K fluorescence properties of freshly isolated chloroplasts before proceeding with a discussion of experimental results.

Most of the light energy absorbed by Chls dissolved in organic solvents, is dissipated as fluorescence. At the temperature of liquid N_2 (77 °K) MV Chl *a* dissolved in diethyl ether coordinates to two solvent molecules (*i. e* the central Mg atom becomes hexacoordinated by axial coordination to two lewis bases) (19, 20). It exhibits a major red emission maximum at 674 nm [Q_y (0'-0) transition], a minor maximum at 725 nm [Q_y (0'-1) transition], and Soret excitation maxima at 447 nm [B_y, B_x (0-0') transition] (20, 21), 422 nm (η_1 transition) and 400 nm (η_2 transition) (22, 23). Under the same conditions, MV Chl *b* is also hexacoordinated and exhibits a major red emission maximum at 659 nm [Q_y (0'-0) transition], a minor maximum at 722 nm [Q_y (0'-1) transition], and Soret excitation maxima at 475 nm [B_y (0-0') transition], 449 nm (η_1 transition) and 427 nm (η_2 transition) (24). The *eta* (η) transitions are forbidden in unsubstituted porphyrins, but become allowed in reduced porphyrins or when there is a conjugated carbonyl substituent as in the Chls (23).

In the chloroplast, MV Chl *a* and *b* are non-covalently associated with various thylakoid polypeptides. This special pigment-protein environment changes drastically the population and energy levels of various electronic transitions and results in different spectroscopic properties than in ether. As a consequence the spectroscopic properties of a given Chl-polypeptide complex, depends on the specific Chl-protein interactions within the complex. This picture is complicated further by the fact that not all Chl-protein complexes are capable of fluorescence. Depending on the structural proximity of various complexes, some Chl-polypeptides transfer their excitation energy to other fluorescing complexes, instead of emitting their excitation energy as fluorescence. These non-fluorescing Chl-polypeptides may become fluorescent only when their structural relationship to other Chl-poplypeptides is disrupted. For example Chl *b* does not fluoresce in healthy thylakoid membranes because it transfers its excitation energy to Chl *a*. It becomes fluorescent when its structural organization is disrupted.

A fraction of the light energy absorbed by chloroplast membranes is converted to chemical energy via the process of photosynthesis. Another fraction of that energy is dissipated via several mechanisms including fluorescence. At 77 °K, freshly isolated chloroplasts exhibit a deceptively simple three banded fluorescence emission spectrum with emission maxima at 683-686 nm (F686), 693-696 nm (F696) and 735-740 nm (F740) (26, 27) (Fig. 1A). It is believed that the fluorescence emitted at F686 nm arises from the Chl *a* of the light-harvesting Chl-protein complexes (LHCII and LHCI-680), that emitted at F696 nm originates mainly from the Photosystem (PS) II antenna Chl *a* (CP47 and/or CP29), and that emitted at F740 nm originates primarily from the PS I antenna Chl *a* (LHCI-730) (26, 27). Under the same experimental conditions, each fluorescence excitation spectrum recorded at emission wavelengths of 685 (LHCII and LHCI-680), 695 (CP47 and/or CP29) or 740 nm (LHCI-730) exhibit four excitation bands with maxima at 415-417, 440 nm, 475 nm and 485 nm (Fig. 1B). The excitation band with a maximum at 415-417 nm is probably caused by the η_1 transition of Chl *a*, while the 440 nm band corresponds to the bulk of light absorption by Chl *a* in the Soret region. The excitation bands with maxima at 475 and 485 nm are excitation energy transfer bands and correspond to light absorbed by Chl *b* and carotenoids in the Soret region. In healthy chloroplasts the photons absorbed at these wavelength by Chl *b* and by carotenoids, are transferred to Chl *a* where they are converted to chemical energy or wasted as Chl *a* fluorescence.

As mentioned above, this simple picture of the fluorescence properties of thylakoid membranes is rather deceptive, since thylakoid membranes contain several Chl *a* and *b*-binding polypeptides which may not fluorescence until their structural organization is disrupted. In this context, the ratio of emission at 739-740 nm relative to that at 685 nm (F740/F686), as well as F740/F696 have been used to determine changes in the relative distribution of excitation energy between PSI and PSII which is mediated mainly by LHCII (28). The magnitude and blue shift of these fluorescence ratios have also been used to study the onset of chloroplast degradation which disrupts the normal distribution of excitation energy between the photosystems and results in a steady decrease in the F740/F696 and F740/F686 fluorescence emission ratios (29). Furthermore disorganization of the chloroplast structure results in a blue shift of the emission and excitation maxima to shorter wavelength and eventual disappearance of the emission peaks between 680 and 740 nm, and the excitation bands between 470 and 490 nm.

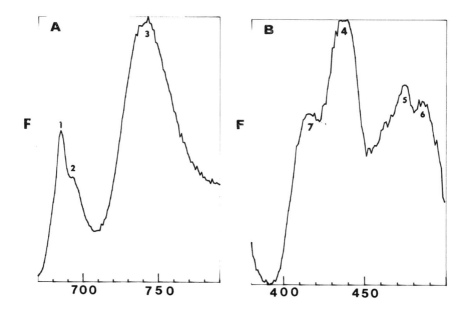

Fig. 1. Emission and Excitation Fluorescence Spectra at 77 °K of Percoll-Purified Cucumber Chloroplast at the Beginning of Incubation. F = relative fluorescence amplitude.   A: Fluorescence emission spectrum elicited by excitation at 420 nm. Abscissa = 670 to 790 nm. 1 = 685 nm, 2 = 694 nm, 3 = 740 nm. B: Fluorescence excitation spectrum recorded at an emission wavelength of 685 nm. Abscissa = 380 to 500 nm. 4 = 440 nm, 5 = 475 nm, 6 = 485 nm, 7 = 416 nm.

Effect of Light on Percoll-Purified Chloroplasts Incubated for 2 Hours in the Light in the Absence of Added Tetrapyrroles. After 2 hr of incubation in the light, Percoll-purified chloroplasts exhibited significant net synthesis of Chl a and b and Chlide a and b (Table II).   Under similar incubation conditions, in the presence of added ALA, isolated etiochloroplast were capable of massive Chl a biosynthesis and accumulation and formation of massive grana (30).   The observed rate of pigment biosynthesis in the absence of added substrates was rather unexpected since to our knowledge endogenous substrates cannot account for these synthetic rates.   The slight increase in the amount of pheophytin a and Pheophorbide a may originate from demetalation of Chl a and dephytylation of pheophytin a.

Information about the state of the thylakoid pigment-protein complexes was derived from 77 °K spectroscopy of the incubated plastids.   In 6 out of 15 experiments, no change was observed in the 77 °K profile, after 2 hr incubation of the isolated chloroplasts in the light, in the absence of added substrates.   In 9 out of 16 experiments, the emission maxima at 685 nm and 740 nm and the excitation maximum at 440 nm underwent blue shifts to 683, 733 and 437 nm respectively

(Fig 2). This in turn, indicated a certain change in the native state of the pigment-protein complexes of LHCII, LHCI-680, and LHCI-730 . The most dramatic effect was manifested by the appearance of a new emission maximum at 702-703 nm which obscured the detection of the 695 nm peak. The latter was now observed only as a weak shoulder on the long wavelength tail of the 702-703 nm band (Fig. 2). A similar emission maximum at 703 nm has been attributed to the fluorescence of a minor LHCI complex (27). All these changes may be

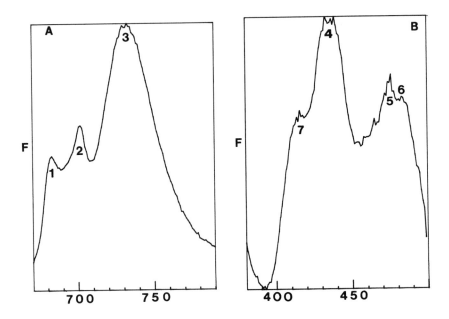

Fig. 2. **Emission and Excitation Fluorescence Spectra at 77 °K of Percoll-Purified Cucumber Chloroplast After 2 hr Incubation in the Presence of Light.** F = relative fluorescence amplitude. A: Fluorescence emission spectrum elicited by excitation at 420 nm. Abscissa = 670 to 790 nm. 1 = 683 nm, 2 = 702 nm, 3 = 733 nm. B: Fluorescence excitation spectrum recorded at an emission wavelength of 702 nm. Abscissa = 380 to 500 nm. 4 = 437 nm, 5 = 476 nm, 6 = 485 nm, 7 = 414 nm.

Table II.   Metabolic Profile of Chloroplasts Incubated in the Absence of Added Tetrapyrroles for 2 hr in the Light. Values are means of 15 Experiments. Pheo a = pheophytin a Pheobide a = pheophorbide a.

Incubation	Change in tetrapyrrole content after 2 hr incubation (nmol/100 mg protein/2 hr)					
	Chl a	Chlide a	Chl b	Chlide b	Pheo a	Pheobide a
0 hr	9271	13	3490	5	528	2
2 hr	11810	74	4495	20	695	24
Significance level (%)[a]	0.03	1.34	0.19	1.18	5.85	0.29

[a] Refers to the probability for a greater absolute value of the Student's T value to occur by chance.

attributed to a slight disorganization of the native pigment-protein complexes, and an imperfect association of newly formed Chl a and b with apoproteins. It is worth noting that the Chl b-carotenoid energy transfer bands at 476-490 nm were not affected by the 2 hr incubation.

Effect of Exogenous DV Proto on Photodynamic Damage in Isolated Cucumber Chloroplasts. Divinyl protoporphyrin IX (I) is the precursor of DV Mg-Proto. (II). In its native state DV Proto is bound to the plastid membranes (31). It is formed by oxidation of DV protoporphyrinogen IX (Protogen), the hexahydro reduction product of DV Proto, by protoporphyrinogen oxidase (Protox). Divinyl Protogen is a highly mobile metabolite. It moves readily from

Table III.   Tetrapyrrole Metabolic Profile of Chloroplasts Incubated in the Absence and Presence of 12881 nmol DV Proto per 100 mg Plastid Protein for 2 hr in the Light. Values are means of three experiments. Proto = DV Proto; Pheo a = pheophytin a Pheobide a = pheophorbide a ns = not significant.

Substrate	Change in tetrapyrrole content after 2 hr incubation (nmol/100 mg protein/2 hr)					
	Chl a	Chlide a	Chl b	Chlide b	Pheo a	Pheobide a
None	8206	19	3188	5	1299	_[a]
Proto	8260	98	2780	35	300	_[a]
Significance level (%)[b]	ns	3.64	ns	2.53	ns	_[a]

[a] DV Proto interfered with the determination of pheobide a.

[b] Refers to the probability for a greater absolute value of the Student's T value to occur by chance.

one cellular compartment to another, where it is rapidly converted to DV Proto by Protox (31). Diphenyl ethers, belong to a family of potent herbicides, which act via Protox inhibition in the chloroplast. Protogen which can no longer be converted to Proto in the chloroplast, diffuses out of the chloroplast to various subcellular compartments. There, it causes considerable photodynamic damage after conversion to DV Proto by Protoxes which are resistant to inhibition by diphenyl ethers (31). A large number of photodynamic herbicide modulators also result in the accumulation of DV Proto in the chloroplast when plants are treated with modulators and ALA (3-7).

After 2 hr of incubation of isolated chloroplasts with Proto in the light, about 90% of the added Proto disappeared. This is compatible with earlier observations that indicated a rapid photodestruction of ALA + modulator-induced tetrapyrroles in treated plants, in the light (1). Proto photosensitization exerted negligible effects, however, on other pigments. Except for a very modest increase in Chlide a and b content, Proto had essentially no effects on other pigment pools (Table III).

Further insight into the effects of DV Proto photosensitization on isolated chloroplasts was derived from 77 °K spectroscopy. Proto photosensitization appeared to affect mainly the structure of CP47 and/or CP29. CP47 is an internal Chl a antenna which is in direct contact with PSII. In that polypeptide, 14 conserved histidine residues bind 9-10 Chl a molecules by coordination of the central Mg atom to the imidazole N. CP29 is a minor PSII antenna which consists of a 31 KDa polypeptide. It is located close to the PSII reaction center and is confined to grana membranes (26). Each 31 KDa polypeptide binds about 4-12 Chls with an a/b ratio of 2.8-3.0 (26). The disruption of these pigment-protein complexes by exogenous Proto in the light is based on the following observations (Fig. 3): (a) reduction in the magnitude of the emission at 695 nm which originates in CP47 and/or CP29 (24), (b) near disappearance of the 440 nm excitation maximum, blue-shift and considerable reduction in the magnitude of the Chl b-carotenoids energy transfer bands at 470-490 nm, in the excitation spectrum of CP47 and/or CP29 (recorded at F695 nm), which is a further indication of structural disruption of the CP complexes, (c) appearance of a 674 nm emission peak which is identical to the 674 nm $[Q_y (0'-0)]$ fluorescence transition of hexacoordinated Chl a at 77 °K (20), and (d) appearance of a pronounced excitation maximum at 411 nm. This Soret excitation maximum is that of DV Proto. It is detectable in the excitation spectrum recorded at F696 nm because Proto has been able to transfer its excitation energy to CP47 and/or CP29. This in turn indicates that the Proto substrate has positioned itself close enough to CP47 and/or CP29 to cause efficient excitation energy transfer to these complexes. The confinement of the effects of DV Proto to CP47 and/or CP29 is further indicated by (a) the lack of effects on the 685 nm (LHCII and LHCI-680 complexes) and 739 nm emission (LHCI-730 complex), and (b) lack of effects on the 440 nm Chl a Soret and Chl b-carotenoids energy transfer bands of the LHCII, LHCI-680 and LHCI-730 complexes, (i. e. in excitation spectra recorded at F685 and F739 nm) (data not shown).

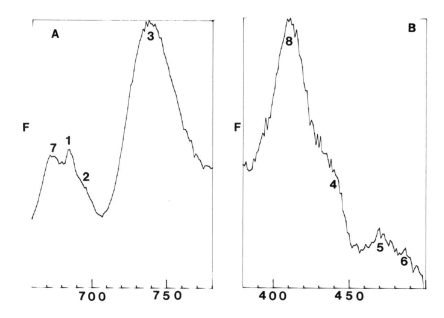

Fig. 3. **Emission and Excitation Fluorescence Spectra at 77 °K of Percoll-Purified Cucumber Chloroplasts After 2 hr Incubation in the Light with 12757 nmol of Proto per 100 mg Plastid Protein**   F = relative fluorescence amplitude.   A: Fluorescence emission spectrum elicited by excitation at 420 nm.   Abscissa = 660 to 780 nm.   1 = 685 nm, 2 = 696 nm, 3 = 739 nm, 7 = 675 nm.   B: Fluorescence excitation spectrum recorded at an emission wavelength of 696 nm.   Abscissa = 380 to 500 nm.   4 = 440 nm, 5 = 470 nm, 6 = 485 nm, 8 = 411 nm.

Effect of Exogenous DV Mg-Proto on Photodynamic Damage in Isolated Cucumber Chloroplasts.   Divinyl Mg-Proto (II) is the precursor of Mpe (III).   In its native state it is bound to the plastid membranes (31).   After 2 hr of incubation with isolated chloroplasts in the light, 87 % of the added Mg-Proto disappeared probably as a consequence of photodestruction (1).   In the light Mg-Proto had no significant effects on the pigment pools of incubated chloroplasts, except for a modest increase in the Chlide *a* and *b* content (Table IV).

In three replicates, the 77 °K fluorescence emission and excitation profiles after 2 hr of incubation in the absence of added DV Mg-Proto, were indistinguishable from those of freshly isolated chloroplasts (Fig. 1).   After incubation with 7289 nmoles of Mg-Proto per 100 mg plastid protein for 2 hr in the light, the 77 °K

Table IV. Tetrapyrrole Metabolic Profile of Chloroplasts Incubated in the Absence and Presence of 7289 nmol DV Mg-Proto per 100mg Plastid Protein for 2 hr in the Light. Values are means of three experiments. Proto = DV Proto, Pheo *a* = pheophytin *a*, Pheobide *a* = pheophorbide *a*, ns = not significant.

Substrate	Change in tetrapyrrole content after 2 hr incubation (nmol/100 mg protein/2 hr)					
	Chl a	Chlide a	Chl b	Chlide b	Pheo a	Pheobide a
None	10135	26	3687	6	1560	13
Mg-Proto	12832	46	4490	36	458	356
Significance level[a] (%)	ns	2.25	ns	8.32	ns	5.9

[a] Refers to the probability for a greater absolute value of the Student's T value to occur by chance.

fluorescence emission profile, and the excitation profile recorded at F739 nm (LHCI-730 complex) were also indistinguishable from those of freshly isolated chloroplasts (Fig. 1,4A). However, slight differences in the low temperature fluorescence excitation profile recorded at F685 (LHCII and LHCI-680 complexes) (Fig 4B) and F696 nm (CP47 and/or CP29 complexes, data not shown) became apparent. It consisted in the appearance of a 418 nm excitation peak, a 430 nm excitation shoulder, and a 2-3 nm blue shift of the remaining excitation maxima to 437, 473 and 483 nm respectively (Fig. 4B). The 2-3 nm blue shifts of these peaks indicated a slight disorganization of the LHCII and LHCI-680 light-harvesting, and the CP47 and/or CP29 Chl-protein complexes. The 418 nm peak corresponds to the Soret excitation of pentacoordinated Mg-Proto in a semi-aqueous environment, such as aqueous acetone, at room temperature (20, 33). Nevertheless in this environment DV Mg-Proto was able to transfer its excitation energy to the LHCII, LHCI-680 and the CP47 and/or CP29 Chl-protein complexes, as evidenced by the presence of the 418 nm peak in the excitation spectra recorded at an emission maximum of 685 nm (Fig. 4B) and 695 nm (data not shown). Because of the slight disorganization of the LHCII and LHCI-680 light-harvesting, and the CP47 and/or CP29 Chl-protein complexes the 430 nm excitation shoulder may be a degradation product of these complexes.

Effect of Exogenous DV Mpe on Photodynamic Damage in Isolated Cucumber Chloroplasts. Divinyl Mpe (III) is the precursor of DV Pchlide *a* (IV). It differs from DV Mg-Proto (II) by methyl esterification of the propionic acid residue at position 13 of the macrocycle.   In its native state it is bound to the plastid membranes (31). Membrane-bound Mpe exhibits an emission maximum at 598-600 nm and a Soret excitation maximum at 424-425 nm at room temperature (33).

After 2 hr of incubation with isolated chloroplasts in the light, 79 % of the

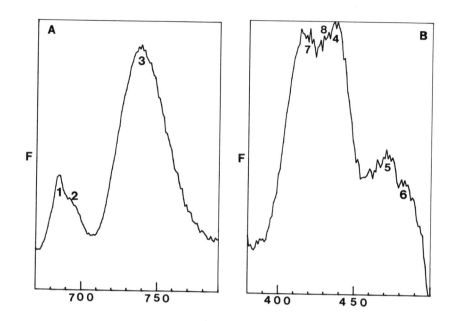

Fig. 4. Emission and Excitation Fluorescence Spectra at 77 °K of Percoll-Purified Cucumber Chloroplasts After 2 hr Incubation in the Light with 6808 nmol DV Mg-Proto per 100 mg Plastid Protein. F = relative fluorescence amplitude.   A: Fluorescence emission spectrum elicited by excitation at 420 nm. Abscissa = 660 to 780 nm. 1 = 685 nm, 2 = 695 nm, 3 = 739 nm. B: Fluorescence excitation spectrum recorded at an emission wavelength of 685 nm. Abscissa = 380 to 500 nm. 4 = 437 nm, 5 = 473 nm, 6 = 483 nm, 7 = 418 nm, 8 = 430 nm.

Table V.   Tetrapyrrole Metabolic Profile of Chloroplasts Incubated in the Absence and Presence of 6698 nmol DV Mpe per 100 mg Plastid Protein for   2 hr in the Light.   Values are means of three experiments.   Proto = DV Proto, Pheo *a* = pheophytin *a*, Pheobide *a* = pheophorbide *a*, ns = not significant.

	Change in tetrapyrrole content after 2 hr incubation					
			(nmol/100 mg protein/2 hr)			
Substrate	*Chl a*	*Chlide a*	*Chl b*	*Chlide b*	*Pheo a*	*Pheobide a*
None	11775	84	4687	22	79	23
Mpe	8468	506	3547	131	203	360
Significance level[a] (%)	3.91	3.70	11.69	ns	13.78	12.60

[a] Refers to the probability for a greater absolute value of the Student's T value to occur by chance.

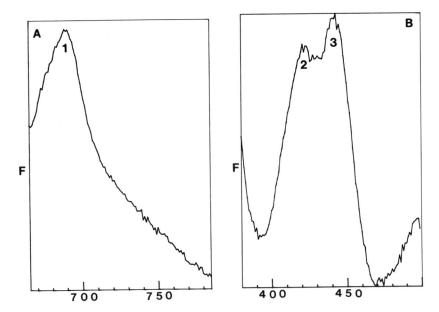

Fig. 5. Emission and Excitation Fluorescence Spectra at 77 °K of Percoll-Purified Cucumber Chloroplasts After 2 hr Incubation in the Light with 6808 nmol DV Mpe per 100 mg Plastid Protein. F = relative fluorescence amplitude. A: Fluorescence emission spectrum elicited by excitation at 420 nm. Abscissa = 665 to 785 nm. 1 = 690 nm. B: Fluorescence excitation spectrum recorded at an emission wavelength of 685 nm. Abscissa = 380 to 500 nm. 2 = 424 nm, 3 = 443 nm.

added DV Mpe disappeared probably as a consequence of photodestruction (1). It resulted in the destruction of MV Chl a and b and the formation of MV Chlide a, probably by hydrolysis of MV Chl a (Table V).

In three replicates, the 77 °K fluorescence emission and excitation profiles after 2 hr of incubation in the absence of added DV Mpe, were indistinguishable from those reported in (Fig. 2). After incubation with 6698 nmoles of DV Mpe per 100 mg plastid protein for 2 hr in the light, the 77 °K fluorescence emission and excitation profiles underwent dramatic changes (Fig. 5). Essentially the organized structure of the chloroplast was completely destroyed. This was evidenced by disappearance of the three-banded emission profile and the appearance of only one fluorescence emission maximum at 690 nm (Fig. 5A). This in turn indicated the complete disorganization of the LHCII, LHCI-680, CP47 and/or CP29, and LHCI-730 complexes. Excitation at 472 nm $i.~e.$ close to the Soret excitation maximum of MV Chl b, elicited a Chl b emission maximum at 660 nm (data not shown). Appearance of Chl b fluorescence is usually an indication of a certain degree of disruption of the structural relationship of Chl a and b in the thylakoid membranes, as Chl *b* in healthy thylakoids, does not fluoresce but transfers its excitation energy to Chl *a*.

Further evidence of disorganization of the aforementioned complexes, was evidenced by the 77 °K fluorescence excitation spectra. In three excitation spectra, recorded at F685, 696 and 739 nm, the normal three-banded fluorescence excitation profile with maxima at 440, 475 and 485 nm was replaced by one Soret excitation maximum at 443 nm (Fig. 5B). This Soret excitation maximum corresponds to MV Chl *a* coordinated to two small ligands such as pyridine at room temperature. The Soret excitation maximum at 424 nm is that of membrane. bound DV Mpe which transfers its excitation energy to the remnants of the LHCII, LHCI-680, Chl *a* CP47 and/or CP29, and LHCI-730 complexes, evidenced by the sloping tail between 685 and 740 nm (Fig. 5A).

Effect of Exogenous DV Pchlide *a* on Photodynamic Damage in Isolated Cucumber Chloroplasts. Divinyl Pchlide *a* (IV) is the precursor of DV Chlide *a*. It differs from DV Mpe (III) by the presence of a fifth ring, the cyclopentanone ring at position 13 and 15 of the macrocycle. In its native state it is bound to the plastid membranes (31). Membrane-bound DV Pchlide *a* exhibits emission maxima between 629 and 658 nm, depending on its state of aggregation and the stage of greening of the tissue (17).

After 2 hr of incubation with isolated chloroplasts in the light, 20 % of the added DV Pchlide *a* disappeared probably as a consequence of photodestruction (1). It caused considerable destruction of MV Chl *a* and *b*, which was accompanied by the formation of significant amounts of Chlide *a* and *b* probably by hydrolysis of the corresponding Chls (Table VI). In three replicates, the 77 °K fluorescence emission and excitation profiles after 2 hr of incubation in the absence of added DV Pchlide *a*, were indistinguishable from those reported in (Fig. 2). After incubation with 5493 nmoles of DV Pchlide *a* per 100 mg plastid protein for 2 hr in the light, the 77 °K fluorescence emission and excitation profiles underwent profound changes (Fig. 6). Essentially the organized structure of the chloroplast was profoundly disrupted. The 740 nm fluorescence emission decreased considerably in magnitude thus indicating disruption of the LHCI-730 protein-pigment complex (Fig. 6A). The fluorescence emission maximum at 696 nm disappeared completely, thus indicating the complete disorganization of the CP47 and/or CP29 complex. The 686 nm fluorescence emission peak underwent a 3 nm blue-shift which also indicated a certain degree of disorganization of the LHCII and LHCI-680 complexes (Fig. 6A).

Evidence of strong disorganization of the aforementioned complexes was provided by changes in the 77 °K fluorescence excitation spectra. In excitation spectra recorded at F685, 696 and 739 nm, the normal three-banded fluorescence excitation profile with maxima at 416, 440, 475 and 485 nm disappeared and was replaced by one Soret excitation maximum at 447 nm and an excitation shoulder at 421 nm (Fig. 6B). The Soret excitation maximum at 447 nm belongs to MV Chl *a* hexacoordinated to a small ligand such as diethyl ether at 77 °K (20). The Soret excitation maximum at 421 nm probably corresponds to the η_1 transition of Chl *a*.

Table VI. **Tetrapyrrole Metabolic Profile of Chloroplasts Incubated in the Absence and Presence of 4986 nmol DV Pchlide _a_ per 100 mg Plastid Protein for 2 hr in the Light.** Values are means of three experiments. Proto = DV Proto, Pheo _a_ = pheophytin _a_, Pheobide _a_ = pheophorbide _a_, ns = not significant.

| | _Change in tetrapyrrole content after 2 hr incubation_ | | | | | |
| | (_nmol/100 mg protein/2 hr_) | | | | | |
Substrate	_Chl a_	_Chlide a_	_Chl b_	_Chlide b_	_Pheo a_	_Pheobide a_
None	18411	212	7217	49	235	61
DV Pchlide _a_	11345	2183	4433	929	135	10
Significance level[a] (%)	3.52	0.88	4.89	2.28	ns	ns

[a] Refers to the probability for a greater absolute value of the Student's T value to occur by chance.

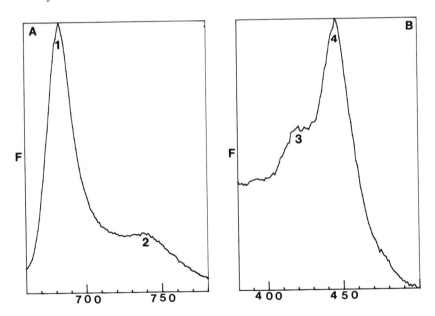

Fig. 6. Emission and Excitation Fluorescence Spectra at 77 °K of Percoll-Purified Cucumber Chloroplasts After 2 hr Incubation in the Light with 5493 nmol DV Pchlide _a_ per 100 mg Plastid Protein. F = relative fluorescence amplitude. A: Fluorescence emission spectrum elicited by excitation at 420 nm. Abscissa = 660 to 780 nm. 1 = 683 nm, 2 =739 nm. B: Fluorescence excitation spectrum recorded at an emission wavelength of 685 nm. Abscissa = 380 to 500 nm. 3 = 421 nm, 4 = 447 nm.

Effect of Exogenous MV Pchlide *a* on Photodynamic Damage in Isolated Cucumber Chloroplasts. Monovinyl Pchlide *a* (V) is the precursor of MV Chlide *a*. It differs from DV Pchlide *a* (IV) by the presence of an ethyl instead of a vinyl group at position 8 of the macrocycle. In its native state it is bound to the plastid membranes (31). Membrane-bound MV Pchlide *a* exhibits emission maxima between 629 and 658 nm, depending on its state of aggregation and the stage of greening of the tissue (17, 30).

After 2 hr of incubation with isolated chloroplasts in the light, 12 % of the added MV Pchlide *a* disappeared probably as a consequence of photodestruction (1). During incubation in the light, monovinyl Pchlide *a* exerted negligible effects on the pigment pools of the chloroplast (Table VII). Nevertheless it did cause disruption of various pigment-protein complexes as evidenced by 77 °K fluorescence spectroscopy. In one of three replicates, the 77 °K fluorescence

Table VII. Tetrapyrrole Metabolic Profile of Chloroplasts Incubated in the Absence and Presence of 3877 nmol MV Pchlide *a* per 100 mg Plastid Protein for 2 hr in the Light. Values are means of three experiments. Proto = DV Proto, Pheo *a* = pheophytin *a*, Pheobide *a* = pheophorbide *a*, ns = not significant.

Substrate	Change in tetrapyrrole content after 2 hr incubation (nmol/100 mg protein/2 hr)					
	Chl *a*	Chlide *a*	Chl *b*	Chlide *b*	Pheo *a*	Pheobide *a*
None	10525	28	3696	19	301	11
MV Pchlide *a*	10354	694	3670	27	466	0
Significance level[a] (%)	ns	14.03	ns	ns	ns	0.34

[a] Refers to the probability for a greater absolute value of the Student's T value to occur by chance.

emission and excitation profiles after 2 hr of incubation in the absence of added MV Pchlide *a*, were indistinguishable from those reported in (Fig. 1), and in two of three replicates, the 77 °K fluorescence emission and excitation profiles, were indistinguishable from those reported in (Fig. 2).

After incubation with 5997 nmoles of MV Pchlide *a* per 100 mg plastid protein for 2 hr in the light, the 77 °K fluorescence emission and excitation profiles underwent profound changes (Fig. 7). Essentially the organized structure of the chloroplast was strongly disrupted. The 740 nm fluorescence emission disappeared and was replaced by a long wavelength emission at 747 nm, thus indicating disruption of the LHCI-730 protein-pigment complex (Fig. 7A). The fluorescence emission maxima at 686 and 696 nm also disappeared and were replaced by a single emission maximum at 690 nm. This also indicated a certain disorganization of the LHCII, LHCI-680 and CP47 and/or CP29 complexes (Fig. 7A). Excitation at 472 nm *i. e.* close to the Soret excitation maximum of MV Chl *b*, elicited an emission maximum at 660 nm (data not shown). Appearance of Chl *b* fluorescence is usually an indication of a certain degree of disruption of the structural relationship of Chl *a* and *b* in the thylakoid membranes, as Chl *b* in healthy thylakoids, does not fluoresce but transfers its excitation energy to Chl *a*. The 77 °K fluorescence excitation recorded at F685, 696 and 739 nm, were similar. They exhibited excitation maxima at 418, 443, and 476 nm and a 485 nm excitation shoulder. The presence of the 476 and 485 nm excitation bands indicated that the disruption of the Chl *b*-carotenoid association was not as complete as in the case of DV Pchlide *a*. The Soret excitation maximum at 443 nm is equivalent to that of MV Chl *a* coordinated to two small ligands such as pyridine at room temperature. The 418 nm excitation peak probably corresponds to the η_1 transition of Chl *a*.

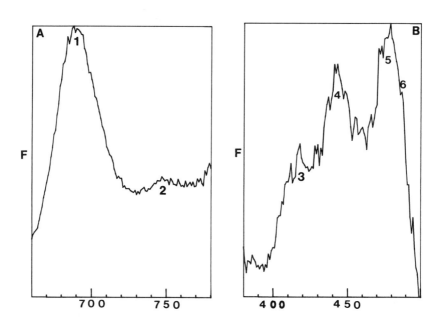

Fig. 7. Emission and Excitation Fluorescence Spectra at 77 °K of Percoll-Purified Cucumber Chloroplasts After 2 hr Incubation in the Light with 5977 nmol MV Pchlide *a* per 100 mg Plastid Protein. F = relative fluorescence amplitude. A: Fluorescence emission spectrum elicited by excitation at 420 nm. Abscissa = 660 to 780 nm. 1 = 690 nm, 2 =747 nm. B: Fluorescence excitation spectrum recorded at an emission wavelength of 696 nm. Abscissa = 380 to 500 nm. 3 = 418 nm, 4 = 443 nm, 5 = 476 nm, 6 = 485 nm.

Distribution of Exclusive Negative Charge Binding (B) and Repelling (R) Electrostatic Volumes Around DV Proto (Not Present in DV Mg-Proto) and Around DV Mg-Proto (Not Present in Proto).   In an effort to understand the molecular mechanisms underlying the differential effects of the various tetrapyrroles on chloroplast structure, the electrostatic fields of the tetrapyrroles were compared.   It was conjectured that different tetrapyrroles exhibit different effects on thylakoid pigment-protein complexes because of their specific binding to specific pigment-protein complexes.   In that case tetrapyrrole pigment-protein binding is likely to be controlled by negative charge binding (*i. e.* positive charge repelling) and/or negative charge repelling (*i. e.* positive charge binding) volumes surrounding each tetrapyrrole.   For example it can be readily visualized how a tetrapyrrole with a positively charged electrostatic volume (*i. e.* with a negative charge-binding volume) surrounding part of its structure may bind to a region of the apoprotein with a negatively-charged volume (*i. e.* with a positive charge-binding volume) such as histidine, arginine, lysine, serine, threonine, tyrosine, proline or OH-proline-rich regions.

Table VIII.   **Distribution of Exclusive Negative Charge Binding (B) and Repelling (R) Electrostatic Volumes Around DV Proto (Not Present in DV Mg-Proto) and Around DV Mg-Proto (Not Present in Proto).**

	Exclusive Volumes (Å³)[a]													
Tet	C1 C15 C31 C32	C3 C11 C24 C25	C5 C20 C21 C23	C10 C26	C21 C25	Mg	N3 C14 C20	N3 N4 C31	O2 C29 C33	O3 C29	O3	O3 O5	O4 O5	O5 C31
DV Proto			R24		B14	R6	B14		B24		R24		R18	
DV Mg-Proto	R33	R33		B6	B7				B14			R33		B7

[a] Exclusive volumes, in cubic angstroms, are on either side or orthogonal to the molecular plane, at sites around the molecule indicated by atom numbers (*i. e.* O3), or at sites delineated by a range of atoms (*i. e.* C1C15C31C32). The atom numbering is the same as in (I). Tet = tetrapyrrole.

Electrostatic field forces can be visualized by examination of the exclusive electrostatic volumes peculiar to each tetrapyrrole.   For example, in comparison to DV Mg-Proto, DV Proto exhibited negative charge-binding volumes at C21C25, N3N4C20, and O2C29C31 (Table VIII).   Large negative charge-repelling volumes surrounded the carboxylic group of the propionic acid residue at position 13 (O3), at position 17 (O4-O5) and at the northern pole of the DV Proto macrocycle (I) (Table VIII)

Insertion of Mg into DV Proto had profound effects on the electrostatic field

240 LIGHT-ACTIVATED PEST CONTROL

of the resulting DV Mg-Proto molecule, which rendered the substituted molecule an ineffective photosensitizer (Figs. 3, 4) . It resulted in the appearance of negative charge-repelling volumes at C1C15C3C32 and C3C11C24C25 (Table VIII). The largest negative charge-binding volume appeared in the southern pole of the DV Mg-Proto molecule at O3C29C33. The exclusive negative charge-binding volume surrounding the Mg atom, which is responsible for axial coordination to Lewis bases, amounted to about 7 Å^3. It may be recalled that DV Proto photosensitization affected mainly CP47 and/or CP29, while the effect of DV Mg-Proto photosensitization on these complexes was minimal (Fig. 3,4). Evidence of energy transfer from DV Proto to CP47 and/or CP29 (Fig. 3B) and from DV Mg-Proto to LHCII and LHCI-680 (Fig 4B) also indicates that DV Mg-Proto was positioned in a different thylakoid environment than DV Proto. On this basis it is logical to assume that the aforementioned changes in the distribution of various electrostatic volumes around the two molecules, changed their binding properties, and was also responsible for the observed differences in the disorganization of the pigment-protein complexes.

Distribution of Exclusive Negative Charge Binding (B) and Repelling (R) Electrostatic Volumes Around DV Mg-Proto (Not Present in DV Mpe) and Around DV Mpe (Not Present in DV Mg-Proto). Methyl esterification of the propionic acid residue at position 13 of the DV Mg-Proto macrocycle which results in the formation of DV Mpe drastically altered the electrostatic properties of the molecule and its photosensitizing properties. As reported above, incubation of chloroplasts with Mpe resulted in complete disorganization of the LHCII, LHCI-680, CP47 and/or CP29, and LHCI-730 complexes (Fig. 5). Energy transfer from Mpe to all the aforementioned pigment-protein complexes (Fig. 5B), indicated that esterification of the propionic acid residue at position 13 of the

Table IX. **Distribution of Exclusive Negative Charge Binding (B) and Repelling (R) Electrostatic Volumes Around DV Mg-Proto (Not Present in DV Mpe) and Around DV Mpe (Not Present in DV Mg-Proto).**

Tet	C2 C18 C32	C2 C22 C27	C3 C23	C11 C15 C18 C30	C25 C26	C25 C28	C31 C35	C33 C34	N3 C14	O2 O4	O3 C29	O4 C31
DV Mg Proto	R6		R4		R4	B2			B1	R21	B7	
DV Mpe		R7		R10	B2			B3				B3 R16

Exclusive Volumes (Å^3)ᵃ

ᵃ Exclusive volumes, in cubic angstroms, and all abbreviations are as defined in Table VIII. Tet = tetrapyrrole.

macrocycle allowed Mpe to bind to all these complexes. The electrostatic changes brought about by esterification consisted mainly in the appearance of negative charge binding volumes at the southern (C33-C34, O4-C31) and northern (C25-C26) poles of the molecule (Table IX). Large negative charge repelling volumes appeared at C11C15C18C30 and O4C31 (Table IX).

Distribution of Exclusive Negative Charge Binding (B) and Repelling (R) Electrostatic Volumes Around DV Mpe (Not Present in DV Pchlide *a*) and Around DV Pchlide *a* (Not Present in DV Mpe). Cyclization of the propionic methyl ester group at position 13 of the DV Mpe macrocycle (III), results in the formation of DV Pchlide *a* (IV). The difference in photosensitization between DV Mpe and DV Pchlide *a* resided mainly in their different effects on the pigment content of the various Chl-protein complexes as Chl destruction was more pronounced during DV Pchlide *a* incubation (Tables V, VI). The differences in electrostatic properties between the two molecules that may account for these effects are described in Table (X). Formation of the electron-rich cyclopentanone ring created a very large negative charge-repelling volume around that ring at

Table X.   Distribution of Exclusive Negative Charge Binding (B) and Repelling (R) Electrostatic Volumes Around DV Pchlide *a* (Not Present in DV Mpe) and Around DV Mpe (Not Present in Pchlide *a*).

						Exclusive Volumes ($Å^3$)ᵃ								
Tet	C1 C2 C27	C3 C29	C3 C5 C25 C26	C8 C11 C27	C10 C28 C30 C33	C19 C21	C20	C26	Mg	O1 C28 C12 C16	O1 O4	O3 C31 C33	O3 C34	O4 O5 C35
DV Mpe	R19	B2		R13	R21	B2				B13			R17	B8 R19
DV Pchlide *a*			B11	B3	B3				B10		R89	B14		

ᵃ Exclusive volumes, in cubic angstroms, and all abbreviations are as defined in Table VIII. Atom numbering is the same as in (I) and (IV). Tet = tetrapyrrole.

O1O4C12C16. Also negative charge binding volumes appeared in the equatorial plane of the DV Pchlide *a* molecule at C10, C20C21 and the Mg atom as well as at C3C5C25C26 and O3C31 C33. The larger negative charge binding volume surrounding and the Mg atom of DV Pchlide *a* in comparison to DV Mpe is compatible with the stronger hexacoordination properties of DV Pchlide *a* at 77 °K (20).

Distribution of Exclusive Negative Charge Binding (B) and Repelling (R) Electrostatic Volumes Around DV Pchlide _a_ (Not Present in MV Pchlide _a_) and Around MV Pchlide _a_ (Not Present in MV Pchlide _a_). Reduction of the vinyl group at position 8 of the DV Pchlide _a_ macrocycle (IV), converts the vinyl group to ethyl, and results in the formation of MV Pchlide _a_ (V). Incubation of chloroplasts with MV Pchlide _a_ for two hours in the light resulted in the disruption of the organized structure of all the monitored pigment-protein complexes (Fig. 7). However in comparison to DV Pchlide _a_ the effects on LHCII, LHCI-680, CP47 and CP29, appeared to be less severe as evidenced by a lesser effect on the MV Chl _b_-carotenoid excitation bands for all pigment-protein complexes (Fig. 7B). The most pronounced electrostatic effects that may account for this change in the

Table XI. Distribution of Exclusive Negative Charge Binding (B) and Repelling (R) Electrostatic Volumes Around MV Pchlide _a_ (Not Present in DV Pchlide _a_) and Around DV Pchlide _a_ (Not Present in MV Pchlide _a_).

Exclusive Volumes (Å^3)[a]

Tet	C6 / C9	C16	C23 C24 / C25 / C26	C25 C25 / C26	C28 / C29	C30 C33 / C31 C34	N1 / N3	O4	O4 / O5	O4 / O5 / C31	O5
DV Pchlide _a_		B1	B9	R3	R1	B3	B3	R3			B4 R1
MV Pchlide _a_	R3	R1		B1		B2			R5	B7	

[a] Exclusive volumes, in cubic angstroms, and all abbreviations are as defined in Table VIII. Atom numbering is the same as in (I) and (IV). Tet = tetrapyrrole.

photosensitization behavior of MV Pchlide _a_ may be the appearance of negative charge-repelling volumes at C6C9 and O4O5 and negative charge-binding volumes at C33C34 and O4O5C31 (Table XI).

Discussion

The effects of the various metabolic tetrapyrroles used in this study on the thylakoid pigment-protein complexes are summarized in Table XII.

Table XII. Summary of the Effects of Various Metabolic Tetrapyrroles on the Thylakoid Pigment-Protein Complexes. N = none, W = weak, M = medium, S = strong, Y = yes. The Chl destruction values are calculated from Tables III-VII.

Tetrapyrrole	Substrate disappearance (%)	Chl destruction (%)	Chl b fluorescence	LHCI-680 (F686)	LHCI-730 (F740)	LHCII (F686)	CP47 (F696)	CP29 (F696)
DV Proto	90	3	N	N	N	N	S	S
DV Mg-Proto	87	0	N	W	N	W	W	W
DV Mpe	79	27	Y	S	S	S	S	S
DV Pchlide a	20	38	N	S	S	S	S	S
MV Pchlide a	12	1	Y	M	S	M	M	M

It appears that added tetrapyrroles were metabolized at different rates during incubation in the light. In one extreme, added DV Proto disappeared nearly completely during incubation (Table XII). In another extreme, only 12 % of the added MV Pchlide *a* disappeared after 2 h of incubation (Table II). No consistent trend was observed, however, between added substrate disappearance and photodynamic damage to the thylakoid membranes.

In some cases, structural disorganization of thylakoid pigment-protein complexes, was not necessarily accompanied by Chl *a* and *b* destruction. For example, DV Proto exhibited strong disorganization effects on the CP47 and/or CP29 complexes, and little effects on Chl disappearance (Table XII). It should be emphasized that the photodynamic effects of DV Proto appear to be much more devastating, however, outside than inside the chloroplast. For example upon treatment with diphenyl ethers, Protoporphyrinogen IX migrates out of the chloroplast to unprotected subcellular compartments, where the generated DV Proto causes extensive photodynamic damage at very low concentrations (8, 31). In general, strong thylakoid pigment-protein disorganization was accompanied by Chl destruction as evidenced by the effects of DV Mpe and DV Pchlide *a* (Table XII). This in turn indicates that during chloroplast destruction, thorough disorganization of the thylakoid pigment-protein complexes, probably precedes, Chl *a* and *b* destruction.

In the thylakoid membranes, most but not all Chl is associated with apoproteins by coordination of the central Mg atom to the imidazole N of histidyl residues. It has been suggested that other coordination types or weak bonding besides Mg coordination to histidyl residues may be operative (26). Whatever the precise nature of Chl-apoprotein associations, it certainly is a beneficial one, as it allows the harvesting of light energy by antenna Chl, and its conveyance to reaction centers where it is converted to chemical energy. Association of the various exogenous tetrapyrroles with the thylakoid pigment-protein complexes was more destructive, and was probably mediated by electrostatic forces. Indeed at physiological pH, most of the thylakoid proteins exhibit negative electrostatic fields generated by basic amino acids such as lysine, arginine and histidine, by hydroxylated amino acids such as serine, threonine tyrosine and OH-proline, as well as by some heterocyclic amino acids such as proline. As depicted in Tables VIII-XI, the metabolic terapyrroles used in this study exhibited negative charge-binding and repelling electrostatic volumes around the macrocycle. These forces were probably instrumental in causing the associations of a particular tetrapyrrole with specific pigment protein complexes. These associations did not appear to be random, however, as evidenced by specific effects on specific pigment-protein complexes.

An examination of Table XII, reveals that in the light, all exogenous tetrapyrroles affected one or more pigment-protein complexes, except DV Mg-Proto. It is not clear at this stage whether the poor performance of DV Mg-Proto is due to the inherent poor photosensitizing properties of DV Mg-Proto or to poor access to pigment-protein complexes. One thing is clear however: pigment-protein complexes are susceptible to photosensitization by exogenous metabolic tetrapyrroles irrespective of their extrinsic or intrinsic location in the thylakoid membranes.

Abbreviations.   ALA, δ-aminolevulinic acid; Chlide, chlorophyllide; Protogen, protoporphyrinogen IX; Protox, protoporphyrinogen IX oxidase; Proto, protoporphyrin IX; DV, divinyl; MV, monovinyl; Pchlide, protochlorophyllide; Mpe, Mg-protoporphyrin IX monomethyl ester; TDPH, tetrapyrrole-dependent photodynamic herbicide; PS, photosystem; LHCI-680, light harvesting Chl *a* antenna of PSI with an emission maximum at 685 nm at 77 °K; LHCI-730, light harvesting Chl *a* antenna of PSI with an emission maximum at 739-740 nm at 77 °K; LHCII, Chl *a* and *b*- light harvesting protein complex; CP47 and CP29, PSII Chl *a* antennae.

* To whom correspondence should be addressed.

[1] Unless preceded by MV or DV, tetrapyrroles are used generically to designate metabolic pools that may consist of MV and DV components.

Acknowledgments.   Supported by funds from the Illinois Agricultural Experimental Station and by the John P. Trebellas Photobiotechnology Research Endowment to C.A.R.

Literature Cited

1. Rebeiz, C. A.; Montazer-Zouhoor, A.; Hopen, H. J.; Wu, S.-M. *Enzyme Microb. Technol.* **1984**, *5*, 390-401.
2. Rebeiz, C. A.; Montazer-Zouhoor, A.; Mayasich, J. M.; Tripathy, B. C.; Wu, S.-M.; Rebeiz, C. C. *Amer. Chem. Soc. Symp. Ser.* **1987**, *339*, 295-328.
3. Rebeiz, C. A.; Montazer-Zouhoor, A.; Mayasich, J. M.; Tripathy, B. C.; Wu, S.-M.; Rebeiz, C. C. *Crit. Rev. Plant Sci.* **1988**, *6*, 385-436.

4. Rebeiz, C. A.; Reddy, K. N.; Nandihalli, U. B. Velu, *Photochem. Photobiol.* **1990**, *52*, 1099-227.

5. Rebeiz, C. A.; Nandihalli, U. B.; Reddy, K. N. In *Topics in Photosynthesis - Volume 10, Herbicides*; Baker, N. R.; Percival, M. P., Eds.; Elsevier, Amsterdam, **1991**; pp. 173-208.

6. Rebeiz, C. A., In *Active Oxygen/Oxidative Stress and Plant Metabolism*; Pell, E, Steffen K., Eds.; Am. Soc. Plant Phsysiol. Maryland, **1991**; pp. 193-203.

7. Rebeiz, C. A.; Amindari, S., Reddy, K. N., Nandihalli, U. B., Moubarak, M. B., Velu, J. *Amer. Chem. Soc. Symp. Ser.* **1994**, *559*, 48-64.

8. Duke, S. O., Rebeiz, C. A. *Amer. Chem. Soc. Symp. Ser.* **1994**, *559*, 1-16.

9. Rebeiz, C. A. Ever Green. *The Sciences.* **1987,** 27, 40-45.

10. Ioannides, I. M., Fasoula, D. A., Robertson K. R., Rebeiz, C. A. *Biochem. System. Ecol.* **1994**, 22, 211-220.

11. Rebeiz, C. A., Parham R., Fasoula, D. A., Ioannides, I. M. *Ciba Foundation Symposium 180*, Chadwick D. J., Ackril, K., Eds. Wiley, New York, **1994,** pp. 177-193.

12. Duggan, J. X., Rebeiz, C. A. *Plant Sci. Lett.* **1982**, *24*, 27-37.

13. Tripathy, B. C., Rebeiz, C. A. *J. Biol. Chem.* **1986**, *261*, 13556-13564.

14. Bazzaz, M. B., Rebeiz, C. A. *Photochem. Photobiol.* **1979**, *30*, 709-721.

15. Rebeiz, C. A., Mattheis, J. R., Smith, B. B., Rebeiz, C. C., Dayton, D. F. *Arch. Biochem. Biophys.* **1975**, *171*, 549-567.

16. Tripathy, B. C., Rebeiz, C. A. *Anal. Biochem.* **1985**, *149*, 43-61.

17. Cohen, C. E., Rebeiz, C. A. *Plant Physiol.* **1978**, *61*, 824-829.

18. Smith, P. K., Krohn, R. I., Hermanson, G. T., Nadlia, A. K., Gardner, F. H., Provenzano, M. D., Fujimoto, E. K., Goeke, N. M,\., Olson B. J., Klenk, D. C. *Anal. Biochem.* **1985**, *150,* 76-85.

19. Rebeiz C. A., Belanger F. C. *Spectrochim. Acta.* **1984**, *40A*, 793-806.

20. Belanger F. C., Rebeiz C. A. *Spectrochim. Acta.* **1984**, *40A*, 807-827.

21. Belanger F. C., Duggan, J. X., Rebeiz, C. A. *J. Biol. Chem.*, **1982**, *257*, 4849-4858.

22. Weiss, C. *Ann. N. Y. Acd. Sci.* **1975**, *244*, 204-213.

23. Weiss, C. In *The Porphyrins*, Dolphin D., Ed.; Academic Press New York, **1978**; pp. 211-223.

24. Duggan J. X., Rebeiz, C. A. *Biochim. Biophys. Acta.*, **1982**, *714*, 248-260.

25. Goedheer J. C., *Biochim. Biophys. Acta.*, **1964**, *88*, 304-317.

26. Bassi, R., Rigoni, F., Giacometti, G. M. *Photochem. Photobiol.*, **1990**, *52*, 1187-1206.

27. Butler W. L., Kilajima, M. *Biochim. Biophys. Acta.* **1975**, *396*, 72-85.

28. Hipkins, M. F., Baker, N. R., *Photosynthesis Energy Transduction* IRL Press, **1986**, pp. 1-7.

29. Rebeiz, C. A., Bazzaz, M. B. *Biotech. Bioeng. Symposium.* **1978**, *No. 8*, 453-471.

30. Rebeiz, C. A., Montazer-Zouhoor, A., Daniell, H. *Isr. J. Bot.* **1984**, *33*, 225-235.

31. Smith B. B., Rebeiz, C. A. *Plant Physiol.* **1979**, *63*, 227-231.

32. Lee, H. J., Duke, M. V., Duke, S. O. *Plant Physiol.* **1993**, *102*, 881-889.
33. Rebeiz, C. A., Mattheis, J. R., Smith, B. B., Rebeiz, C. C., Dayton, D. F. *Arch. Biochem. Biophys.* **1975**, *166*, 446-465.

RECEIVED August 8, 1995

Author Index

Affiliation Index

Subject Index

Production: Amie Jackowski & Charlotte McNaughton
Indexing: Deborah H. Steiner
Acquisition: Michelle D. Althuis
Cover design: Alan Kahan

Printed and bound by Maple Press, York, PA

Supplement

Responses of the Mexican Fruit Fly (Diptera: Tephritidae) to Two Hydrolyzed Proteins and Incorporation of Phloxine B To Kill Adults

Daniel S. Moreno and Robert L. Mangan

Crop Quality and Fruit Insects Research, Subtropical Agriculture Research Laboratory, Agricultural Research Service, U.S. Department of Agriculture, 2301 South International Boulevard, Weslaco, TX 78596

In fruit fly eradication programs, insecticide-baits are one of the essential control strategies. The bait normally used is NuLure in combination with the insecticide malathion. However, social pressures demand that the use of insecticides be reduced to an absolute minimum. We found in the field that the hydrolyzed protein Mazoferm is more attractive to Mexican fruit flies than NuLure. Mazoferm also stimulates feeding more so than NuLure in the Mexican fruit fly. In independent tests Mazoferm with fructose stimulated fly feeding beyond that of fructose alone; whereas NuLure consumption was increased by adding fructose but flies consumed less than fructose alone. In choice tests a Mazoferm-fructose formulation was consumed more by flies than a NuLure-fructose formulation. The photoactive dye phloxine B mixed with Mazoferm-fructose (dye-bait) was acceptable to flies as measured by their consumption. When these flies were exposed to sunlight, they died in less than 1 h at concentrations of 2 and 4%. The effective kill of the dye-bait was increased by adding the adjuvants SM-9 or Kinetic and further increased by adding soybean oil. The attraction of the dye-bait to flies with all its additives did not diminish below that of Mazoferm-fructose alone and this attraction was significantly increased by adding acetic acid.

Some fruit flies in the family Tephritidae are the most injurious pests in the world. The most important one is the Mediterranean fruit fly, *Ceratitis capitata* (Wiedemann), which is found in nearly every continent and has a host list exceeding 300. This species is followed in importance by: oriental fruit fly, *Bactrocera dorsalis* Hendel; melon fruit fly, *B. cucurbitae* Coquillett; Queensland fruit fly, *B. tryoni* (Froggatt); olive fruit fly, *B. oleae* (Gmelin); apple maggot fly, *Rhagoletis pomonella* (Walsh); cherry fruit fly, *R. cerasi* (Linnaeus); Mexican fruit fly, *Anastrepha ludens* (Loew); West Indian fruit fly, *A. obliqua* Macquart; zapote fruit fly, *A. serpentina* Wiedemann; South American fruit fly, *A. fraterculus* Wiedemann, and guava fruit fly, *A. striata* Schiner. These fruit flies oviposit beneath the exocarp of fruits, larvae develop in their edible tissues, the fruit decays, fruit dehisces, and mature larvae burrow into the soil where they pupate. After a transformation period, adult flies emerge; and, if adults are not controlled before mating, the cycle is repeated and flies can destroy entire crops. The fruit fly immatures, for the most part, are inaccessible for control. Almost all

NOTE: This supplement has not been indexed.

control strategies are currently directed at the adults because they are the most vulnerable targets.

Bait and insecticide sprays to control fruit flies have been used since the beginning of the 20th century (1). Initially, inorganic insecticides were recommended (2, 3). After the second world war, chlorinated hydrocarbon insecticides replaced the inorganic insecticides (4, 5) to be replaced by the organophosphorous insecticides which are used at present. As recorded by Back & Pemberton (1), baits used for fruit fly control were first recommended by Mally in South Africa for the control of the Mediterranean fruit fly in 1908-1909 and by Berlese in Italy for the control of the olive fruit fly. The methods were further developed by Lounsboury in South Africa in 1912 for the control of the Mediterranean fruit fly and by Newman during 1913-1914 in Australia for the control of the Queensland fruit fly. In 1910, Marsh (6) used low-volume insecticide applications against the melon fly in Hawaii. Thereafter, a number of investigators adopted the low-volume approach to control fruit flies. Whenever baits were used, they incorporated carbohydrates and fermenting substances such as sugars, molasses, syrups, or fruit juices. In the 1930's McPhail (7) in his work with attractants found that sugar-yeast solutions attracted flies and in 1939 found that protein lures were attractive to *Anastrepha* species, especially to the guava fruit fly (8). It was not until 1952, however, when Steiner demonstrated the use of hydrolyzed proteins and partially hydrolyzed yeast in combination with organophosphates to control fruit flies, that investigators began to test the idea of attracticides to kill flies. Protein-hydrolyzate baits were first used in Hawaii for the control of the oriental fruit fly. The spray mixture contained protein hydrolyzate, sugar, and parathion (9). In 1956, the combination of malathion (a contact, stomach poison) and protein hydrolyzate bait spray was used to eradicate the Mediterranean fruit fly from Florida (10). A combination of technical grade malathion and NuLure (previously known as Protein Insecticide Bait 7 = PIB-7) at a ratio of 1 : 4 (malathion to NuLure) was used to eradicate the Mediterranean fruit fly from Brownsville, Texas (11). Lopez et al. (12) also used malathion-bait combination at the same ratio to control the Mexican fruit fly at discrete locations; however, they did not elucidate the rationale for their 1 : 4 ratio. This demonstration was left to Harris et al. (13) who showed that the effectiveness of the malathion in killing fruit flies was directly proportional to the ratio of the concentration of toxicant to NuLure with a ratio of 1 : 4 being more effective than other ratios tested, when the amount of formulation tested was the same. Thus, flies responding to the attracticide needed only to walk on the mixture, taste it, or ingest it, whereupon they died in a short time.

The main reason for the wide use of malathion is its low mammalian toxicity as well as its low price. However, ultralow-volume sprays contain concentrations of technical malathion which can be as high as 20% AI.

Recent studies have shown that a number of insect species exhibit photooxidative reactions when exposed to certain dyes (14). Among these are artic white TX, eosin Y, indigocarmine, erythrosine B, sunset yellow, methylene blue, and sodium fluorescein (15). Some of these dyes, such as erythrosine B and phloxine B are food additives which have been found to have light-induced toxicity to certain species of insects (14). Several economic dipteran species such as the house fly, *Musca domestica* L., and the face fly, *M. autumnalis* De Geer have shown susceptibility to photoactive dyes (16, 17, 18, 19). More recently Krasnoff et al. (20) showed that erythrosine B had a light-induced toxicity to the apple maggot. Erythrosine B has an oral LD_{50} in rats of ≈ 6700 mg/kg as compared to malathion of ≈ 1375 mg/kg which indicates a much greater safety margin in using this dye. Malathion is an organophosphate contact and stomach poison whereas most of the photoactive dyes with insecticidal properties could be classified as stomach poisons; yet, they are not poisons per se. This means the dye has to be delivered in an acceptable manner to the target species and it has to be ingested. Moreno et al. (21) encountered problems with food consumption with the Mexican fruit fly in their studies with cyromazine. These problems were largely overcome when 20% fructose was added to the formulation. They hypothesized that the addition of fructose increased the phagostimulatory properties of the NuLure-cyromazine mixture.

Therefore, the formulation in the delivery system should attract the target species and induce it to consume the given product.   Also, the formulation should fall within conventional practices of handling similar products, it should be safe to mammals, and should be relatively inexpensive.

Our purpose in these studies was to examine: (1) the acceptability of NuLure to the Mexican fruit fly; (2) develop an acceptable formulation which would attract flies and induce them to feed; (3) test the light-induced toxicity of phloxine B on the Mexican fruit fly; and (4) find ways of increasing the toxicity of phloxine B.

General Approaches Used in Tests

All the tests (except one) that involved feeding were done with 3 to 7-d old flies. The flies were from a culture established in 1987 and larvae reared on an artificial diet of torula yeast, carrot powder, sugar, corn cob grits, preservatives, and water (22). The flies were acquired as pupae; and, when adults emerged, they were fed 4% sugar water during the pretesting period. The day before feeding flies, these were separated by aspirating and placing them inside cubed (20 cm per side) screened aluminum cages. The screen was made of fine, blue, polyester netting (10 mesh per centimeter). Isolated flies were given only water until test food was provided. One milliliter of aqueous suspension was placed in 60 mm plastic dishes. The suspension was swirled to spread it on the surface of the dish, then, covered with 1 mm mesh fluortex monofilament screen (Tetko, Inc., Monterey Park, CA). The screen opening permitted flies to eat uninhibited and at the same time flies could walk above the food source without getting stuck. Generally flies were without food for 16 to 18 h. Test food was left in the presence of flies for 24 h. They could eat food *ad libitum*; however, flies do not eat during scotophase (10 h). Light was provided by overhead VitaLite (DURO-TEST CORPORATION, Fairlfield, NJ) fluorescent lamps which provided a maximum energy intensity of 8 μmol s^{-1} m^{-2} (measured with a LI-COR quantum meter, Model LI-185B; LI-COR, Inc., Lincoln, NB) against the inside top screen of the cage at the working level. The test room was kept at 25 ± 1 °C and 65 ± 5% RH. The experimental design for these tests was a completely randomized block.

The laboratory tests for attractiveness were cage top bioassays. We used an aluminum screened (6 x 7 mesh/cm) cage measuring 155 x 155 x 84 cm. The top surface of the cage was subdivided equally, such that 30, 9-cm Petri dishes could be placed equidistant from each other on its surface. Each dish was placed on ≈ 24 cm centers from each other. A treatment was applied to each 12.7 mm diam filter-paper disc attached to a Petri dish bottom. After application of the test solution, the dish was inverted and placed on top of the cage containing ≈ 1000 flies of both sexes. Fly attractiveness responses were determined from counts taken every 5 min for 30 min of flies found directly under the area of the Petri dishes. In this fashion, six treatments of five replicates each could be tested simultaneously. Tests were repeated three times and the means of the three tests analyzed. A randomized complete block design was used for this type of testing. The test room was kept at 25 ± 1 °C and 70 ± 5% RH. Illumination was provided by four banks of four Power Twist, 40 watt Vitalite bulbs (DURO-TEST) each, mounted on two walls, two meters above the floor.   Light intensity impinging on top of the cage was from 900 to 1200 lux.

All data were analyzed using SuperAnova AGLM (Abacus Concepts, Berkeley, CA) one factor analyses of variance (ANOVA). Significant differences ($P = < 0.05$) among means were separated using Fisher's protected LSD.

Responses of Adult Mexican Fruit Fly

Attractants. Based on the problems that Moreno et al. (21) encountered in feeding flies cyromazine in NuLure and considering that NuLure is commonly used as the bait in malathion-bait spray applications, and, knowing that a 10% NuLure or a compressed torula yeast-borax tablet is commonly used as a fly attractant in McPhail traps, it appeared essential to evaluate these two materials against other potentially attractive proteinaceous sources. The selected sources were torula yeast, casein hydrolyzate, yeast hydrolyzate, and Mazoferm. All these products were developed for human or animal consumption and therefore are considered safe. In addition, torula yeast, yeast hydrolyzate, or casein have or are being used in insect artificial rearing of flies.

For laboratory and field tests the following treatments were made in deionized water: 2.22% torula yeast (equivalent amount in a torula yeast tablet; $\approx$ 50% protein), 2.22% casein hydrolyzate ($\approx$ 87% protein) 2.22% yeast hydrolyzate ($\approx$ 50% protein), 10% Mazoferm 802 ($\approx$ 22% protein), and 10% NuLure ($\approx$22% protein). In addition, 0.01% Triton X-100 was added to break water surface tension. The pH of all the suspensions was adjusted to eight with dibasic sodium phosphate, and in addition, potassium hydroxide in the case of Mazoferm and NuLure which had received 4% dibasic sodium phosphate. These concentrations were chosen based on conventional use of torula yeast and NuLure in the field. The field test was conducted with released irradiated flies in a citrus orchard comprising 'Rio Red' grapefruit planted in 1990. Pupae were irradiated according to the procedures established by USDA-APHIS, i.e., 1-2 days before adult eclosion, pupae were exposed to 70-92 Gy emitted at a rate of 23.95 Gy per minute from a cesium[137] source, M-001 irradiator. After irradiation, pupae were dyed with Day-Glo fluorescent signal green color at a rate of 3 g dye per liter of pupae. Then, 150 pupae were placed in each of 36, 0.5 liter cartons with screened tops. Flies emerging in the carton were provided with a 6% sugar-water solution and maintained in the laboratory at 25 °C and 70% RH. Flies ($\approx$ 5000 total) from cartons were released in trees diagonal to the trap tree the evening before trap set-up. The following morning, plastic McPhail-type traps were set out in a 6 x 6 Latin Square experimental design and traps placed midway up the trees on the northeast corner. Traps were placed in every other tree within a row and in every other row with a distance of $\approx$ 9.1 x 14.6 m from each other. Each trap contained 200 ml of test solution. Traps were left in the field for two full days, then, were collected and taken to the laboratory for all fly-counts. Data were subjected to analyses of variance for Latin Square design. Data were also subjected to log normal transformation based on an analysis of the residuals. Means were retransformed for presentation.

Laboratory results (Figure 1) indicate that the hydrolyzed proteins of casein, yeast, Mazoferm, and NuLure were significantly (F = 25.93; df = 5; p = < 0.001) more attractive than the unhydrolyzed torula yeast or water. This, indicates that any of these could be used as attractive sources for *A. ludens*. However, when this same test was conducted in the field, torula yeast, yeast hydrolyzate, casein hydrolyzate, and Mazoferm were significantly (F = 16.26; df = 5; p = < 0.001) more attractive than water or NuLure (Figure 2). The surprise in the field test was the result of the NuLure which had been significantly more attractive than torula yeast or water in the previous test. However, due to cost considerations of torula yeast, yeast hydrolyzate, and casein and because Mazoferm is relatively inexpensive and had shown to be very attractive in both tests, we decided to test it further against the long-standing standard NuLure.

As the malathion-protein hydrolyzate mixture in ultralow-volume applications had been used since $\approx$ 1966 in the ratio of 1 : 4, we decided to test the relative attractiveness of low (10%), medium (40%), and high (70%) concentrations of Mazoferm against NuLure in the laboratory. The 70% concentration approached that of NuLure ($\approx$ 80%) used in aerial ultralow-volume applications. We wanted to determine if increasing concentrations of hydrolyzed protein would enhance attraction to flies. Results (Figure 3) of the tests indicate that higher concentrations of Mazoferm enhance attraction but NuLure does not. The attraction profile of NuLure indicates a slight descending order whereas that of Mazoferm is ascending. The attraction of Mazoferm and NuLure at a concentration of 10% is about equal but at concentrations of 40 and 70%, Mazoferm is

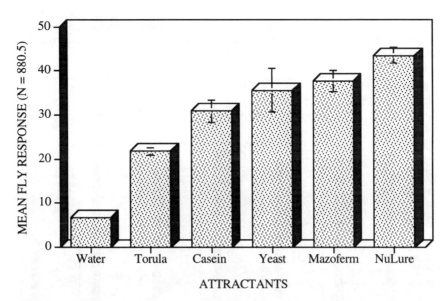

Figure 1. Response in the laboratory of *A. ludens* to water homogenates of 2.22% torula yeast, 2.22% casein hydrolyzate, 2.22% yeast hydrolyzate, 10% Mazoferm 802, and 10% NuLure. All attractants were adjusted to pH 8.

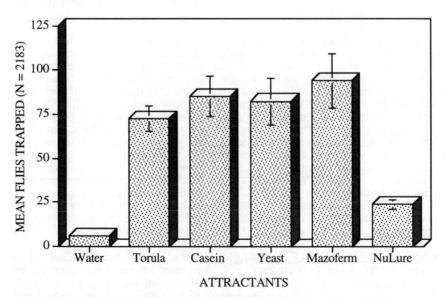

Figure 2. Response of *A. ludens* to water homogenates of, 2.22% torula yeast, 2.22% casein hydrolyzate, 2.22% yeast hydrolyzate, 10% Mazoferm 802, and 10% NuLure in the field. All attractants were adjusted to pH 8 and placed in plastic McPhail type traps.

significantly (F = 9.14; df = 5; p = < 0.001) more attractive than NuLure. It clearly appears that Mazoferm would be the ingredient of choice among these two hydrolyzed proteins and that a medium to high concentration of Mazoferm is preferred over lower concentrations.

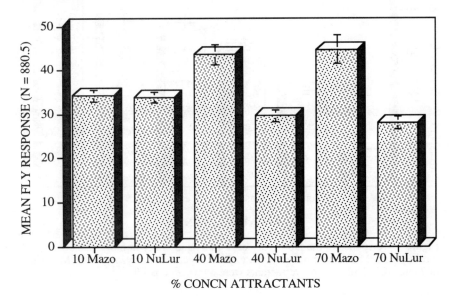

Figure 3. Response in the laboratory of *A. ludens* to various percentage concentrations of Mazoferm 802 or NuLure in water.

The previous tests indicate that any one of the attractants could be used to lure flies in ultralow-volume spray applications but because torula yeast, yeast hydrolyzate, and casein are costly, the best candidates are Mazoferm and NuLure. The former three are produced for human consumption and for that reason should be considered safe. The latter two are corn products and are used in animal food mixes because of their high protein content. Mazoferm is produced in the wet milling process in which the corn is soaked in a warm dilute sulfurous acid solution, then, the extract undergoes a mild lactic acid fermentation from naturally occurring microorganisms. Whereas NuLure is the product of corn gluten meal hydrolyzed with strong sulfuric acid. The difference in the method of hydrolysis eventually differentiates few characteristics among the products. The approximate typical (information provided by the producers) composition (%) of Mazoferm:NuLure is protein 22:22 (with very similar amino acid profiles differing only in the presence of cystine and tyrosine in Mazoferm), sugars 6:14, fat 0.2:2, salt 0.15:10.4, and solids 49:48, and with pH 3.7:4, respectively. However, NuLure has been the standard in ultralow-volume aerial spray applications and for that reason further evaluation of the two products was necessary.

Consumption: No Choice and Choice Tests. These tests were set up to determine the acceptance of Mazoferm or NuLure based on the consumption of the products by adult Mexican fruit flies. In the no choice tests, the consumption of each of the products was measured against the consumption of fructose, a simple sugar that insects quickly utilize and was found to be highly accepted by the house fly, *Musca domestica* L., (23). In a separate study by Baker et al. (8), it was found that fructose and sucrose promoted maximum survival of adult Mexican fruit flies as compared to other sugars (8). The treatments for each of the hydrolyzates were 0, 4, 8, 12, 16, 20, and 24% fructose in a 60% Mazoferm- or NuLure-base compared to the consumption of fructose crystals. Fructose was added on a weight to volume basis using deionized water as the diluent and 0.01% Triton X-100 (SIGMA Chemical Company, St. Louis, MO) as an aid in the emulsification of the products. Consumption of food was measured by weighing a 16 mm plastic dish, a screen which served as a walking platform for flies, and the formulated food before and after exposure to flies. Each of the dishes was placed next to the screen inside the cage; a similarly prepared dish was placed on the outside of the screen tangential to the dish inside the cage. Net consumption of food by flies in replicated groups was corrected to changes occurring in the control (outside dish). In the no choice tests we used groups of 40 flies (20 females and 20 males) per replicate. Thus, the results (Table I) for the no-choice tests indicate that for Mazoferm increasing amounts of fructose stimulate greater consumption of food. Consumption at the 16% or higher concentrations of fructose was significantly ($F = 28.81$; df = 7; p = < 0.001) higher than that of fructose alone. This indicates that the flies found the food more palatable with the addition of fructose as compared to no fructose. However, the addition of 4% fructose to Mazoferm does not significantly increase food consumption over the base formulation. Thus, whatever factors the sweetness of fructose needs to overcome in Mazoferm at 4%, the amount is not sufficient to do so. In contrast, a 4% addition of fructose to NuLure significantly ($F = 23.19$; df = 7; p < 0.001) increased consumption. However, increased consumption flattens at a concentration of 16% fructose but at no concentration did consumption exceed that of fructose alone. The data for NuLure tends to suggest, as compared to Mazoferm, that something in the NuLure-base formulation limits feeding.

Table I.   Consumption of Fructose in a 60% Mazoferm 802 or NuLure Aqueous Suspension by *A. ludens*, in 24 h, in Independent No-choice Tests

% Concn fructose	Mean[a] (±SEM) Consumption (mg)	
	Mazoferm	NuLure
0	79.6 ± 8.7	47.7 ± 3.1
4	87.7 ± 1.9	78.0 ± 4.0
8	111.5 ± 7.4	97.6 ± 6.7
12	126.8 ± 4.3	81.5 ± 3.1
16	137.6 ± 5.2	101.2 ± 5.7
20	152.6 ± 10.0	110.3 ± 5.4
24	210.1 ± 12.0	114.3 ± 12.4
100 (fructose only)	109.7 ± 7.1	156.6 ± 6.9

[a]Mean of five replicates, 40 flies per replicate

In the choice tests, we took the same approach as in the no choice tests of placing dishes inside and outside the screen of each test cage. However, in this test, the flies had a choice of consuming either a 60% Mazoferm- or 60% NuLure-base formulation that contained 0, 4, 12, or 20% fructose. We used 80 females per replicate. The results of this test (Table II) clearly show that flies significantly ($F = 60.61$; df = 7; p = < 0.001) preferred a Mazoferm-base formulation with or without fructose to that of NuLure with or without fructose. Flies consumed 6.9 times more Mazoferm without fructose as that of NuLure without fructose. Again, we see that a 4% fructose in the

Mazoferm formulation does not increase consumption significantly over that without fructose but is significantly increased with the addition of 12 or 20% fructose. However, there is no significant difference among these two concentrations. Inasmuch as some consumption of NuLure took place, perhaps sensorial sensitivity became attenuated.

As a follow-up to the previous results, we set up a test to determine the food quality of Mazoferm and NuLure measured by the time that it took 50% of flies to die (LT_{50}). The idea was that the nutritional quality of the food should be reflected in the adult longevity of flies. The treatments for this test were a 60% Mazoferm- or NuLure-base formulation and the addition of 12% fructose to each. We used 80 females per replicate times five replicates per treatment. Deaths were recorded three times a day from the time they were given the formulation to the time 50% of them died. Old food was removed and fresh food was provided every other day (except week ends). The results (Table III) of this test indicate that the Mazoferm-base formulation has significantly ($F = 295$; df = 3; $p < 0.001$) better nutritional qualities than those of the NuLure-base formulation. Inasmuch as basic survival of flies depends on assimilative carbohydrates, the data suggests that Mazoferm has more assimilative carbohydrates than NuLure. This inference is attested by the similarity in survival of flies given Mazoferm or NuLure with 12% fructose. Here, the amount of fructose provided, apparently superseded the differences in assimilative carbohydrates among Mazoferm and NuLure.

Table II.   Consumption of Fructose in a 60% Mazoferm or NuLure Aqueous Suspensions by _A. ludens_ in 24 h, Choice Tests

% Concn fructose	Mean[a] (±SEM) Consumption (mg)	
	Mazoferm	NuLure
0	170.8 ± 9.7	24.7 ± 4.1
4	183.1 ± 13.4	48.5 ± 11.5
12	280.7 ± 4.3	91.5 ± 14.4
20	280.6 ± 13.6	110.3 ± 5.4

[a]Mean of five replicates, 80 females per replicate.

Table III.   Time Lapsed for 50% Female _A. ludens_ to Die When Fed Only Indicated Diets Plus Water.

60% ingredient	Mean[a] (±SEM) LT_{50} Death (h)
Mazoferm 802	106.0 ± 0.0
NuLure	58.0 ± 0.0
Mazoferm + 12% fructose	236.2 ± 5.3
NuLure + 12% fructose	237.8 ± 9.3

[a]Mean of five replicates, 80 females per replicate.

Results of the three previous tests indicate that, in terms of attraction and food consumption, Mazoferm 802 is preferred over NuLure. This important fact was not noticed by the various workers in their protein bait-malathion applications because, though NuLure may not be very attractive, the frequency of a fly encountering a droplet is great in ultralow-volume aerial applications, which deposit ≈ 10 8-μl droplets of the formulation per meter square. In addition, as long as flies walked on, tasted, or ate the malathion bait, flies died. So, from about the time that Steiner (9) introduced the protein-malathion bait it really has not been necessary to examine the delivery system. However, now that researchers such as Moreno et al (21) tried to introduce third generation insecticides to control fruit flies, gustatory behavior patterns of fruit flies have come to the forefront. Acceptability of a food bait is even more important when the effect of a latent toxicant, such as a photoactive dye, depends on total consumption and physiological uptake.

Lethality of Some Photoactive Dyes Alone or With Adjuvants

The following tests were set-up to determine acceptability of the developed formulation with the incorporation of a photoactive dye, phloxine B. Krasnoff et al. (20) had already shown that erythrosine B was an effective compound to kill the apple maggot under laboratory conditions when flies fed in the dark on a 10% sugar solution were then exposed to light. They chose to develop a logarithmic model to account for observed mortality. We chose to take a more practical approach in trying to develop a food bait that would simply replace the malathion-bait currently used in aerial applications for the control of fruit flies. We also chose to feed flies under subdued light (8 μmol s^{-1} m^{-2}) because the Mexican fruit fly does not feed in the dark and in nature feeds shortly after daybreak until it gets hot ($\approx$ 32 °C) or start feeding much later depending on temperature. In our first test we chose to test concentrations of 0 (control, Mazoferm-fructose formulation only), 0.25, 0.5, 1, 2, and 4% phloxine B (92% AI) (Hilton Davies, Cincinnati, OH) because we had no idea about the relative activity of this dye in the Mazoferm-fructose formulation. Twenty female flies per cage were exposed to food treatments (five replicates per treatment) for 24 h and consumption measurements conducted as before. The following day, cages containing the flies were taken outside the laboratory and exposed to natural sunlight for a period of four hours. The number of flies killed during that period was recorded every hour.

The results of this test are shown in Table IV. Temperature at the start of the test was 25 °C and the sky was mostly clear with patchy clouds. Minimum energy reading

Table IV. Sunlight-activated Death (Mostly Clear Sky) of Female *A. ludens* After Exposing Them for 24 h to a Mazoferm Diet Containing Various Concentrations of Phloxine B

% Concn	Mean[a] (±SEM) Percentage Females Dead in Hour:			
	1	2	3	4
0	3 ± 2.0	4 ± 2.9	5 ± 3.0	6 ± 3.7
0.25	11 ± 4.6	28 ± 3.0	62 ± 7.5	79 ± 7.0
0.5	8 ± 2.0	38 ± 7.2	67 ± 2.6	93 ± 2.0
1	37 ± 4.1	74 ± 8.1	87 ± 4.1	98 ± 1.3
2	49 ± 6.2	90 ± 3.5	96 ± 2.5	99 ± 1.0
4	98 ± 1.2	100 ± 0.0	100 ± 0.0	100 ± 0.0

[a]Mean of five replicates, 20 females per replicate.

inside the cage, against the screen where flies rested, was 720 μmol s^{-1} m^{-2} half an hour after the beginning of the test and a maximum of 2100 μmol s^{-1} m^{-2} intermittently between cloud shadows. For the last hour of recording there were no clouds in the sky. As can be seen from the table, the percentage of females killed was dosage dependent. At a concentration of 4% phloxine B, some females were dying within 15 min of exposure to sunlight. At lower concentrations flies took longer to die. It appeared that acute light-activated reactions started to level off after 3 h of exposure to sunlight at concentrations above 0.5% phloxine B. However after 4 h, it also appears that accumulated death at 0.5% concentration was as effective as higher concentrations. Nonetheless, this concentration appears to be at the margins of acute light-activated reactions.

A repeat of the previous test was conducted to determine food consumption (procedure described previously) with the incorporation of dye and exposed flies to slightly different conditions outside. At the beginning of the test temperature was 20 °C, very foggy, and energy readings varied between 60 and 150 μmol s^{-1} m^{-2}. Conditions were similar the second hour with maximum energy recordings of 240 μmol s^{-1} m^{-2}, skies were patchy the third hour and energy readings were at least 700 μmol s^{-1} m^{-2}. Skies were completely clear the fourth hour. Beginning temperature for the third hour was 25 °C. The results in Table V indicate that flies were not deterred from consuming the Mazoferm-fructose formulation regardless of dye content. Fly kill at 2 and 4% phloxine B was much lower in this test the first 2 h than in the previous one. The lower effect can be attributed directly to the much lower light energy levels the first 2 h. Despite the lower energy intensity, by the end of the fourth hour fly kill for 1 to 4% phloxine B was similar to the previous test. However, the effect of 0.5% phloxine B in this test was 23 points lower than in the previous one. Thus, the data confirms that the latter concentration is at the margins of acute light-activated reactions, but also indicates that given enough light exposure, even low concentrations can be effective.

Table V. Food Consumption and Sunlight-activated Death (Mostly Cloudy Sky) of Female _A. ludens_ After Exposing Them for 24 h to a Mazoferm Diet Containing Various Concentrations of Phloxine B

% Concn	Mean[a] (±SEM) Food Consumption (mg)	Mean[a] (±SEM) Percentage Killed by Hour:		
		1	2	4
0	105.7 ± 11.3	0 ± 0.0	0 ± 0.0	3 ± 1.4
0.25	92.9 ± 5.1	0 ± 0.0	10 ± 3.2	53 ± 14.4
0.5	93.5 ± 5.9	0 ± 0.0	7 ± 2.6	70 ± 2.7
1	104.4 ± 11.3	0 ± 0.0	20 ± 2.7	93 ± 1.2
2	98.4 ± 4.1	1 ± 1.0	34 ± 4.3	96 ± 1.0
4	95.8 ± 5.6	18 ± 4.8	71 ± 7.9	99 ± 0.8

[a]Mean of five replicates, 20 females per replicate.

The fact that flies did not reject the Mazoferm-fructose formulation even with relatively high concentrations of phloxine B and the fact that light-activated death occurred in flies after exposing them to sunlight, encouraged us to improve the formulation. Carpenter et al. (24) reported that by incorporating fluorescein to rose bengal, they were able to enhance the toxicity of rose bengal against _Aedes_ larvae. Subsequently they showed (25) very similar results with eight xanthene dyes. We chose to examine the possibility of enhancing fly mortality by incorporating the nonphototoxic uranine (75% AI) (sodium salt of fluorescein) (Hilton Davies, Cincinnati, OH 45237) to our formulation containing 0.5% phloxine B. This concentration was chosen because in the two previous tests it appeared to approximate acute light-activated reactions. To a dye-bait formulation containing 0.5% phloxine B we added 0, 0.1, 0.2, 0.3 and 0.4% uranine and compared the consumption of each concentration to our base formulation of Mazoferm-fructose (check). The results of this tests are shown in Table VI. Again, consumption is very similar for all uranine concentrations, phloxine B only, and check (contains no dyes). Thus, indicating food acceptance by flies. However, percentage fly kill after 3 h of exposure to sunlight (a range of 420 to 1500 μmol s^{-1} m^{-2}) indicates that the addition of uranine, at all concentrations, does not enhance fly death. Quite the contrary, it appears to protect flies from acute, lethal, light-activated reactions. Perhaps this effect can be explained by the fact that uranine is a smaller molecule than phloxine B and as such, penetrates faster and occupies same or similar sites as phloxine B and, when exposed to high energy intensities, the light absorbing properties of uranine can cause quenching interactions owing to a crossing-over of excited and ground state energy surfaces of phloxine B. Thus, in a sense, reducing lethal phloxine B photo-active reactions. However, these results do not negate the possible synergism that may occur between uranine and phloxine B under subdued energy conditions such as the closed canopies of grapefruit or orange trees that the Mexican fruit fly occupies during sunny days or the shady understory coffee plantations that the Mediterranean fruit fly occupies seasonally. However, in a test conducted earlier using exactly the same procedures, no fly mortality was observed after exposing flies to 5 h of sunlight. The day of the test was cloudy, drizzling on and off, and cool (15 to 18 °C). The energy flux recorded for the day was from 10 to 60 μmol s^{-1} m^{-2}. After the 5 h of light exposure, flies were fed sugar water and the following morning an additional death count was made. The mean death count for each treatment of check, 0, 0.1, 0.2, 0.3, and 0.4% uranine was 2, 17, 16, 12, 6, and 7%, respectively. These death percentages follow the general profile of results presented in Table VI but the effect was minimal and do not validate or invalidate the previous argument. The data does suggest a latent, lethal, light-activated reaction despite low energy levels.

Table VI.   Food Consumption and Sunlight-activated Death of Female *A. ludens* After Exposing Them to a Mazoferm Diet Containing 0.5% (AI) Phloxine B and Varying Amounts of Uranine

% Concn Uranine	Mean[a] (±SEM)	
	Consumption (mg)	% Females Killed (3 h)
Check	135.1 ± 18.7	0 ± 0.0
0.0	134.1 ± 15.0	99 ± 1.0
0.1	140.3 ± 3.7	82 ± 6.0
0.2	137.9 ± 10.4	50 ± 4.5
0.3	159.2 ± 10.9	31 ± 2.5
0.4	140.5 ± 12.2	21 ± 2.9

[a]Mean of five replicates, 20 females per replicate.

In a continued effort to improve onset of insect mortality and increase toxicity at lower dosages of phloxine B, we hypothesized that if we could somehow increase gut membrane permeability, the effectiveness of the dye could be increased. Biological membranes are composed of lipid bilayers, integral proteins, and peripheral protein molecules held together by noncovalent interactions. Membranes are highly selective permeability barriers. High amounts of lipids, particularly phospholipids, glycolipids, and sterols are characteristic of biological membranes; a common feature of these membrane lipids is that they are amphipathic molecules (26). They spontaneously form extensive bimolecular sheets in aqueous solutions. These lipid bilayers are highly impermeable to ions and most polar molecules, yet they are quite fluid, which enables them to act as a solvent for membrane proteins (26). Thus when these membranes come in contact with a water soluble food, such as that offered to flies, an interfacial tension is created. If this interfacial tension can be partially or completely broken, penetration of the membrane by hydrophilic molecules and ions can be increased. For such an effect we turned to the adjuvants. Adjuvants are commonly used in conjunction with insecticide, fungicide, herbicide, or nutrient applications. According to Ebeling (4), adjuvant or helper substance is a material which, while not itself toxic, is added to an insecticide to improve its physical or chemical characteristics. Adjuvants are now broadly classified (27) as spreaders-stickers-buffers-foliar nutrients, penetrants-crop oil concentrates-extenders, drift control agents-deposition and retention agents, defoamers, foam markers-dyes, soil wetting agents-soil conditioners and compatibility agents. Further, adjuvants that are formulated to break interfacial tension are classified as anionic, cationic and nonionic agents. The former two because of their ionic charge can react with other compounds and can potentially cause precipitation. Respicio et al. (28) and Carpenter (29) used a dispersible formulation (containing sodium dodecyl sulfate) of erythrosine B against the mosquito larvae of *Culex pipiens quinquefascitus* Say and found it to be more toxic than its nondispersible counterpart. This probably occurred due to the better distribution of erythrosine B molecules. However, we inferred that we could in fact improve upon our newly derived formulation. With the number of variables that we had in our formulation, plus the physiological conditions of the fly gut, we chose to test a nonionic adjuvant, SM-9 (proprietary blend of linear secondary alcohols reacted with ethylene oxide 99.97% AI; SMI, Valdosta, GA). The nonionic adjuvants are chemically inert and highly efficient as surface-active agents (4).

As in the previous test, we selected a test concentration of 0.5% phloxine B to which SM-9 was added. The treatments were check (Mazoferm-fructose formulation with 0.1% SM-9 as an emulsifier), and the concentrations of 0.1, 0.25, 0.5, 1, and 2% SM-9 in the phloxine B formulation. The test was conducted with young and mated female flies. Young and mated females were maintained with a sugar solution only. Mated females were 13 days old (at 26.7, °C they normally mate in eight days). One test with young females was conducted as previously described but for a second test the females were allowed to feed only once. To accomplish this, groups of 1000 flies in each of six cages were deprived of food for 20 h, after which treated food was given to them. As soon as flies were observed to leave the feeding dishes with marked abdomens, these were aspirated from the various feeding cages and groups of 20 females transferred into each of five cages. From the time that food was first provided to the time that all transfers were completed, took 1.5 h. Then they were immediately taken outside and exposed to sunlight. The day of sunlight exposure for the first test with young females was completely sunny, and partially cloudy for mated females and singly-fed females. The energy flux during sunlight exposure for the latter two groups was at least 500 µmol s^{-1} m^{-2} under partial cloud cover. Temperature during the testing period for all three days was 25 to 30 °C.

Data shown in Table VII suggest that the addition of SM-9 greatly enhances the lethal effect of phloxine B on flies. In the first test (sunny day) with young females at concentrations of 1 and 2%, females started dying within five minutes of exposure to sunlight. All flies were killed in 1 h at the two highest concentrations. By the end of the second hour, nearly 100% of flies were killed at concentrations of 0.5% SM-9 and above. The rate of fly kill for the check and 0.25% SM-9 was about the same for the second and third hour. No flies were killed in the control during the entire test period. Compared to the concentration of 0.1% SM-9, fly kill was definitely accelerated by increasing amounts of SM-9. The same can be said for results obtained with mated females and singly-fed young females. A 100% kill was not observed in the first hour in the latter two tests because the initial energy impinging on these test flies was not as high (at least 1100 µmol s^{-1} m^{-2}) as that impinging on test insects of the first test. However, by the end of the second hour fly kill was over 90% at concentrations of 1 and 2%. There appears to be a slight lag with singly-fed flies but this lag may be attributed to the degree of metabolized dye in insect tissues rather than energy intensity.

Table VII. Sunlight-activated Death of Female A. *ludens* 24 hours (Except Single Feeding) After Exposing Them to a Mazoferm-Fructose Diet Containing 0.5% Phloxine B and Varying Amounts of the Adjuvant SM-9

Mean[a] (±SEM) Cumulative Percentage Killed by Indicated Hour:

% Concn SM-9[c]	Young Females			Mated Females			Single Feeding[b] of Young Females		
	1	2	3	1	2	4	1	2	4
Check	0 ± 0.0	0 ± 0.0	0 ± 0.0	0 ± 0.0	0 ± 0.0	0 ± 0.0	0 ± 0.0	0 ± 0.0	0 ± 0.0
0.1	11 ± 1.0	81 ± 3.3	100 ± 0.0	4 ± 1.9	55 ± 6.9	95 ± 5.0	0 ± 0.0	4 ± 1.9	74 ± 1.9
0.25	29 ± 6.0	82 ± 6.0	98 ± 0.0	3 ± 1.2	63 ± 6.2	99 ± 1.0	2 ± 1.2	31 ± 6.2	85 ± 4.5
0.5	67 ± 6.0	99 ± 1.0	100 ± 0.0	42 ± 5.6	95 ± 2.7	99 ± 1.0	3 ± 1.2	71 ± 4.0	91 ± 4.0
1	100 ± 0.0	100 ± 0.0	100 ± 0.0	66 ± 4.8	100 ± 0.0	100 ± 0.0	44 ± 5.3	91 ± 4.0	99 ± 1.0
2	100 ± 0.0	100 ± 0.0	100 ± 0.0	81 ± 4.3	100 ± 0.0	100 ± 0.0	75 ± 4.7	97 ± 2.0	100 ± 1.9

[a]Mean of five replicates, 20 females per replicate.
[b]Females were allowed to feed once, transferred to test cages, and 1 h after observed first feeding taken out to sunlight.
[c]Check contained 0.1% SM-9 but no dye.

At a concentration of 0.1% of singly-fed flies there was a lag of four hours before substantial kill was observed.

A concern arose about the possibility that SM-9 at high concentrations could be killing flies. So we set-up a test where adult flies were maintained on a Mazoferm-fructose base formulation from the first day of adult life. All insects were provided with water on the side and new food was provided three times a week. The treatments were check (base formulation), 0.25, 0.5, 1, 2, and 4% SM-9 in the base formulation. Flies were fed on their respective diets for 10 days. Death rate was 4, 3, 5, 0, 3 and 5%, for the respective concentrations in 10 days. Thus indicating that SM-9 is not innately toxic.

The data suggest that the adjuvant SM-9 noticeably enhances the phototoxicity effect of phloxine B. At this point we do not know if SM-9 is aiding by increasing the dissolution of phloxine B or aiding in membrane penetration of phloxine B molecules within the gut and within cellular target sites. What we see is the integrated effect of phloxine B and SM-9 in the form of greater and more rapid kill of flies at the same concentration of phloxine B. If a rapid kill of flies is desirable then the data suggest that one should use at least 0.5% SM-9.

As the use of SM-9 enhanced the phototoxic effect of phloxine B and because we wished to make further evaluations of other potential photoactive dyes and other adjuvants, we decided to standardize the concentration of phloxine B such that when mixed with a Mazoferm-fructose-SM-9 (1%) formulation, the dye-bait would kill ≈ 50% of flies. We selected 1% SM-9 because we wanted to incorporate a rapid-kill component. We conducted a series of tests to obtain the final concentrations that would provide us with an approximate linear regression fit. The selected phloxine B concentrations were 0, 0.01, 0.02, 0.04, 0.06 and 0.08% AI. Flies were handled as previously described. The day of sunlight exposure was partly cloudy and irradiation varied from 210 to 640 μmol s^{-1} m^{-2}. Temperature during the testing period was from 21 to 25 °C. Data gathered during the day was submitted to an analysis of variance and to regression analysis using StatView (Abacus Concepts, Inc., Berkeley, CA) statistical program.

There was no significant fly kill the first hour of the test. Significant ($F = 21.2$; df = 5; $p = < 0.001$) kill from the control occurred in the second hour of exposure at a concentration of 0.08% phloxine B. At the end of the fourth hour significant ($F = 172$; df = 5; $p = < 0.001$) kill over the control occurred at 0.02% concentration and above. Mean ($\pm$SEM) percentage kill was 0 (0), 4 (1.9), 17 (2.6), 67 (7.2), 88 (2.0), and 98 (1.2) for concentrations of 0, 0.01, 0.02, 0.04, 0.06 and 0.08%, respectively. Percentage fly kill obtained during the fourth hour was regressed against phloxine B concentration which provided a reasonable linear fit (Figure 4). From this regression we were able to compute an estimated phloxine B lethal concentration = 0.0382% that would kill ≈ 50% (LC_{50}) of flies under the same conditions. Thus, the 0.0382% concentration phloxine B became our standard to make comparisons with other candidate photoactive dyes or adjuvants.

By generating the previous data, it allowed us to test other potential photoactive compounds and compare their lethality to that of our standard 0.0382% phloxine B. Thus, the standard test formulation contained 70% Mazoferm, 20% fructose, 1% SM-9, and 0.0382% phloxine B in water which equilibrated at pH 3.8. Candidate photoactive test dyes contained the same formulation except that phloxine B was substituted with equivalent concentration of candidate dye. By taking this approach we could only see phototoxicity in those compounds that were as active or more so than phloxine B. Thus, in two different tests we evaluated the phototoxicity of methylene blue, rhodamine B, erythrosine B, and mercurochrome (SIGMA), brilliant blue BMA and eosin Y (Allied Chemical, Morristown, NJ), fast green (Eastman Organic Chemicals, Rochester, NY), and rose bengal (ICN Biochemicals, Cleveland, OH). Procedures in these tests were the same as those previously described.

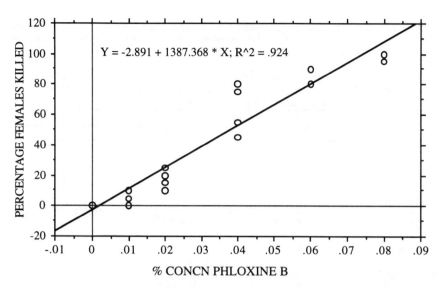

Figure 4. Regression and curve fit for fourth hour percentage fly kill on percent concentration of phloxine B. Predicted LC_{50} occurred at a concentration of 0.0382% phloxine B.

The results of these tests (Table VIII) show that rose bengal, erythrosine B, and eosin Y are sufficiently photoactive to kill flies at the concentrations tested. Erythrosine B is the most phototoxic, rose bengal about as active as phloxine B and eosin Y about half as active as phloxine B. Flies in the first test received at least 120 μmol s^{-1} m^{-2} while in the second test they received 270 μmol s^{-1} m^{-2} at the onset of the experiments. In either test the maximum energy flux was 1500 μmol s^{-1} m^{-2}. These results are not surprising as Heitz and Downum (14) had already reported on their activity. The compounds are structurally related and their activity appears to be influenced by the presence of chlorine in the phthalic ring. Phloxine B and eosin Y have tetrabrominated xanthene rings with no chlorine in the latter, yet, it is half as active as phloxine B. Rose bengal and erythrosine B have tetraiodized xanthene rings with no chlorine in the latter, yet, it is almost twice as active as the former or phloxine B. Perhaps their activity is related to the specific halogen on the xanthene ring or a combined effect with that of the phthalic ring. We did not see any activity on the Mexican fruit fly with rhodamine B or methylene blue as Yoho et al. (30) reported for the house fly but it could be that our concentration was too low and exposure time to short to detect obvious adverse photoactive reactions with these dyes.

We took the same approach as before to the testing of adjuvants, and used our previous formulation containing 0.0382% phloxine B. Where indicated, we substituted the 1% AI SM-9 with the appropriate test adjuvant at equivalent concentration. In addition, one test was conducted using 1% soybean oil in conjunction with some of the adjuvants to see if we could enhance the lethality of phloxine B. Therefore, we tested the following adjuvants: MICRO (a laboratory cleaner, International Products Corporation, Burlington, NJ), Silwet L-77 (99% AI, copolymer surfactant, Union Carbide Chemicals & Plastic Co. Inc., Danbury, CT), Triton X-100 (99% AI, polyoxyethylene ethers and other surface active compounds, SIGMA), Kinetic (99% AI, blend of polyalkyleneoxide modified polydimethylsiloxane) and Citrufilm (99% AI, paraffin base petroleum oil) (Helena Chemical Company, Memphis, TN), Latron AG-

98 (80% AI, alkyl aryl polyoxyethylene, Rohm and Haas Company, Philadelphia, PA), and NuFilm P (96% AI, poly-1-p-menthene) and NuFilm 17 (96% AI, di-1-p-menthene) (Miller Chemical & Fertilizer Corporation, Hanover, PA).

Table VIII. Sunlight-activated Death of Female *A. ludens* After Exposing Them 24 h to a Mazoferm Diet Containing 0.0382% Concentration of Various Dyes

Dye	C.I. Number	Mean[a] (±SEM) Percentage Killed	
		Test 1	Test 2
Check	--	0 ± 0.0 a	0 ± 0.0 a
Phloxine B (Stand.)	45410	53 ± 6.3 b	98 ± 1.2 c
Methylene Blue	52015	0 ± 0.0 a	--
Rhodamine B	45170	0 ± 0.0 a	--
Rose Bengal	45440	51 ± 8.9 b	--
Erythrosine B	45430	94 ± 1.9 c	--
Brilliant Blue BMA	--	--	0 ± 0.0 a
Fast Green	42053	--	0 ± 0.0 a
Mercurochrome	--	--	3 ± 1.2 a
Eosin Y	45380	--	41 ± 4.0 b

[a]Mean of five replicates, 20 females per replicate. Means in same column followed by the same letter are not significantly different (p = 0.05; Fisher's Protected LSD)

The results in Table IX show that in the first test (conducted during a sunny day), SM-9, Silwet L-77, Triton X-100 and Kinetic were about equally as effective in increasing fly mortality. However, during the first hour only Kinetic significantly (F = 31.87; df = 5; p = < 0.001) killed more adults (67 ± 8.9%) than any of the other adjuvants over the check (0%). The second test was conducted during a cloudy, hazy day with an energy flux of 80 to 170 μmol s^{-1} m^{-2} for the first three hours of the experiment and the last hour turning sunny. Nonetheless, in this test Latron AG-98 and Citrufilm were significantly (F = 27.1; df = 5; p = < 0.001) more effective than the standard SM-9. For the third test each of the most promising adjuvants was combined with soybean oil in Mazoferm-fructose formulation and fed to flies. Twenty-four hours later fly exposure to sunlight was conducted during a cloudy day. The energy flux during the exposure period was from 100 to 300 μmol s^{-1} m^{-2}. The results show that, in fact, soybean oil enhances the phototoxicity of phloxine B. The addition of soybean oil to formulations with SM-9, Kinetic, and Latron AG-98 significantly (F = 159.9; df = 5; p = < 0.001) killed more flies in comparison to the standard (with 1% SM-9). However, the addition of soybean oil to Citrufilm adversely affected the proven additive effects of Citrufilm. This implies that the active ingredients (polyoxyethylated polyol fatty acid esters, 99%) of Citrufilm were incompatible with soybean oil (a vegetable oil) but we need to remember that the data were taken at different conditions. The total amount of energy is about the same but the dose-rate and irradiation time are quite different.

Table IX.   Effect of 1% Adjuvants on Sunlight-activated Death of Female *A. ludens* After Exposing Them for 24 h to a Mazoferm-Fructose Formulation Containing 0.0382% Phloxine B

Adjuvant	Mean[a] (±SEM) Percentage Killed in:		
	Test 1, 2 h	Test 2, 4 h	Test 3[b] 2 h
Check	3 ± 0.0 a	4 ± 2.5 a	2 ± 1.2 d
SM-9 (Standard)	91 ± 4.6 c	64 ± 7.8 c	77 ± 4.1 c
SM-9, Soybean Oil			87 ± 2.0 b
Micro	35 ± 5.0 b	--	--
Silwet L-77	74 ± 13.3 c	--	--
Triton X-100	88 ± 4.4 c	--	--
Kinetic	99 ± 1.0 c	--	100. ± 0.0 a
NuFilm 17		10 ± 5.2 a	--
NuFilm P		38 ± 12.0 b	--
Latron AG-98		83 ± 5.8 d	95 ± 2.2 ab
Citrufilm		86 ± 2.9 d	73 ± 4.6 c

[a]Mean of five replicates, 20 females per replicate. Means in same column followed by the same letter are not significantly different (p = 0.05; Fisher's Protected LSD).
[b]In addition to 1% adjuvant, treatments contain 1% soybean oil, except the standard.

The addition of phloxine B with/without uranine to the base Mazoferm formulation did not deter the Mexican fruit fly from feeding according to our results. This is significant because Fondren et al. (17) reported that the housefly fed dietary concentrations of $5X10^{-3}M$ (= 0.42% AI) had less mortality than a lower concentration $(3X10^{-3}M = 0.25\%$ AI) and was routinely observed at high dye concentrations. The plausible explanation offered was that of feeding inhibition at high dye levels. This was not observed in our test even though our highest concentration (4%) was ≈10X higher than theirs. Our fly kill was higher and faster at the high concentration during the first hour of exposure to sunlight. The difference may be in the Mazoferm-fructose diet vs their milk-sugar diet or a difference in sensorial discrimination among the two species. However, the addition of adjuvants dramatically increased fly mortality in our formulated preparation. The addition of adjuvants, so far as we know, has never been reported with the use of malathion-bait sprays. We used the adjuvants to specifically facilitate penetration of the dye from the gut tract into the body tissues of the insect. The results confirmed that our idea was correct. Respicio et al. (28) and Carpenter et al.(29) used what they termed a dispersant with their dye formulation to increase the dispersion of their test material in water but no one else has reported the use of such an approach with other dye formulations. The sensitivity of the Mexican fruit fly to the photooxidants erythrosine B, phloxine B, rose bengal, and eosin Y using LC_{50} equivalent concentrations showed that erythrosine B is the most effective followed by phloxine B and rose bengal being nearly equal, and eosin Y about half as active as phloxine B. Fondren et al. (17) showed that in their case rose bengal was the most effective followed by erythrosine B, phloxine B and eosin Y in terms of LT_{50}. Also the study with adjuvants showed that fly kill could be improved, at the same dye concentration, and that with the addition of an appropriate vegetable oil this kill can be further enhanced. Yet, it appeared that we could still improve the formulation in the area of attraction.

Improvements on Attractiveness of Formulation

There is always a concern that with chemical additions the desirable qualities of the product may change. In our case we had already established (Figures 1-3) that Mazoferm and NuLure have attractive qualities to the flies. Mazoferm was found to be more attractive than NuLure in two of the tests (Figures 1, 2) but with the addition of 0.5% phloxine B we did not know whether this had changed. We set-up a test to determine the relative attraction of 30% Mazoferm or NuLure with and without dye and at 70% with dye. Procedures were the same as those previously described for laboratory attraction tests. The results (Table X) show that though Mazoferm with dye did not lose any attraction in relation to Mazoferm without dye, there was a significant (F = 8.35; df = 5; p = < 0.001) decrease in attraction in NuLure when dye was added to it. There was no reason for this change that we could detect; even pH range was fairly stable (4 to 4.1) for NuLure. The pH range for Mazoferm was from 3.78 to 3.83. Mazoferm may be more attractive than NuLure at lower pHs.

Table X. Response of *A. ludens* to Mazoferm 802 or NuLure With or Without 0.5% (AI) Phloxine B in Laboratory Tests

% Concn	Mean[a] (±SEM) Fly Response	
	Mazoferm	NuLure
30 Without	60.0 ± 4.3	53.6 ± 2.6
30 With	59.3 ± 2.6	46.2 ± 2.5
70 With	64.0 ± 2.3	41.3 ± 3.5

[a]Mean of two tests, five replicates per treatment in each test

Mazoferm was shown to be a more acceptable food to flies than NuLure and ostensibly its gustatory appeal and attraction to flies is not lost with the addition of phloxine B; however, we questioned this status when the adjuvant SM-9 and soybean oil were added. Therefore, we conducted an attractants test with the Mazoferm-base formulation at concentrations of 10, 40, and 70% Mazoferm against same concentrations with the addition of 0.5% phloxine B, 1% SM-9, and 1% soybean oil. Procedures were the same as already described. The results of the test indicated that there was no loss in attraction by the addition of phloxine B, SM-9, and soybean oil. In fact, there was a significant (F = 4.64; df = 5; p = 0.004) increase in attraction of Mazoferm-base with additives over that with none at the 10% Mazoferm concentration. The mean (±SEM) fly response was 37.2 (1.2), 43.2 (2.7), and 41.6 (2.2) with and 30.1 (1.5), 35.7 (2.1), and 37 (2.8) without additives for concentrations of 10, 40, and 70% Mazoferm, respectively. The response profile indicates a slight increase in attraction of Mazoferm with additives over that with none but it could be that under the conditions of the test the added attraction can be attributed to differences in color of the treatments. The color of the Mazoferm-fructose formulation is amber and with 0.5% phloxine B is dark red. However Robacker et al. (31), when using plastic panels, found that black, red, blue, and white were not more attractive than colorless control panels. Thus, indicating that color is not a confounding factor. The important aspect of this test is that no loss in attraction was observed despite all the additives.

The idea of using ultralow-volume (≈ 1 liter/hectare) bait sprays with large droplets (≈ 1-2 mm diam) is that each droplet is a center of attraction to which flies will respond. They should find the source phagostimulatory and feed *ad libitum*. In this manner, one can deliver the intended toxicant without repelling the flies. To this end, one should be able to increase the attractiveness of the food source. Bateman and

Morton (32) combined ammonium carbonate with a protein hydrolyzate and reportedly increased the attractiveness of the hydrolyzate. Hedstrom and Jimenez (33) attracted the West Indian fruit fly and the guava fruit fly with ammonium acetate and Robacker and Warfield (34) used a four-component mixture containing ammonium bicarbonate to attract the Mexican fruit fly. Therefore, we tried to improve our formulation with the addition of an ammonium compound. Laboratory tests had shown some ammonium salts attractive to the Mexican fruit fly. For this test we selected two acids and two ammonium salts. The treatments were water control, Mazoferm-bait (includes 0.5% phloxine B, 1% SM-9, and 1% soybean oil) check, to which we added 0.6% propionic acid, 1% ammonium acetate, 2.5% ammonium citrate, or 0.6% acetic acid. Test procedures were those previously described. The results in Table XI showed that propionic acid tended to reduce the attraction of the Mazoferm-bait but acetic acid was

Table XI. Response of *A. ludens* to Volatiles from Various Attractants in a Mazoferm-Bait in the Laboratory[a]

Attractant	% Concn	Mean[b] (±SEM) Response
Water	0.0	6.7 ± 0.4 a
Mazoferm-Bait Check	0.0	29.3 ± 1.5 bc
Propionic Acid	0.6	25.2 ± 0.9 b
Ammonium Acetate	1.0	31.9 ± 1.7 c
Ammonium Citrate	2.5	34.0 ± 2.3 c
Acetic Acid	0.6	39.8 ± 2.8 d

[a]Mazoferm-bait includes 0.5% AI phloxine B, 1% SM-9, and 1% soybean oil.

[b]Mean of three tests, five replicates per treatment in each test. Means in same column followed by the same letter are not significantly different (p = 0.05; Fisher's Protected LSD).

the only test compound that significantly ($F = 40.84$; df = 5; $p = < 0.001$) increased the attraction of the Mazoferm-bait. The pH ranged from 3.66 for the Mazoferm-bait to 3.98 for the Mazoferm-bait plus ammonium acetate formulation. Apparently the slight increase in pH was not a factor in the attraction of the formulation nor did the propionic acid made the formulation more acidic but reduced the attractiveness of the formulation. We ran a series of acetic acid dilutions (0.075, 0.15, 0.3, 0.6, and 1.2%) to determine if we had selected the best concentration for the test and compared them to the attraction of the Mazoferm-bait check. The results showed no significant ($F = 1.69$; df = 5; $p = 0.176$) difference among means. Then we chose to run a test with 0.6% acetic acid and propionic acid at their approximate innate pH mixtures (≈ 3.7) and at pH 4 in a Mazoferm-bait formulation and 2% SM-9 and compared fly responses to that of a water control and check (Mazoferm-bait formulation). The results of three tests showed significant ($F = 50.3$; df = 5; $p = < 0.001$) differences among means (see Figure 5). The results of this test confirm those seen in the ammonium salts test, i.e., the tendency of propionic acid to reduce attractiveness and the increase in attractiveness with acetic acid mixed with the Mazoferm-bait.

These results show that it is possible to increase the attraction of a food bait; but not necessarily in the expected manner. Based on published information, it would appear that ammonium acetate should have been a good candidate to increase the attractiveness of the bait, but was not. This may be due to the dynamics of the mixture, i.e., after the dissociation of ammonium acetate the cation is tied up and not released to the environment and the anion reverts back to acetic acid but is not enough to significantly increase attraction or it combines with other cations to form other acetates. This area warrants further exploration including raising the pH of the Mazoferm-bait and taking into consideration the dynamic changes that may occur with additional additives.

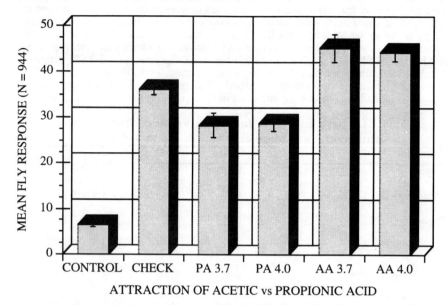

ATTRACTION OF ACETIC vs PROPIONIC ACID

Figure 5.   Mean (three tests; five replicates per treatment) fly response to 0.6% propionic (PA) or acetic acid (AA), at pH 3.7 and 4.0 each, in 0.5% phloxine B-bait as compared to a check (phloxine B-bait) and water control.

Conclusions

The role of protein hydrolyzates in fruit fly control have been important since the early 1950's. For example, partially hydrolyzed yeast, among others, was laced with an insecticide to lure flies and induce them to feed on the poisonous mixture. The hydrolyzed protein NuLure mixed with malathion prevailed and is currently used. We evaluated NuLure against other proteins for its attractiveness in the laboratory and field. In our trials we found that (1) torula yeast, (2) yeast hydrolyzate, (3) casein hydrolyzate, and (4) Mazoferm were better attractants than NuLure.   Moreover, because of cost, we chose to further study the least expensive Mazoferm against NuLure. In these tests we found that (1) in no choice feeding studies, flies consumed more of a Mazoferm-fructose mixture over fructose alone; by contrast, flies consumed more fructose crystals than a mixture of NuLure-fructose; (2) in choice tests, flies preferred Mazoferm significantly over NuLure; and (3) Mazoferm-fed flies had twice the longevity as those fed NuLure. In these studies flies were more attracted to and consumed more Mazoferm and lived longer than flies provided with NuLure. Thus, indicating that Mazoferm is an outstanding candidate to replace NuLure in a bait.

    The studies with the photoactive dye phloxine B as a replacement for malathion show that indeed flies can be killed if they consume the dye. Fly kill over time was relatively dosage dependent. The highest dye concentrations killed flies the fastest. There was no statistical difference in food consumption among concentrations as compared to the check, even when uranine was added. Fly kill rate was increased dramatically when the adjuvant SM-9 was added to the dye-bait formulation. This adjuvant was added with the intent of increasing membrane permeability in the gut. We do not know that this happened but the addition of SM-9 definitely enhanced the lethal effects of phloxine B. However, we were able to lower lethal concentrations. The LC_{50} was determined to be 0.0382% AI, under the conditions of the tests, and as a consequence we were able to find other potential phototoxic dyes which could be used by themselves or in combination with phloxine B. The most lethal dye was erythrosine B > phloxine B ≥ rose bengal > eosin Y. Other dyes such as rhodamine B and methylene blue (previously shown to be phototoxic against the housefly) did not induce mortality in the Mexican fruit fly, but dye concentrations may have been to low. We also found the adjuvants, Kinetic, Latron AG-98, and Citrufilm, to be as good or better than SM-9 in inducing fly mortality in conjunction with phloxine B. The addition of soybean oil further enhanced the activity of SM-9, Kinetic, and Latron AG-98. Of the three, Kinetic appears to be the most promising because it induces the most rapid kill.

Our final tests indicated that with the addition of phloxine B, SM-9, and soybean oil to the Mazoferm-fructose base no attraction is lost. Flies respond about the same to a formulation with or without additives. Even so, we were still able to significantly increase the attraction of the entire mixture with the addition of acetic acid. This simply indicates that there is room for improvement and that the quality of the Mazoferm-fructose mixture is quite stable and remains so despite the other additions.

We started with the idea of developing a bait that could be used conventionally as NuLure has been in the past for ultralow-volume aerial applications. Based on our results, we have a much better bait than NuLure. The idea of using 70% Mazoferm was to approximate that used with the malathion-bait formulation. However, our results show that we can go as low as 40% Mazoferm without losing the attractive qualities of the Mazoferm. Conventional ground spray equipment can be used to apply the lower 40% concentration of Mazoferm. If the formulation needs to be thickened as we did for our field-cage work (shown elsewhere in this book, Mangan and Moreno) it can be done with ease with the addition of xanthan gum. This gum dissolves readily in water at room temperature.

Therefore, we propose that the formulation of 70% Mazoferm 802, 20% fructose or invert sugar, 0.1 to 1% phloxine B, 1% SM-9 or equivalent adjuvant, 1% soybean oil, 0.6% acetic acid, and optionally 0.4 to 1% xanthan gum, volume to volume or weight to volume in water makes an attractive, phagostimulatory dye-bait that can be used not only for the Mexican fruit fly but for other fruit flies such as the Mediterranean fruit fly (discussed by Liquido et al. elsewhere in this book). We do not presume this to be the final formulation but rather a foundation from which one can start. In addition, it is an environmentally safe formulation and most of the products (excepting SM-9), are already used for human or animal consumption. The linear alcohols in SM-9 are generally safe and for that reason are used as adjuvants. The residual activity of phloxine B is not perceived to be long, as it is degraded rapidly by sunlight and will not affect nontarget organisms unless attracted to and consume the dye-bait formulation. There is no toxic contact activity with phloxine B as is the case with malathion.

Acknowledgments

We thank the technical assistance of Robert Rivas for conducting feeding and photoactive dye bioassays, Ismael Saenz for conducting attractiveness tests in the laboratory and in the field and for assisting in the design and modification of cages and Ruben Garcia for providing flies in a timely manner.

Literature Cited

1. Back, E.A.; Pemberton, C.E. *The Mediterranean fruit fly in Hawaii*; USDA Bull 538, 1918; p. 118.

2. Bodenheimer, F.S. *Citrus Entomology in the Middle East*; Dr. W. Junk,'s-Gravenhage, The Netherlands, 1951; p. 663.
3. Woglum, R.S. *The Mediterranean Fruit Fly*; Bull No. 6, Citrus Fruit Growers Exchange, Calif., 1929; pp. 33-48.
4. Ebeling, W. *Subtropical Fruit Pests*; Univ. Calif., 1959; p. 436.
5, Talhouk, A.M.S. *Insects and Mites Injurious to Crops in the Middle Eastern Countries*; Monog. Angew. Entomol., Verlag P. Parey, Hamburg-Berlin, 1969; p. 239.
6. Marsh, H.O. *Report of the Assistant Entomologist*, Board Commissioners Agr. Forest. Hawaii, 1910; pp. 152-159.
7. McPhail, M. *J. Econ. Entomol.* **1937**, 30, 793-799.
8. Baker, A.C.; Stone, W.E.; Plummer, C.C.; McPhail, M. *A Review of Studies on the Mexican Fruit Fly and Related Mexican Species*, USDA Misc. Publ. 521, 1944; p. 155.
9. Steiner, L.F. *J. Econ. Entomol.* **1952**, 45, 838-843.
10. Steiner, L.F.; Rohwer, G.G.; Ayers, E.L.; Christenson, L.D. *Fla. J. Econ. Entomol.* **1961**, 54 (1), 30-35.
11. Stephenson, B.C.; McClung, B.B. *Bull. Entomol. Soc Am.*, **1966**, 12, 374.
12. Lopez-D, F.; Chambers, D.L.; Sanchez-R, M.; Kamasaki, H., *J. Econ. Entomol.* **1969**, 62, 1255-1257.
13. Harris, E.J.; Chambers, D.L.; Steiner, L.F.; Kamakahi, D.C.; Komura, M. *J. Econ Entomol.*, **1971**, 64, 1213-1216.
14. *Light Activated Pesticides*, Heitz, J.R.; Downum, K.R., Eds., ACS Symposium Series 339; American Chemical Society: Washington, D.C., 1987.
15. Lemke, L.A.; Koehler, P.G.; Patterson, R.S.; Feger, M.B.; Eickhoff, T. In *Light-Activated Pesticides*, Heitz, J.R.; Downum, K.R., Eds., ACS Symposium Series 339; American Chemical Society: Washington, D.C., 1987, Chapter 10.
16. Fondren, J.E., Jr.; Heitz, J. R. *Environ. Entomol.* **1978**, 7, 843-846.
17. Fondren, J.E., Jr.; Norment, B.R.; Heitz, J.R. *Environ Entomol.* **1978**, 7, 205-208.
18. Burg, J.G.; Webb, J.D.; Knapp, F.W.; Cantor, A.H. *J. Econ. Entomol.* 1989, 82, 171-174.
19. Pimprikar, G.D.; Fondren, J.E., Jr.; Heitz, J.R. *Environ. Entomol.* **1980**, 9, 53-58.
20. Krasnoff, S.B.; Sawyer, A.J.; Chapple, M.; Chock, S.; Reissig, W.H. *Environ. Entomol.* **1994**, 23, 738-743.
21. Moreno, D.S.; Martinez, A.J.; Sanchez Rivello, M. J. Econ. Entomol. **1994**, 87, 202-211.
22. Rhode, R.M.; Spishakoff, L.M. II Memoria de dia del Parasitologo (1964), Departamento de Parasitologia, Escuela Nacional de Agricultura, Chapingo, Mexico, **1965**; p 23-28.
23. Dethier, V.G. *The Hungry Fly*, Harvard University, 1976; p 489.
24. Carpenter, T.L; Mundie, T.G.; Ross, J.H.; Heitz, J.R.   Environ. Entomol. **1981**, 10, 953-955.
25. Carpenter, T.L.; Johnson, L.H.; Mundie, T.G. Heitz, J.R. J. Econ. Entomol. **1984**, 77, 308-312.
26. Kleinsmith, L.J; Kish, V.M. *Principles of Cell and Molecular Biology*, HarperCollins, 1995; p. 1089.
27. Thomson Harvey, L. *Agricultural Spray Adjuvants*; Thomson Publications, 1992-1993 ed; p. 244.
28. Respicio, N.C.; Carpenter, T.L.; Heitz, J.R. *J. Econ. Entomol.* **1985**, 78, 30-34.
29. Carpenter, T.L.; Respicio, N.C.; Heitz, J.R. *J. Econ. Entomol.* **1985**, 78, 232-237.

30.   Yoho, T.P.; Weaver, J.E.; Butler, L. *Environ. Entomol.* **1973**, 2, 1092-1096.
31.   Robacker, D.C; Moreno, D.S.; Wolferbarger, D.A. *J. Econ. Entomol.* **1990**, 83, 412-419.
32.   Bateman, M.A.; Morton, T.C. *Aust. J. Agric. Res.* **1981**,31, 883-903.
33.   Hedstrom, I.; Jimenez, J. *Revista Brasiliera Entomologia.* **1988**, 32, 319-322.
34.   Robacker, D.C.; Warfield, W.C. *J. Chem. Ecol.* **1993**, 19, 2999-3016.

RECEIVED October 31, 1995

Bestsellers from ACS Books

The ACS Style Guide: A Manual for Authors and Editors
Edited by Janet S. Dodd
264 pp; clothbound ISBN 0–8412–0917–0; paperback ISBN 0–8412–0943–X

Understanding Chemical Patents: A Guide for the Inventor
By John T. Maynard and Howard M. Peters
184 pp; clothbound ISBN 0–8412–1997–4; paperback ISBN 0–8412–1998–2

Chemical Activities (student and teacher editions)
By Christie L. Borgford and Lee R. Summerlin
330 pp; spiralbound ISBN 0–8412–1417–4; teacher ed. ISBN 0–8412–1416–6

Chemical Demonstrations: A Sourcebook for Teachers,
Volumes 1 and 2, Second Edition
Volume 1 by Lee R. Summerlin and James L. Ealy, Jr.;
Vol. 1, 198 pp; spiralbound ISBN 0–8412–1481–6;
Volume 2 by Lee R. Summerlin, Christie L. Borgford, and Julie B. Ealy
Vol. 2, 234 pp; spiralbound ISBN 0–8412–1535–9

Chemistry and Crime: From Sherlock Holmes to Today's Courtroom
Edited by Samuel M. Gerber
135 pp; clothbound ISBN 0–8412–0784–4; paperback ISBN 0–8412–0785–2

Writing the Laboratory Notebook
By Howard M. Kanare
145 pp; clothbound ISBN 0–8412–0906–5; paperback ISBN 0–8412–0933–2

Developing a Chemical Hygiene Plan
By Jay A. Young, Warren K. Kingsley, and George H. Wahl, Jr.
paperback ISBN 0–8412–1876–5

Introduction to Microwave Sample Preparation: Theory and Practice
Edited by H. M. Kingston and Lois B. Jassie
263 pp; clothbound ISBN 0–8412–1450–6

Principles of Environmental Sampling
Edited by Lawrence H. Keith
ACS Professional Reference Book; 458 pp;
clothbound ISBN 0–8412–1173–6; paperback ISBN 0–8412–1437–9

Biotechnology and Materials Science: Chemistry for the Future
Edited by Mary L. Good (Jacqueline K. Barton, Associate Editor)
135 pp; clothbound ISBN 0–8412–1472–7; paperback ISBN 0–8412–1473–5

For further information and a free catalog of ACS books, contact:
American Chemical Society
Product Services Office
1155 16th Street, NW, Washington, DC 20036
Telephone 800–227–5558

WORK IN THE 1980s